AF361853

Economic and Societal Losses Due to Environmental Impacts on Forestry Productivity

Economic and Societal Losses Due to Environmental Impacts on Forestry Productivity

Editors

Noriko Sato
Tetsuhiko Yoshimura

Basel • Beijing • Wuhan • Barcelona • Belgrade • Novi Sad • Cluj • Manchester

Editors
Noriko Sato Tetsuhiko Yoshimura
Kyushu University Shimane University
Fukuoka, Japan Matsue, Japan

Editorial Office
MDPI
St. Alban-Anlage 66
4052 Basel, Switzerland

This is a reprint of articles from the Special Issue published online in the open access journal *Forests* (ISSN 1999-4907) (available at: https://www.mdpi.com/journal/forests/special_issues/Economic_Forestry_Productivity).

For citation purposes, cite each article independently as indicated on the article page online and as indicated below:

Lastname, A.A.; Lastname, B.B. Article Title. *Journal Name* **Year**, *Volume Number*, Page Range.

ISBN 978-3-0365-8936-7 (Hbk)
ISBN 978-3-0365-8937-4 (PDF)
doi.org/10.3390/books978-3-0365-8937-4

Cover image courtesy of Ai Ichinose

Contents

About the Editors

Noriko Sato

Professor Noriko Sato received her Ph.D. from Kyushu University in 1989. She is currently a professor at the Faculty of Agriculture, Kyushu University, and an adjunct lecturer at the University of Tokyo in forest sociology. She is the president of the Japanese Forest Economic Society and the president of the Non-Profit Organization Kyushu Forest Network. Her research interests include forest policy, small-scale forestry management, and rural development. Recently, she has focused on the feasibility of transitioning to Continuous Cover Forestry as a sustainable, disaster-resistant forest management, while she is increasingly interested in the role of dogs in wildlife management.

Tetsuhiko Yoshimura

Professor Tetsuhiko Yoshimura received his Ph.D. from Kyoto University in 1997. He was a visiting scholar at the Department of Biological and Agricultural Engineering, University of California, Davis, in 2006 and 2007. He is currently a full professor at the Faculty of Life and Environmental Science, Shimane University. He is also a board member of the Japan Forest Engineering Society. His research interests include forest engineering, forest mechanization, forest roads, GNSS technology in forestry, forest education and tourism. In recent years, he has focused on studying the productivity of forest harvesting operations in steep terrain, and is increasingly interested in mountain tourism worldwide.

Preface

This Special Issue presents 13 noteworthy articles on the economic and societal losses due to environmental impacts on for-estry productivity after restoration. The aim of this issue is to provide a comprehensive overview of recent advances in forest conservation, forest engineering, disaster sociology, and forest policy in various fields. The papers featured in this issue aim to promote a discussion in disaster-resilient forestry between the natural and social sciences. The stark contrast between the four articles from China and the nine articles from Japan is worthy of note. The articles from China focus mainly on the successful reforestation efforts since the 1990s, the relationship between forests and economic development, and the public perception of forests as portrayed in the media. Meanwhile, the papers from Japan focus on the forest management issues that arise during the utilization phase of the forest plantations and on the challenges of managing these plantations. Extensive research has been conducted on forestry operations, optimization of forest resources, and the transformation of mountain communities in re-sponse to climate change and more frequent torrential rain disasters. These disasters need to be analyzed to understand the full scope of their impacts. The economic and social impacts of these disasters are dependent on natural factors such as climate and topography, as well as social structures. As urban areas continue to grow economically, rural areas suffer from depopulation and aging, resulting in significant societal changes that exacerbate the impact of natural disasters. Although they may appear to be two extreme cases, the editorial situates the papers on China and Japan within East Asia, highlighting one country in the early stages of successful government-led afforestation projects and the other country in the mature stage of plantation forests. This provides insight into the forest management challenges facing East Asia, where forest cover has increased.

We would like to thank the authors for their valuable contributions to this Special Issue and all the reviewers for their con-structive comments and suggestions on the submitted manuscripts.

Noriko Sato and Tetsuhiko Yoshimura
Editors

Editorial

Post-Restoration Forest Management Issues in East Asia under Climate Change: Based on the Special Issue "Economic and Societal Losses Due to Environmental Impacts on Forestry Productivity"

Noriko Sato [1,*] **and Tetsuhiko Yoshimura** [2]

1   Faculty of Agriculture, Kyushu University, Fukuoka 819-0395, Japan
2   Faculty of Life and Environmental Sciences, Shimane University, Matsue 690-8504, Japan; t_yoshimura@life.shimane-u.ac.jp
*   Correspondence: sato.noriko.842@m.kyushu-u.ac.jp

Abstract: Forests provide diverse ecosystem services to people. Consequently, initiatives have been undertaken to restore deforested areas. In East Asian countries, particularly those within the Asian Monsoon region, deforestation has contributed to natural disasters such as sediment run-off, landslides, and flooding, which are exacerbated by torrential rainfall. Restoring forest cover is a critical aspect of national land conservation. To achieve this goal, state-led afforestation initiatives have been launched. Successful afforestation efforts have also been considered an indicator of economic development. However, Japan, which implemented afforestation projects successfully in the 1950s and 1970s, has experienced the under-utilization of its forests due to significant changes in economic and societal conditions since afforestation took place. During the 2010s, the Japanese government promoted the industrialization of forestry, encouraging final felling and reforestation. However, there have been issues with immature forest operation methods and low forestry productivity. Furthermore, in the context of intensifying climate change, heavy rainfall-induced disasters have become more intense, with an increased threat to human safety. Research efforts from the natural and social science fields in Japan have helped identify issues that need to be addressed concerning forests where plantation trees are now utilizable. There is a need to identify improved methods of forestry practice that reduce the risk of climate change-related disasters and establish related forest policies.

Keywords: afforestation; under-utilization; forestry operation; clearcutting; economic losses; societal losses; rural community; disaster-resilient forestry; East Asia

Citation: Sato, N.; Yoshimura, T. Post-Restoration Forest Management Issues in East Asia under Climate Change: Based on the Special Issue "Economic and Societal Losses Due to Environmental Impacts on Forestry Productivity". *Forests* **2023**, *14*, 1845. https://doi.org/10.3390/f14091845

Received: 9 August 2023
Revised: 25 August 2023
Accepted: 8 September 2023
Published: 11 September 2023

1. Introduction

The Food and Agricultural Organization's (FAO's) Forest Resources Survey shows that global deforestation has continued but at a slower rate than that in the 1990s. Although deforestation has continued in Africa and South America, recent increases in forest cover in Asia have attracted increasing attention. Among Asian countries, China [1] and Vietnam [2] have experienced significant increases in forest cover since the 1990s. The change from deforestation to recovery efforts is correlated with economic development and an increase in gross domestic product (GDP). Both countries have initiated government-led efforts to promote natural forest conservation and plantation afforestation. Furthermore, although their political systems and timing of afforestation projects differ, Japan [3] and South Korea [4] have both implemented state-led afforestation projects during their economic growth periods to increase their forest cover. Afforestation projects were implemented in Japan between the 1950s and 1970s and in South Korea between the 1970s and 1990s.

East Asian countries have been successful in restoring forest cover as their economies have grown and are in a region where the Environmental Kuznets Curve (EKC) is applicable [5,6]. These countries fall into two groups: China and Vietnam, where forest cover is still recovering, and Japan and South Korea, where forest cover is above 60% and planted forests have reached the utilization stage. Historically, the implementation of afforestation projects was triggered by severe forest degradation and consequent extensive human and material damage caused by heavy rainfall and extensive flooding, such as during Typhoon Kathleen (1947) in Japan [7] and the Yangtze River flooding in China (1998) [8]. The importance of forest restoration from a land conservation perspective has been widely recognized. In addition, the abundance of labor available for afforestation and forestry in rural villages during the early stages of economic development has been identified as a factor in the success of afforestation projects [9].

Despite these government-led afforestation projects and successful forest restoration in East Asian countries, various challenges in the context of climate change remain, and solutions are being sought to achieve sustainable forest management. With these considerations in mind, we wrote this article about the Special Issue titled "Economic and Societal Losses Due to Environmental Impacts on Forestry Productivity". This Special Issue comprised 13 noteworthy papers; four of which focused on China and nine on Japan, highlighting a striking contrast. Therefore, the purpose of this paper is to provide a summary of the post-reconstruction challenges in East Asia by reviewing the literature published in the Special Issue, along with some additional references.

2. Literature Review on Current Forestry Issues in China and Japan

Although forest cover is rapidly increasing in China (i.e., from 9% in 1990 to 23.3% in 2020) and the total area of plantation forests is expanding, there are areas where afforestation projects have failed. In addition, different regions of China are at different stages of economic development. For this reason, attempts have been made to apply the EKC theory, which has been used to make comparisons between countries and regions within China. For example, Zhang, Q. et al. [10] analyzed the relationship between regional GDP and forest cover in 41 cities in the Yangtze River Basin and showed that the EKC is valid within China. They concluded that economic development allows alternative income generation and forest cover recovery, which can reciprocally contribute to the economic development of the region. Forestry eco-efficiency (FECO) has been developed as an integrated indicator of resource, economic, and environmental strength. A literature review comparing China as a whole, by provinces and cities, found that there is a positive relationship between economic development and FECO in the economically developed eastern region; however, this relationship is not significant elsewhere [11].

In China, forests are owned by the state, while forest management units are divided into state-owned forests and collective forests. In a case study of Heilongjiang Province, China, where there are many state-owned forest areas, the correlation between the conservation of natural forest resources in state-owned forest areas and local economic development was significant until the early 2010s; however, coordinated environmental and economic development was limited after 2015 [12]. The authors also found that the free exercise of forest rights under the collective forest rights reform, which was implemented in 2003 and transferred the right to manage and control forests to local people, had a generally positive impact on management incentives [13]. A significant issue for the future will be comparing forest conservation and the promotion of forestry in China, specifically focusing on state-owned forests directly managed by state-led organizations and collective forests where management is decentralized to the local people. Determining whether successful afforestation can be linked to appropriate utilization in the area is important for the evaluation of afforestation projects and forest resource management. In addition, researchers have begun analyzing how the disclosure of these topics by forest companies and governments (i.e., CSR reports) to their stakeholders (i.e., shareholders, customers, and employees) is being evaluated on social media to make environmental investments in the forest sector more favorable [14].

In Japan, afforestation progressed from the 1950s to the 1970s, mainly in coniferous cedar and cypress plantations. By 1980, the forest cover had reached 68%, with planted forests accounting for 40% of the forested area. As a result, there are no studies comparing forest cover and economic development across regions. It has been noted that the success of afforestation projects in Japan led to a gradual increase in forest accumulation, resulting in a significant reduction in the high incidence of land cover loss throughout the country in the 1950s [15]. However, resource under-utilization has become a major forest policy issue in Japan, as more than 50 years have passed since the afforestation projects were implemented. Forest stocks have increased to a stage where they can be harvested and utilized; however, economic and social factors have prevented the harvesting and utilization of these resources [16]. Although the problem of over-exploitation due to deforestation has been resolved by successful afforestation projects, these projects have now created under-utilization issues.

Economic factors contributing to these under-utilization issues include the high cost of forestry due to the rising cost of labor associated with economic development and Japan's steep topography, and in particular, the strong exchange rate of the Yen resulting from the Plaza Accord of 1985. This puts the country at a disadvantage when competing in the market against overseas forest products. In addition, social factors including the small-scale ownership of private forests and the depopulation and aging of mountainous areas are barriers to utilization. In Japan, approximately 55% of forests are privately owned, and the micro-dispersed ownership structure hinders efficient resource use. A previous study [17] revealed that as the generation involved in afforestation ages and retires, management succession fails, forest management and boundaries become unclear, and the level of forest management declines, particularly in micro-owned areas. Even in many forest producers' cooperatives and authorized neighborhood associations that were established by reorganizing the remaining historical forest commons (Iriai Forest), management is becoming increasingly difficult owing to low timber prices, a decreased number of members, and tax burdens [18]. Ota also emphasized that the members had maintained an attachment to and responsibility for the forests of Iriai origin and a sense of public contribution.

In response to these socio-economic challenges, the Japanese government has worked to overcome under-utilization by clarifying ownership boundaries that had become unclear, creating forest plans by grouping small private forests into estates, promoting forestry mechanization by outsourcing to large forestry companies, and reducing production costs. The government has also been zoning environmental and production forests and promoting final felling through clear cutting in production forests [19]. To promote these plans, a forest environment tax was introduced in 2018 [20]. Despite these initiatives, a recent report [21] highlighted that the road networks in forests, which are an essential production base for increasing forestry productivity, are inadequate, implying that more timber removal processes are required in Japan than in Central European countries with the same sloping forestry, and therefore, no immediate economic benefits can be expected.

In addition, biomass power plants are being built through a feed-in tariff for renewable energy to promote resource use (i.e., to utilize construction and sawmill waste, as well as materials generated from thinning) and CO_2 reduction as renewable energy. However, a previous study [22] conducted an inter-industry analysis of the economic and environmental effects of a forest biomass power plant. The study revealed that the economic effects were estimated in terms of investment and job creation; however, the plant may not be effective in contributing to the global environment in terms of CO_2 reduction, as stated in the forest policy.

In recent years, climate change has caused frequent heavy torrential rainfall disasters during the typhoon and the rainy season in Japan, with an increasing number of days where the hourly rainfall level exceeds 50 mm [15]. Landslides and floods caused by heavy rains have damaged mountain villages, leading to the rapid depopulation of aging villages [23]. In addition, many forest roads have been damaged in various regions, and research on how to rebuild forest roads and make them more disaster-resistant is urgently needed [24]. Suzuki, Y. et al. [25] suggest that, in Japan, where the terrain is steep and the geology is complex, special precautions, such as extensively excavating the soil beneath the roadbed on the cut slope side before compacting the roadbed, are needed while constructing spur roads to ensure that they have sufficient strength.

These issues related to under-utilization, such as low resource utilization and inadequate infrastructure, along with problems associated with the monocultural structure of conifer plantations are highlighted. The spread of cedar pollinosis, intensification of driftwood catastrophes caused by landslides, and forest damage caused by wildlife are becoming more serious, especially in reforested areas [26]. In short, Japan, which was one of the first countries to successfully restore its forests, needs to develop resilient forests that can withstand disaster risks. While torrential rain disasters have the potential to cause the disappearance of mountain villages, some villages are seeking to recover from disasters through activities to create new forest landscapes, with the cooperation of external organizations and former residents. The recovery of these communities will require residents to form their own initiatives. A previous study [23] has meticulously described the recovery process and presented the nature of resident-led recovery in communities affected by disasters.

3. Future Research Issues for Sustainable Forest Management and Disaster-Resilient Forestry in East Asia

The above discussion shows that the pertinent issues for forest science research differ in China and Japan, which are at different stages of forest restoration. Due to differences in land area and political systems, it is unlikely that the challenges faced in Japan will manifest themselves in other countries. However, East Asia has a high population density, and there are similarities in the transformation of the social structure: the shift of the population from rural to urban areas (creating a working population that supported rapid industrial development), agriculture and forestry becoming comparatively minor industries and East Asia becoming a major importer of round wood and forest products, and the declining birthrate and increasingly aging population [27]. There are also similarities in the natural environment, such as the Asian monsoon climate and the high rates of plant growth during the summer, which necessitate intensive weeding work to preserve the success of artificial afforestation and natural regeneration. Therefore, the forestry challenges that arose in Japan after the success of its early afforestation projects may spread to other East Asian countries in the future. Comparative research is needed not only between China and Japan but also with other East Asian countries.

We found that a deeper understanding of current forest problems, which are exacerbated by climate change, can be gained by reviewing not only empirical social science research on Japan (based on field studies) but also natural science research on mountain disasters and forestry operations. To promote disaster-resilient forestry in the future and successfully revitalize rural villages through diverse forest use (not just for timber production), it is necessary to move from a monocultural forest plantation structure ("far from natural forests") to "closer-to-nature forests" [28] suitable for East Asia. This points to the need to consider the creation of "close-to-nature forests" that incorporate research findings related to silviculture and forest management theory. Our paper is limited in its discussion of these theories. The EU Commission's Forest Strategy 2030 indicated in 2021 [29] that production forests also need to explore close-to-nature forestry, with an emphasis on biodiversity conservation, to prevent the consequences of climate change.

4. Conclusions

This paper discusses forest management issues that have arisen after successful afforestation efforts in East Asia, using case studies from China and Japan. While sustainable forest management issues have been discussed in Europe based on comparative studies of forests and forestry in different countries, there has been little comprehensive forest science research based on the similarities and differences between East Asian countries. In Asia, where torrential rain damage occurs yearly (and is exacerbated by climate change), there is a need for collaborative research and discussion on measures to address nature-based forest conservation and resource use issues.

Author Contributions: Conceptualization, N.S. and T.Y.; Writing and editing, N.S. and T.Y.; Funding acquisition, N.S. All authors have read and agreed to the published version of the manuscript.

Funding: This study was supported by the MEXT/JSPS KAKENHI Grant Numbers JP18H04152 and JP25292090.

Conflicts of Interest: The authors declare no conflict of interest.

References

1. Guo, J.; Gong, P.; Dronova, I.; Zhu, Z.L. Forest Cover Change in China from 2000 to 2016. *Int. J. Remote Sens.* **2022**, *43*, 593–606. [CrossRef]
2. Khuc, Q.V.; Le, T.A.T.; Nguyen, T.H.; Nong, D.; Tran, B.Q.; Meyfroidt, P.; Tran, T.; Duong, P.B.; Nguyen, T.T.; Tran, T.; et al. Forest Cover Change, Households' Livelihoods, Trade-Offs, and Constraints Associated with Plantation Forests in Poor Upland-Rural Landscapes: Evidence from North Central Vietnam. *Forests* **2020**, *11*, 548. [CrossRef]
3. Iida, S. *Afforestation—The History and Current Situation*; Forestry Management Series 10; Forestry Management Research Institute: Tokyo, Japan, 1975. Available online: https://iss.ndl.go.jp/books/R100000039-I003347613-00 (accessed on 7 September 2023). (In Japanese, Translated Title by Authors)
4. An, K.W.; Kang, H.M.; Okamori, A. Development of Korea Type Forest Resources Policy and the Role of Government. *J. Fac. Agric. Kyushu Univ.* **2001**, *46*, 153–163. [CrossRef] [PubMed]
5. Mather, A.S.; Needle, C.L.; Fairbairn, J. Environmental kuznets curves and forest trends. *Geography* **1999**, *84*, 55–65.
6. Caravaggio, N. Economic Growth and the Forest Development Path: A Theoretical Re-Assessment of the Environmental Kuznets Curve for Deforestation. *For. Policy Econ.* **2020**, *118*, 102259. [CrossRef]
7. Okada, T.; McAneney, K.J.; Chen, K. Estimating Insured Residential Losses from Large Flood Scenarios on the Tone River, Japan—A Data Integration Approach. *Nat. Hazard Earth Syst.* **2011**, *11*, 3373–3382. [CrossRef]
8. Heng, L.; Xu, Z.K. The 1998 Floods on the Yangtze, China. *Nat. Hazards* **2000**, *22*, 165–184. [CrossRef]
9. Yagi, T. The Composition and Organization of the Forestry Labor Force under Rural Overpopulation Restructuring (in Japanese, translated title by authors). *Jpn. For. Econ. Soc.* **1980**, *98*, 65–71. [CrossRef]
10. Zhang, Q.; Tang, D.C.; Boamah, V. Exploring the Role of Forest Resources Abundance on Economic Development in the Yangtze River Delta Region: Application of Spatial Durbin SDM Model. *Forests* **2022**, *13*, 1605. [CrossRef]
11. Tan, J.L.; Su, X.; Wang, R. Exploring the Measurement of Regional Forestry Eco-Efficiency and Influencing Factors in China Based on the Super-Efficient DEA-Tobit Two Stage Model. *Forests* **2023**, *14*, 300. [CrossRef]
12. Liu, X.Y.; Arif, M.; Wan, Z.F.; Zhu, Z.F. Dynamic Evaluation of Coupling and Coordinating Development of Environments and Economic Development in Key State-Owned Forests in Heilongjiang Province, China. *Forests* **2022**, *13*, 2069. [CrossRef]
13. Wu, C.; Nagata, S.; Furuido, H.; Shibasaki, S.; Haga, K. Farmers' Subjective Comprehension of Forest Tenure Reform and Its Implication on Forest Management Incentives: A Case Study in Jinggangshan, Jiangxi Province, China. *For. Econ.* **2021**, *74-1*, 1–16. (In Japanese) [CrossRef]
14. Zhong, M.; Lu, F.F.; Zhu, Y.F.; Chen, J.R. What Corporate Social Responsibility (CSR) Disclosures Do Chinese Forestry Firms Make on Social Media? Evidence from WeChat. *Forests* **2022**, *13*, 1842. [CrossRef]
15. Sato, T.; Shuin, Y. Impact of National-Scale Changes in Forest Cover on Floods and Rainfall-Induced Sediment-Related Disasters in Japan. *J. For. Res.-Jpn.* **2023**, *28*, 106–110. [CrossRef]
16. Forestry Agency, Japan. *Annual Report on Forest and Forestry in Japan (Summary)*; Forestry Agency, Ministry of Agriculture, Forestry and Fisheries, Japan: Tokyo, Japan, 2021; p. 46. Available online: https://www.rinya.maff.go.jp/j/kikaku/hakusyo/r3hakusyo/attach/pdf/index-2.pdf (accessed on 8 August 2023).
17. Onda, N.; Ochi, S.; Tsuzuki, N. Examination of Social Factors Affecting Private Forest Owners' Future Intentions for Forest Management in Miyazaki Prefecture: A Comparison of Regional Characteristics by Forest Ownership Size. *Forests* **2023**, *14*, 309. [CrossRef]
18. Ota, M. Current Status and Challenges for Forest Commons (Iriai Forest) Management in Japan: A Focus on Forest Producers' Cooperatives and Authorized Neighborhood Associations. *Forests* **2023**, *14*, 572. [CrossRef]
19. Forestry Agency, Japan. *Forest and Forestry Basic Plan*; Forestry Agency, Ministry of Agriculture, Forestry and Fisheries, Japan: Tokyo, Japan, 2021; p. 38. Available online: https://www.rinya.maff.go.jp/j/kikaku/plan/attach/pdf/index-10.pdf (accessed on 8 August 2023). (In Japanese)
20. Ishizaki, R.; Matsuda, S. Message for Solidarity: A Japanese Perspective on the Payment for Forest Ecosystem Services Developed over Centuries of History. *Sustainability* **2021**, *13*, 12846. [CrossRef]
21. Yoshimura, T.; Suzuki, Y.; Sato, N. Assessing the Productivity of Forest Harvesting Systems Using a Combination of Forestry Machines in Steep Terrain. *Forests* **2023**, *14*, 1430. [CrossRef]
22. Fujino, M.; Hashimoto, M. Economic and Environmental Analysis of Woody Biomass Power Generation Using Forest Residues and Demolition Debris in Japan without Assuming Carbon Neutrality. *Forests* **2023**, *14*, 148. [CrossRef]

23. Harada, Y.; Ichinose, A.; Owake, T.; Asahiro, K.; Sato, N.; Fujiwara, T. The Process and Challenges of Resident-Led Reconstruction in a Mountain Community Damaged by the Northern Kyushu Torrential Rain Disaster: A Case Study of the Hiraenoki Community, Asakura City, Fukuoka Prefecture, Japan. *Forests* **2023**, *14*, 664. [CrossRef]

24. Hasegawa, H.; Sujaswara, A.A.; Kanemoto, T.; Tsubota, K. Possibilities of Using UAV for Estimating Earthwork Volumes during Process of Repairing a Small-Scale Forest Road, Case Study from Kyoto Prefecture, Japan. *Forests* **2023**, *14*, 677. [CrossRef]

25. Suzuki, Y.; Hashimoto, S.; Aoki, H.; Katayama, I.; Yoshimura, T. Verification of Structural Strength of Spur Roads Constructed Using a Locally Developed Method for Mountainous Areas: A Case Study in Kochi University Forest, Japan. *Forests* **2023**, *14*, 380. [CrossRef]

26. Takahashi, T.; de Jong, W.; Kakizawa, H.; Kawase, M.; Matsushita, K.; Sato, N.; Takayanagi, A. New Frontiers in Japanese Forest Policy: Addressing Ecosystem Disservices in the 21st Century. *Ambio* **2021**, *50*, 2272–2285. [CrossRef] [PubMed]

27. Kato, K. *Aporia of Global East Asian Capitalism: Thinking from the "Rural" Realm of Japan, Korea, China and Taiwan*; Otsukishoten: Tokyo, Japan, 2020; p. 470. Available online: https://iss.ndl.go.jp/books/R100000002-I030730643-00 (accessed on 8 August 2023). (In Japanese, Translated Title by Authors)

28. Larsen, J.B.; Angelstam, P.; Bauhus, J.; Carvalho, J.F.; Diaci, J.; Dobrowolska, D.; Gazda, A.; Gustafsson, L.; Krumm, F.; Knoke, T.; et al. *Closer-to-Nature Forest Management. From Science to Policy 12*; Pülzl, H., Ed.; European Forest Institute: Joensuu, Finland, 2022.

29. Commission, E. New EU Forest Strategy for 2030. Available online: https://eur-lex.europa.eu/resource.html?uri=cellar:0d918e07-e610-11eb-a1a5-01aa75ed71a1.0001.02/DOC_1&format=PDF (accessed on 30 July 2023).

forests

Article

Exploring the Role of Forest Resources Abundance on Economic Development in the Yangtze River Delta Region: Application of Spatial Durbin SDM Model

Qian Zhang [1,2,*], Decai Tang [3,4] and Valentina Boamah [4]

1   Business School, Nanjing Xiaozhuang University, Nanjing 211171, China
2   School of Economics and Management, Nanjing Forestry University, Nanjing 210037, China
3   School of Law and Business, Sanjiang University, Nanjing 210012, China
4   School of Management Science and Engineering, Nanjing University of Information Science and Technology, Nanjing 210044, China
*   Correspondence: zhangqian@njxzc.edu.cn

Abstract: With the data of 41 cities, including urban and rural areas in the Yangtze River Delta (YRD) region from 2007 to 2019, this paper mainly uses the spatial econometric method to analyze the impact of forest resource abundance in the YRD region on economic development under the background of carbon neutrality. Direct effects, indirect effects, and total effects are further decomposed. The main conclusions are as follows. (1) The abundance of forest resources in the YRD has a U-shaped non-linear effect on economic development, and the curse of forest resources will gradually form forest resource welfare with economic improvement. (2) The phenomenon of economic convergence exists in the YRD region. (3) The spatial effect of forest resource abundance on economic development is non-linear, and the increase in greenery and carbon reduction should be moderately reasonable. (4) The abundance of forest resources can also promote the development of green total factor productivity. The research in this paper complements the existing literature and provides a reference for policymakers.

Keywords: Yangtze River Delta; forest resources abundance; economic development

Citation: Zhang, Q.; Tang, D.; Boamah, V. Exploring the Role of Forest Resources Abundance on Economic Development in the Yangtze River Delta Region: Application of Spatial Durbin SDM Model. *Forests* **2022**, *13*, 1605. https://doi.org/10.3390/f13101605

Academic Editors: Noriko Sato and Tetsuhiko Yoshimura

Received: 31 August 2022
Accepted: 28 September 2022
Published: 30 September 2022

Publisher's Note: MDPI stays neutral with regard to jurisdictional claims in published maps and institutional affiliations.

1. Introduction

Forestry development is related to the sustainable development of China's social economy. Forest resources are pivotal in building ecosystems and improving carbon sink capacity. The role of forest resources in sustainable economic development has attracted more attention from scholars and policymakers. Forest growth plays a very important role in climate regulation [1]. In 2017, the State Forestry Administration issued the "13th Five-Year Plan for Forestry Development" to promote the modernization of China's forestry. In 2021, the "14th Five-Year Plan for Forestry and Grassland Protection and Development Plan" was released. By 2025, China's forest coverage rate will be increased to 24.1%, and the forest stock volume will be increased to 19 billion m^3. The Yangtze River Delta (YRD) region is the hub of China's economic development and has an important strategic position in the overall regional development pattern. In June 2016, the "Yangtze River Delta Urban Agglomeration Development Plan" was promulgated. Boosting the construction of a green and low-carbon ecological city has become one of the important goals of the integrated development of the YRD. In 2018, the integrated development of the YRD was officially elevated to a national strategy. From 2000 to 2018, the forest coverage rate in the YRD region increased from 22.29% to 29.33%, much higher than the national average. However, the distribution of forest resources is uneven, and the structure is unreasonable. During the same period, the economic development of the YRD region was higher than the national average level and showed obvious regional differences. In December 2019, an Outline of the Yangtze River Delta Regional Integrated Development Plan was issued to prioritize ecological protection and strive to build a beautiful Yangtze River Delta. In the 14th Five-Year Plan of each province and city, increasing forest coverage has become one of the common goals for all the cities in the YRD. The YRD region is a cluster of 41 cities in 1 municipality (Shanghai) and 3 Provinces (Jiangsu, Zhejiang, and Anhui) [2,3]. Figure 1 shows the location of the YRD region.

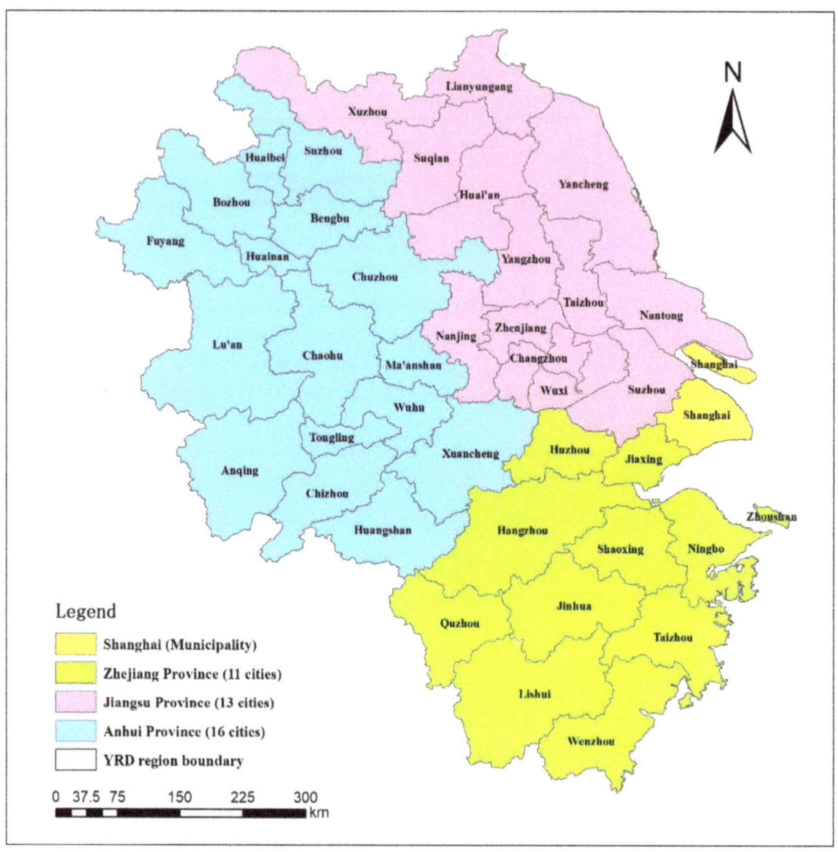

Figure 1. Location of the YRD region [4]. Reproduced from [4], with permission from Scientific Research Publishing, 2020.

There may be a phenomenon of "tragedy of the commons" for forest resources. One of the reasons for this phenomenon is that the beneficiaries of the use of resources and services do not need to pay for it. The costs and benefits are not equal. Unfair allocation of resource property rights will affect the fairness and efficiency of resources and social benefits and may lead to resource rent-seeking and corruption [5]. The resource curse can be turned into welfare if resources are effectively managed [6].

Some scholars have studied whether the environmental Kuznets curve (EKC) exists between forest resources and economic development, but the conclusion is also controversial [7–9]. The EKC assumes an inverted U-shaped relationship between economic growth and environmental quality, which means a country or region's environmental quality will decrease with economic development. Still, when the economy develops to a certain level, the environmental quality will gradually improve [9]. Therefore, this paper's motivation is to empirically explore the impact on the economy at different stages of forest resource development in the YRD region.

This paper mainly studies the following questions: (1) Does the abundance of forest resources in the YRD promote economic development at the city (city in this paper includes urban and rural areas) level? (2) Is there any non-linear relationship between the two? (3) Is there any evidence of the spatial spillover effect? (4) Can the abundance of forest resources improve green total factor productivity? (5) Does economic development affect forest growth, and does the Environmental Kuznets Curve exist? The possible academic contributions of this paper include: (1) Based on the resource curse hypothesis, the impact of natural resources on economic development is discussed from the perspective of renewable resources. (2) Both the quantity and quality of economic development are considered. GDP per capita (although GDP is national and gross regional product (GRP) is regional, GRP is conceptually equivalent to GDP [10]. For readers to understand easily, the term "GDP" is still used in this paper instead of "GRP") and the green total factor productivity are analyzed further to comprehensively measure the economic development level of the YRD region. (3) This paper adopts a spatial econometric model, which not only considers general economic and social factors but also takes the spatial factors into account. The research in this paper complements the existing literature and provides a reference for policymakers.

The paper is structured as follows. Section 2 presents the literature review. Section 3 presents the econometric models, data collection, and categories of the variables. In Section 4, we describe the model estimation's main results, including direct effects, spatial spillover effects, the robustness check, and reverse causation. In Sections 5 and 6, we discuss the results and then draw the main conclusions.

2. Literature Review

2.1. Impacts of Forest Resources on Economic Growth

There are many studies on forest resources and economic development. Whether forest resources can promote economic development is still inconclusive. Some scholars believe that abundant forest resources will promote economic development. The importance of natural capital to economic growth is increasingly recognized, and natural forest capital positively affects national economic growth [11]. Forest resources will contribute to the total economic value through direct use value, option value, existence value, negative externality, etc. [12]. The richer the forest resources, the better the forestry sector, and the more income from harvesting non-precious wood [13]. In addition, forest-rich areas can positively impact the national economy through employment and increased labor income [14]. Rents from forest resources, mineral resources, and oil extraction contribute significantly to economic growth. At the same time, it is necessary to consider the limited availability of natural resources and stimulate the economy by developing policies and utilizing the rents of natural resources to promote development of the business environment [15]. The existence of forest resources is closely related to poverty alleviation. Forest resources and tree systems maintain welfare levels mainly by helping households increase their income and provide food, health, and humanistic values [16]. Based on SWOT and AHP analysis methods, it was found that the rise and development of the Forest Recreation Industry have expanded the regional brand effect, increased income, and enhanced the competitiveness of the industry [17]. The benefits of forest tourism will benefit residents by more than 40%, but urban residents benefit more than rural residents. The infrastructure construction brought by forest resources can also optimize industrial and investment structures [18]. Rural communities that depend on forests experience population decline and economic prosperity often. Van Kooten et al. (2019) explored how communities with rich forest resources as their main economic source stimulate economic development by examining the potential of different forest management regimes to create more jobs and wealth [19].

There is a synergy between forests and sustainable development [20], and forests have a potential role in reducing carbon emissions and poverty [21,22]. There are two main modes in which forest resources can reduce poverty. First, good management patterns can improve the quality of forest resources, thereby improving local economic conditions. Second, increasing forest resources can effectively increase natural capital and ultimately produce economical capital outcomes [23]. However, a review of 242 documents found that although the evidence of forest resource-based poverty reduction is increasing, the results are biased. More comprehensive and robust evidence is also needed to demonstrate and understand differences in outcomes across social backgrounds, social groups, and management objectives [24].

Some scholars believe that abundant forest resources will inhibit economic development. Dependence on forest resources will bring about an obvious resource curse effect [25], mainly because the forestry industry is inefficient and the industrial advantages are not obvious [26]. Another important reason is the existence of the Dutch disease effect. Areas with abundant forest resources mainly rely on the export of forest products to obtain income. However, when the international trade situation is not good, the income of exporting countries will be reduced. For countries with a single source of income, the economy will be greatly affected negatively [27]. Combining the data from 1980 to 2018, using the panel vector autoregression (PVAR) method, it was found that natural resources have an inhibitory effect on economic development in African countries. Economic growth is no longer driven solely by natural resource rents [28].

2.2. Debate of Environmental Kuznets Curve

The EKC between the exploitation and utilization of forests and economic development has always been one of the unsolved problems of environmental economics. Economies make full use of their environmental resources during initial growth. However, when the economy grows beyond a certain level, the environmental recession reaches a tipping point [29]. Hao et al. (2019) tested the relationship between forest resources and economic growth based on the assumption of the EKC using panel data from 30 provinces in China from 2002 to 2015 and the GMM model. The empirical results show that if the economy continues to grow, wood production and afforestation areas will first increase and then decrease after reaching the corresponding inflection point. It proves the existence of EKC between forest resources and economic development [30]. Based on panel data and GMM measurement methods, the empirical test verifies an environmental Kuznets curve between China's provincial-level forest coverage and GDP per capita. Forest resources are already vital to the sustainable development of the economy. The development of urbanization and industrialization has a greater impact on China's forest ecological footprint, while abundant natural resources inhibit economic development. The interaction of urbanization and human capital can alleviate the deterioration of the environment, and human capital has a regulating effect on the sustainable development of the economy [31]. Based on the panel data of 28 provinces in China from 1996 to 2012, the GMM model and the autoregressive distributed lag (ARDL) model were used to confirm the existence of the EKC [32]. However, based on the method of spatial measurement and first-order difference, it was empirically found that there was a U-shaped relationship between the forest coverage rate and GDP per capita in Sichuan, China. Still, there is no evidence to prove the existence of EKC [33].

At the international level, EKC conclusions on forest resources also differ. Combined with the pooled regression model and empirical evidence, it is proved that there is no inverted U-shaped EKC between GDP per capita and forest area in Canada [8]. Furthermore, changes in forest resources have multiple social and economic impacts. Based on the panel data of 111 countries from 1992 to 2015, the cointegration technique was used to analyze the long-term dynamic equilibrium relationship between forest coverage, economic development, agricultural area, and rural population density. Empirical results show that EKC exists in high-income countries and countries in the later stages of forest transition but not in low- and middle-income countries [34]. In addition, many scholars have discussed the EKC of deforestation to observe the impact of deforestation on economic growth [29,35–39]. Among them, through the non-causal investigation of the African heterogeneous panel, it was found that the rational implementation of land policy and trade policy to organize deforestation does not drag down economic growth [38].

2.3. Management of Forest Resources

Economic development will also counteract the development of forest resources. Combining the forest resource input-output model and the forest resource metabolism network model, the forest resources are rationally integrated into the social and economic system. The empirical results show that the primary manufacturing industry consumes more direct wood, and the advanced manufacturing and service industries use wood indirectly. This helps reduce competition between forest industries and sectors. Meanwhile, it is conducive to resource allocation and coordinated development of the economy and ecology [40]. However, increased agricultural productivity and rising wages do not increase local forest cover [41]. The analysis results of the integrated ecological economic model found an N-shaped curve between forest restoration level and economic development, and the forest quality and quantity increase in middle-income countries was the smallest [42]. China's policy interventions in the forest sector have improved environmental and ecological conditions. However, whether forest resources increase depends on issues such as land planning, land practice, and land use rights [43]. China's economic development has driven forest transformation, and economic and population growth has increased demand for forest products and deforestation. However, the government's forest protection program eventually increased forest cover [44].

Regarding forest resource management, China's forest resources have been developing rapidly but still face a very serious situation. For the excessive consumption of forest resources, forest resources have always emphasized efficient management to improve the economy and the effectiveness of forest resource utilization further [32,33]. The continuous increase in forest area and density in China is due to the national afforestation plan and the promotion of forest resources by environmental development [45]. The carrying capacity of forest resources is also affected by urbanization and over-harvesting of forests, especially in Jiangsu and Anhui, where the carrying capacity of forest resources is generally overloaded. Generally, afforestation, energy conservation, and emission reduction measures reduce external pressure on forests [46] by converting forest resource flow into forest resource stock, studying how to reduce dependence on foreign forest resources, adjusting industrial structure, and ultimately improving forest resource utilization structures and efficiency [47].

The purpose of this paper is to construct a spatial econometric model based on the theoretical analysis framework of the impact of forest resource abundance on economic development, and to empirically test the impact of forest resource abundance on economic development at the city level in the YRD region. Finally, this paper puts forward countermeasures and suggestions to promote the high-quality economic development of the YRD region.

3. Methodology

3.1. Theoretical Basis

According to the theory of economic development, in the initial stage, more consideration will be given to allocating limited resources to the industrial sectors with the most productive potential, that is, the greatest linkage effect. Priority is given to developing these sectors while overcoming the bottleneck of economic development. This drives the development of other industries and sectors, and the economic development at this time is unbalanced. However, when the economy enters an advanced stage, from the perspective of industrialization and speeding up economic development, various departments and industries should maintain a certain proportional relationship and development in coordination. The economic development at this time is balanced. The ultimate goal of short-term unbalanced growth is to achieve long-term balanced development. In economic decision-making, resource protection and sustainable resources utilization must be highly valued and considered comprehensively. Improving resource utilization efficiency and economic development are inseparable. For economic development, the excessive use and development of natural resources have caused problems such as resource abuse and environmental pollution, destroying the environment on which people rely, causing further poverty, and ultimately affecting economic development.

The theoretical basis of this paper is mainly analyzed from the perspective of supply and demand. In the early stage of economic development, people's goal was survival. Transportation was inconvenient in places with abundant forest resources, and the industrial structure was single. Farmers mainly lived on cultivated land, and people had more need for basic material life. At the same time, population increase has brought enormous pressure on agricultural land, and the quality of agricultural land is getting lower and lower. This change also gradually reduces the marginal product of labor and lowers incomes [48]. Economic development is inhibited. But with the economic development to a certain extent, the rapid development of industrialization and urbanization has provided many non-agricultural employment opportunities and higher wages [49]. People's lives have improved, and they have begun to pursue a better living environment. The economy has begun to seek green and low-carbon development. A place with abundant forest resources can promote high-quality economic development. The continuous development of science and technology makes the sustainable management of forest resources more efficient. In addition, the public health and environmental functions brought by forest resources also generates corresponding economic value [50].

In Figure 2, in the early stage of economic development, when the forest coverage rate increases from D_1 to D_2, economic growth will decrease from Y_2 to Y_1. Forest resources will affect the allocation of land resources, which may bring about poverty and ultimately affect economic development. Forests have the potential to encroach on agricultural land, and agricultural land change is often used as a proxy for forest cover loss. The increased forest resources have brought about a single industrial structure and inconvenient transportation. Initially, people could only rely on the natural forests or land converted from them for their livelihoods [29]. However, according to the elasticity of demand theory in economics, when the economy expands, people's income levels will increase, and economic prosperity will increase people's demand for green resources. With the advancement of technology, people's incomes are no longer limited to arable land. Abundant forest resources have spawned tourism, led to the gathering of people, and increased employment. Increasing green supply also meets the needs of low-carbon economic development. If the forest cover rises from D_3 to D_4, the level of economic development will rise from Y_3 to Y_4. The curve to the right of the dotted line DE reflects people's demand for forest resources with economic development. Figure 2 reflects the impact of increasing forest resources on economic growth.

Based on the above analysis, this paper proposes four hypotheses:

Hypothesis 1: The impact of forest resource abundance on the level of economic development in the YRD region has a U-shaped non-linear characteristic.

Hypothesis 2: The impact of forest resource abundance on economic development has spatial spillover effects in YRD.

Hypothesis 3: The abundance of forest resources in the YRD region promotes green total factor productivity improvement.

Hypothesis 4: The impact of economic development on the abundance of forest resources presents an environmental Kuznets curve in YRD.

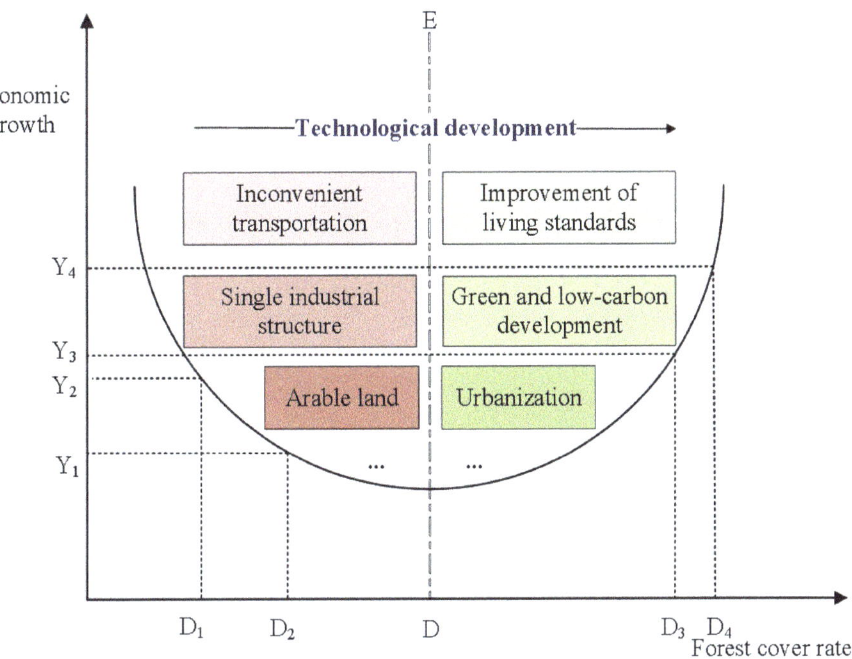

Figure 2. Theoretical framework.

3.2. Spatial Econometric Model

Spatial econometric models usually include the spatial autoregressive model (SAR), spatial error model (SEM), and spatial Durbin model (SDM). Spatial econometric models are gradually being used to verify the relationship between environmental economics studies such as carbon dioxide emissions and economic growth and to demonstrate spatial spillover effects [51].

The steps of regression analysis in this paper are as follows. First, determine whether there is a spatial correlation, and then determine whether a spatial measurement model is available. On the premise of determining the available spatial econometric model, the POLS regression is firstly performed, and the Lagrange multiplier (LM) test of the SAR and SEM models is performed simultaneously. The LM method is applied to test the spatial interaction of the data, specifically the spatial lag and spatial error autocorrelation [52]. The LM test is based on the residuals of non-spatial models with spatial fixed effects, temporal fixed effects, and spatial and temporal double fixed effects, and obeys a chi-square distribution with 1 degree of freedom. If the results reject POLS in favor of the SAR or SEM model, SDM should be selected because SDM model includes both spatially lagged explanatory variables (WY) and spatially lagged explanatory variables (WX). The spatial lag explained variable WY represents the interaction effect between the explained variable and adjacent spatial units. The SDM model can produce better fitting results [53]. The spatial benchmark regression model is constructed as follows.

$$Y_{it} = \rho W_i Y_t + \beta X_{it} + \theta W_i X_{it} + \mu_i + \xi_t + \varepsilon_{it} \tag{1}$$

$$W_i Y_t = \sum_{j=1}^{n} W_{ij} Y_{jt} \tag{2}$$

$$\varepsilon_{it} = \lambda M \varepsilon_t + u_{it} \tag{3}$$

where Y and X denote the explained variable and explanatory variable, respectively. Wi is the i_{th} row of the spatial weight matrix W. μ_i and ζ_t represent the optional individual effect and time effect, respectively. ε represents the disturbance term. $i = 1, 2, \ldots, N$. $t = 1, 2 \ldots, T$. W represents the spatial weighting matrix of the dependent variable, and M is the disturbance term. W_{ij} represents the spatial weight matrices of city i and city j. ρ is the coefficient of the spatial lag term of the explained variable. θ is the spatial autocorrelation coefficient of the explanatory variable. β and θ represent the parameter vector. β reflects the influence of explanatory variables on the explained variables. λ is the coefficient of the error term. Here are two hypotheses: H0: $\theta = 0$; H0: $\theta + \rho\beta = 0$;

If $\theta = 0$, SDM will transform into SAR,

$$Y_{it} = \rho W_i Y_t + X_{it}\beta + \varepsilon_{it} \tag{4}$$

If $\theta = -\rho\beta$, SDM will transform into SEM,

$$Y_{it} = X_{it}\beta + \varepsilon_{it} \tag{5}$$

If $\rho = 0$, $\lambda = 0$, the model will degenerate into POLS.

All the data are normalized to minimize the absolute difference and avoid the influence of extreme values. In order to prevent the bias and endogeneity problems caused by model estimation, this paper constructs SDM, SAR, and a non-spatial panel data model, respectively.

$$GDPPC_{it} = \rho \sum_{j=1}^{n} W_{ij}GDPPC_{jt} + \beta'CORE_{it} + \theta \sum_{j=1}^{n} W_{ij}CORE_{ijt} + \beta''CONT_{it} + \theta \sum_{j=1}^{n} W_{ij}CONT_{ijt} + \mu_i + \zeta_t + \varepsilon_{it} \tag{6}$$

$$GDPPC_{it} = \rho \sum_{j=1}^{n} W_{ij}GDPPC_{jt} + \beta'CORE_{it} + \beta''CONT_{it} + \mu_i + \zeta_t + \varepsilon_{it} \tag{7}$$

$$GDPPC_{it} = \beta'CORE_{it} + \beta''CONT_{it} + \mu_i + \zeta_t + \varepsilon_{it} \tag{8}$$

where Equations (6)–(8) are SDM, SAR, and POLS models, respectively. *GDPPC* is the explained variable GDP per capita, CORE is the core explanatory variable, *CONT* represents the control variables, β' and β'' are the coefficients of the variables, respectively. Other variables and symbols are consistent with the base model.

3.3. Spatial Autocorrelation

This paper constructs the spatial panel data of 41 cities in the YRD region from 2007 to 2019. It uses the spatial panel econometric analysis method to test the impact of forest resource abundance on the economic development level of the YRD region. This paper constructs geospatial weights based on latitude and longitude. The variables are re-estimated based on the spatial distances of all local points in the sample set to the target analysis point. The economic geospatial weighting matrix was not used, mainly because economic conditions change yearly. The distance between two points based on the latitude and longitude distance formula is expressed as follows. The element calculation of the spatial weight matrix is determined by three factors: spatial bandwidth, kernel function, and distance calculation formula. The spatial weight matrix is expressed as follows.

$$W_{\{i\}} = \begin{bmatrix} W_{\{1\}_{h \to i}} & \cdots & 0 \\ \vdots & \ddots & \vdots \\ 0 & \cdots & W_{\{i\}_{h \to i}} \end{bmatrix} \tag{9}$$

The calculation formula is: $W_{\{i\}_{h \to i}} = f\left(D_{\{i\}_{h \to i}}, Bandwidth\right)$. Where $D_{\{i\}_{h \to i}}$ is the distance from all data in the sample set to the local point i. $f(\cdot)$ is the kernel function.

Bandwidth is the spatial bandwidth. One of the commonly used functions is the Gaussian function [54]. The specific representation is as follows.

$$W_{ij} = \exp\left[-\frac{1}{2}\left(d_{ij}/b\right)^2\right] \tag{10}$$

where D_{ij} represents the distance between the centroids of region i and region j. B represents bandwidth. The formula for calculating the distance between two points based on the latitude and longitude distance formula is as follows.

$$D_{\{i\}_{h\to i}} = r_e \times \arccos\left[\sin(v_i\vartheta)\sin\left(v_{\{i\}_h}\vartheta\right) + \cos(v_i\vartheta)\cos\left(v_{\{i\}_h}\vartheta\right)\cos\left(u_i\vartheta - u_{\{i\}_h}\vartheta\right)\right] \tag{11}$$

where ϑ represents an empirical constant and $\vartheta = \pi/180$. r_e represents the radius of the earth. $r_e = 6378.1$km.

Therefore, this paper uses the latitude and longitude data of 41 cities to generate a geospatial weight matrix calculated based on the Gaussian kernel function. According to the distance weight matrix, the first judgment is the spatial autocorrelation of the explained variables. Table 1 provides Moran's Index and Geary's C value of the explained variables from 2007 to 2019. The Moran indices were all significantly greater than 0 and significant at the 1% level. Geary's C is less than 1, proving a positive spatial autocorrelation relationship.

Table 1. Moran's I and Geary's C based on the distance weighting matrix.

Year	Moran's I			Geary's C		
	I	Z Value	*p* Value	C	Z Value	*p* Value
2007	0.431	4.460	0.000	0.547	−3.738	0.000
2008	0.423	4.382	0.000	0.537	−3.829	0.000
2009	0.395	4.100	0.000	0.569	−3.581	0.000
2010	0.355	3.695	0.000	0.596	−3.438	0.001
2011	0.388	4.026	0.000	0.581	−3.559	0.000
2012	0.404	4.185	0.000	0.593	−3.430	0.001
2013	0.383	3.970	0.000	0.602	−3.379	0.001
2014	0.381	3.953	0.000	0.598	−3.402	0.001
2015	0.389	4.032	0.000	0.589	−3.495	0.000
2016	0.423	4.367	0.000	0.535	−3.921	0.000
2017	0.416	4.292	0.000	0.572	−3.613	0.000
2018	0.415	4.280	0.000	0.552	−3.810	0.000
2019	0.401	4.144	0.000	0.568	−3.674	0.000

3.4. Data Sources

This paper constructs an index system from the perspectives of resources, environment, society, economy, and system to verify the influence of the abundance of forest resources in the YRD on economic development (More details in Table 2). The indicators that involve price, such as GDP, per capita consumption level, etc., are deducted from the impact of price.

The raw data are from 2005–2019. Some missing data were filled up by interpolation. There are two main methods for dealing with missing data. One is to use the mean method, which is to take the average value of two adjacent years, such as the CONPC values of Jiaxing in 2014 and 2016; the other is to use the linear interpolation method, that is, the ARIMA imputation method. This method is mainly for the lack of forest resource coverage data in 2019. We then estimate the forest area data and divide it by the land area to calculate the forest cover rate. Missing data is less than 5%. In order to investigate the economic convergence effect and time lag effect of the research city, this paper uses the real GDP per capita data from 2005 to 2019 to generate the first-order difference lag variable of the real GDP per capita. Therefore, this paper uses the 2007–2019 data as the research sample. The descriptive statistics of the data are shown in Table 3.

Table 2. Data source and variables.

Variable	Definition	Measurement	Data source
Dependent variable			
GDPPC	Real GDP per capita	Real GDP divided by the total resident population of each city	SYB
Core independent variable			
FT	Forest cover rate	Forest area divided by land area	SYB, JFB, ZFRMC
FT^2	Square of forest cover rate	Square of (forest area divided by land area)	SYB, JFB, ZFRMC
Control variable			
L.GDPPC	Time lag effect of the level of economic development	The first-order difference lag variable of the GDP per capita	SYB
CSGDP	Carbon sequestration per GDP	Carbon sequestration divided by real GDP	CEADS
ISO2GDP	Industrial sulfur dioxide emissions per GDP	Industrial SO2 emissions divided by GDP	CNKI, CUSY, DEEA
WLI	Water resources carrying capacity	Ratio of per capita domestic water use to total per capita water resources	SYB, WRB
EUGDP	Electricity consumption per capita per GDP	The per capita electricity consumption of the whole society divided by GDP	IFIND
URB	Urbanization rate	Urban population of each city divided by the total permanent population	SYB, SB
SECGDP	Industrial structure	The growth value of the secondary industry as a proportion of GDP	CNKI, SYB
KJ	Science and technology education level	Proportion of education and science and technology expenditures in local budgets to total expenditures	IFIND
GI	Government intervention	The local budget expenditures deduct the remaining expenditures on education and science and technology as a percentage of GDP	IFIND
FDIGDP	Level of foreign investment	The proportion of foreign capital actually utilized by cities in GDP	CNKI, SYB
CONPC	Consumption per capita	The total retail sales of consumer goods in the region divided by the resident population of the region	IFIND

Note: SYB: Statistical yearbooks of provinces and cities over the years; JFB: Jiangsu Forestry Bureau; ZFRMC: Zhejiang Forest Resources Monitoring Center; CEADS: Carbon Emission Accounts & Datasets; CNKI: Platform of China National Knowledge Infrastructure; CUSY: China Urban Statistical Yearbook; DEEA: Department of Ecology and Environment of Anhui Province; WRB: Water Resources Bulletin; IFIND: Financial Data Center; SB: Statistical Bulletins of National Economic and Social Development of All Cities.

Table 3. Descriptive statistics of variables.

Variables	Obs	Mean	Std. Dev.	Min	Max	Unit
GDPPC	533	5.8579	3.6554	0.5515	18.0044	CNY 10,000
L.GDPPC	533	0.4886	0.5572	−3.7511	4.5163	CNY 10,000
FT	533	33.2510	20.3320	3.1700	83.2500	%
CSGDP	533	0.0051	0.0055	0.0002	0.0422	kg/CNY
ISO2GDP	533	2.7067	3.5777	0.0212	34.7532	kg/CNY
WLI	533	0.1069	0.1095	0.0040	1.1268	%
EUGDP	533	0.0877	0.0271	0.0429	0.2099	kWh/CNY
URB	533	58.1388	12.8491	29.0000	89.6000	%
SECGDP	533	48.7669	8.1090	26.9928	74.7346	%
KJ	533	21.0422	3.9938	1.9916	36.9996	%
GI	533	13.5148	5.5856	4.8154	32.1284	%
FDIGDP	533	3.3144	2.1926	0.2114	13.0430	%
CONPC	533	2.2246	1.6116	0.1881	19.4993	CNY 10,000

This paper uses the methods of variance inflation factor (VIF) and tolerance (TOL) to verify the multicollinearity of independent variables. The larger the VIF, the more severe the multicollinearity. Serious multicollinearity exists if the VIF is larger than 10. Tolerance is the inverse of VIF. A collinearity problem exists if TOL is less than 0.1. The VIF of the variables selected in this paper are below 5, and the average is 1.99 (Please see Table 4). In addition, the TOL of each indicator is above 0.1, indicating no multicollinearity.

Table 4. Multicollinearity test of variables.

Variable	VIF	TOL
URB	3.32	0.30
CSGDP	2.56	0.39
SECGDP	2.36	0.42
CONPC	2.30	0.43
EUGDP	2.00	0.50
WLI	1.90	0.53
FT	1.80	0.56
ISO2GDP	1.68	0.59
GI	1.68	0.59
KJ	1.65	0.61
FDIGDP	1.45	0.69
L.GDPPC	1.17	0.85
Average	1.99	0.54

4. Results

4.1. Direct Effects

Table 5 shows the forest resource abundance has a U-shaped non-linear effect on regional economic development. The impact of forest resource abundance on the level of economic development is negative and significant at the 10% level. Still, the relationship between the square of forest coverage and per capita GDP is positive and significant at the 1% level. It shows that in the initial stage of economic development, the abundance of forest resources negatively impacts economic growth. The main reason is that in the early stage of economic development, cities with rich forest resources often have inconvenient transportation and a single industry. The more abundant forest resources, the greater the impact on transportation and arable land. Therefore, the abundance of forest resources will inhibit economic development. However, when the economy develops to a certain extent, abundant forest resources can promote economic level improvement. This is mainly because with technology advancement, improving living standards, and people's changing ideologies, developing the advanced service industry will be faster. Forest resources can provide better spiritual, cultural, and related products and services and improve the ecological environment. The impact of forest resources on economic development mainly includes direct, indirect, and induced effects. The direct effect is reflected in the jobs created by the forest sector, which produces indirect and induced economic added value in other sectors [14].

The first-order lag term of the explained variable is significantly negative, indicating that the phenomenon of economic convergence exists in the YRD region. If the initial state of the economy is relatively underdeveloped, there will be more room for improvement in economic development [55]. This aspect is particularly evident in Anhui Province, which is relatively underdeveloped in the YRD region. Moreover, with the Nanjing metropolitan area's development strategy, Anhui's economic development is relatively fast.

Table 5. Spatial econometric estimation results.

	Dependent Variable: GDP Per Capita			
	Spatial Model			**Non-Spatial Model**
Variable	**SAR**	**SAR**	**SDM**	**POLS**
L.GDPPC	0.0048 ***	−0.0044 **	−0.0051 **	−0.0043 ***
	(−2.8360)	(−2.5635)	(−2.1521)	(−2.7093)
FT	−0.8339 *	−0.7737 *	−0.9200 *	−0.7556 *
	(−1.9476)	(−1.8489)	(−1.6540)	(−1.9437)
FT2	0.1049 ***	0.1049 ***	0.1142 ***	0.1020 ***
	(9.8163)	(9.8842)	(8.3547)	(10.5339)
CSGDP	−0.0239 *	−0.0289 **	−0.0421 **	−0.0290 **
	(−1.6717)	(−1.9894)	(−2.1489)	(−2.1510)
ISO2GDP	0.0031	0.0044	0.0160	0.0048
	(0.3231)	(0.4612)	(1.2394)	(0.5342)
WLI	−0.0017	−0.0012	−0.0016	−0.0011
	(−0.7948)	(−0.5624)	(−0.5204)	(−0.5702)
EUGDP	−0.0564 ***	−0.0541 ***	−0.0585 ***	−0.0524 ***
	(−3.8978)	(−3.6940)	(−2.7759)	(−3.8625)
URB	0.0055 *	0.0059 *	0.0079 *	0.0055 *
	(1.6815)	(1.8217)	(1.7462)	(1.8503)
SECGDP	0.0001	0.0001	0.0001	0.0001
	(−0.6343)	(−0.4708)	(−0.5026)	(−0.4946)
KJ	0.2607 ***	0.2380 ***	0.2282 *	0.2256 ***
	(3.1056)	(2.6675)	(1.7747)	(2.7295)
GI	−0.0081	−0.0062	−0.0166	−0.0062
	(−0.8508)	(−0.6629)	(−1.2419)	(−0.7113)
FDIGDP	0.0001	0.0001	0.0001	0.0001
	(−0.4809)	(−0.2191)	(−0.1173)	(−0.2292)
CONPC	0.0106	0.0103	0.0109	0.0098
	(1.4022)	(1.3795)	(1.0295)	(1.4085)
R^2	0.5224	0.5912	0.5696	0.5442
Obs	533	533	533	533

Note: *** significant at 1% level; ** significant at 5% level; * significant at 10% level; T statistic in brackets.

In terms of carbon sequestration, the impact of carbon sequestration per GDP on the level of economic development is negative and significant. Moreover, carbon sequestration has a certain social cost [56]. Economic development needs to consume a lot of resources and energy, and the production process will cause more carbon dioxide emissions. However, carbon dioxide can be absorbed through carbon sinks such as vegetation and land. The more carbon emissions, the more pollution it brings, which may ultimately inhibit regional economic development. Especially in the Anhui area, it is necessary to do a good job in the strategic layout of the low-carbon economy and circular economy development while undertaking the industrial transfer. In addition, in terms of carbon sequestration methods, there are generally two ways to increase carbon sequestration, one is by technology, and the other is by afforestation. Presently, carbon sequestration through scientific and technological means is costly. However, carbon sequestration through afforestation is currently the most economical and reliable method. Notwithstanding, large-scale afforestation may generate a lot of opportunity costs, such as land use and forestry industry development.

In terms of resources and the environment, the impact of industrial sulfur dioxide emissions per GDP on the level of economic development is positive, but the results are not significant. In addition, the impact of water resource carrying capacity on the level of economic development is negative. The more developed the economic level is, the greater the demand for resources will be and the weaker the resource-carrying capacity will be.

The impact of electricity consumption per GDP on economic development is significantly negative. Electricity consumption can reflect the economic development level of a region. However, electricity consumption and industrial structure are closely related.

Generally, the demand for electricity consumption in primary and tertiary industries is not as large as that of the secondary industry. Although Jiangsu is a large manufacturing province, other industries such as finance, technology, and services are also very developed. In 2020, Shanghai's GDP ranked among the top ten in the country, although Shanghai's economy is relatively developed. Electricity consumption in Shanghai ranks low due to the developed tertiary industry, especially the financial industry. Industrial structure often determines electricity use, but high-tech industries typically generate much more GDP with less energy than low-end manufacturing. Especially after the implementation of power integration development, cities in the YRD gradually developed a mode of sharing and interoperability. The efficiency of energy utilization in economic development is getting higher and higher.

The YRD region is committed to industrial restructuring, optimization, and upgrading. The industry shows a trend of cluster development and can realize the integrated development of modern service and advanced manufacturing industries.

The impact of urbanization on economic development is significantly positive. Improving the urbanization level will help to improve the level of economic development. Urbanization is essentially a process of agglomeration of manpower, capital, and resources. It is also one of the important driving forces for upgrading and transforming the economic structure. The first step is the transformation of the population; that is, the rural population is transformed into an urban population and participates in non-agricultural production activities. The spatial transformation of the population residence was gradually realized after the population transformation. The transfer of rural surplus labor and the construction of cities and towns will gradually promote the development of local enterprises, thereby promoting the continuous improvement of the local economy. However, the way that accompanies economic growth is often extensive. With the continuous improvement of urbanization, higher requirements are put forward on land, culture, society, and other aspects. Ultimately, urbanization will gradually achieve coordinated development with the economy, and economic growth will gradually transit from an extensive development model to an intensive growth model.

Investment in scientific research and education has an obvious role in promoting economic development, and the estimated coefficient of the model is above 0.2. It shows that scientific and technological innovation is very important to realize the economic development of the YRD region.

Government intervention has suppressed the economic development in YRD, but the results are not significant. China's economy has adopted a "government-led market economy" for a long time. The government intervenes deeply in the process of regional economic growth. It suppresses the role of the regional market mechanism, which may lead to imbalances in the economic structure and cannot be adjusted in time. In addition, government intervention may also bring about mistakes in decision-making and improper resources allocation. This will make the economic micro-subjects lose the initiative and vitality of economic activities. Moreover, excessive government intervention can easily lead to rent-seeking. Under modern economic conditions, the government's main role should carry out reasonable macro-intervention and moderate guidance on economic activities through laws and regulations. However, excessive intervention may inhibit economic development.

Foreign direct investment positively impacts economic development. The more foreign direct investment, the higher the level of economic development. First, foreign investment has brought about the application and promotion of advanced green technologies and improved the production efficiency of enterprises. Second, foreign investment will also relatively impact the overall corporate environment positively through production scale expansion, industrial structure adjustment, and talent introduction. Third, by absorbing FDI, the YRD region can accelerate the accumulation of regional capital, accelerate capital formation, and further improve investment level. Meanwhile, it can promote the level of employment in the region.

In terms of consumption, the model results show that the per capita total retail sales of consumer goods positively impact economic development. The average retail sales of social consumption are generally affected by income level, price level, and consumption environment. Only with economic development, continuous improvement of residents' income level, stable price level, and the good consumption environment can the growth of total retail sales of social consumer goods be stimulated. Conversely, the growth of consumer demand will also play a direct and final decisive role in economic growth. Investment demand and aggregate demand depend on consumption demand to some extent. From a medium and long-term perspective, only investment supported by consumer demand is effective, and effective investment and consumption make a greater contribution to economic growth.

4.2. Effects of Decomposition

The effects of spatial econometric models can be divided into direct effects, indirect effects (spatial effects), total effects, and feedback effects. The direct effect is the impact of an independent variable in a certain region on the dependent variable. The feedback effect is the direct effect coefficient minus the regression coefficient of the estimated result. The feedback effect means that the explanatory variables in a certain area will impact the explained variables in the surrounding area, which will affect the explained variables in the local area. Indirect effects are the effects of an explanatory variable on other regions, which is the influence of an explanatory variable in a neighboring area on the explained variable in the local area. The total effect is the sum of the direct and indirect effects.

Table 6 is calculated according to the spatial Durbin model of GDP per capita as the explained variable. The results show that the magnitude and significance of the coefficients of the direct effect are almost consistent with the coefficients and significance of the model estimates. The relationship between forest resource abundance and economic growth always maintains a U-shaped non-linear relationship. Forest resources can inhibit the initial stage of economic development. However, when the economy develops to a certain level, the abundance of forest resources and economic growth will develop together. Moreover, it has a positive pulling effect on the economic development of the YRD region.

Table 6 shows that the spatial effect exists. The impact of the abundance of forest resources in neighboring cities on the region's economic development presents a non-linear inverted U-shaped trend. In the initial state, the more abundant forest resources are in neighboring cities, the more the level of economic development in the region will be promoted. When it reaches a certain level, the excessively abundant forest resources in neighboring regions will inhibit the level of economic development in the region. One reason is that the growth of trees requires material conditions such as land. Efforts to massively increase vegetation cover are increasing to alleviate climate conditions and achieve the grand vision of carbon neutrality. Excessive expansion of forest areas may seriously damage biodiversity. Cutting down old forests and planting new ones may break the original ecological balance. This will eventually crowd out the production and living resources of the region and affect economic development.

From the perspective of the feedback effect, the relationship between the abundance of forest resources of the surrounding areas and the region's economic development again shows a non-linear U-shaped trend. It means that the direct consumption of forest resources has different impacts at different stages of economic development. With the continuous economic improvement, the sustainable development of forest resources has gradually been paid attention to. The ecological, economic, and social values of the forest have been continuously excavated. For example, after the Three Plenary Sessions of the Eleventh Session, Zhejiang entered a period of revitalization and restoration of forestry, which has been vigorously developed. While striving to develop the economy, efforts should be made to realize the sustainable utilization of forest resources. The YRD region gradually realizes the coordinated development of the environment and economy and finally promotes the sustainable development of the economy.

Table 6. Decomposition of direct effects, indirect effects, and total effects.

	Dependent Variable: GDP Per Capita			
	Direct Effect	**Indirect Effect**	**Total Effect**	**Feedback Effect**
L.GDPPC	−0.0048 ***	0.0010 **	−0.0039 ***	0.0001
	(−2.7387)	(2.0410)	(−2.7268)	
FT	−0.8623 **	0.1709 *	−0.6914 *	−0.0284
	(−2.0103)	(1.6379)	(−2.0121)	
FT2	0.1062 ***	−0.0209 ***	0.0853 ***	0.0013
	(9.4994)	(−3.0773)	(8.9238)	
CSGDP	−0.0242 *	0.0048	−0.0195 *	−0.0003
	(−1.7010)	(1.4741)	(−1.6918)	
ISO2GDP	0.0039	−0.0008	0.0031	0.0009
	(0.4115)	(−0.3999)	(0.4084)	
WLI	−0.0017	0.0003	−0.0013	0.0001
	(−0.7504)	(0.7133)	(−0.7466)	
EUGDP	−0.0572 ***	0.0112 **	−0.0459 ***	−0.0008
	(−3.7210)	(2.3800)	(−3.6831)	
URB	0.0056 *	−0.0011	0.0045 *	0.0002
	(1.6479)	(−1.4040)	(1.6441)	
SECGDP	0.0001	0.0001	0.0002	0.0001
	(−0.6463)	(0.6184)	(−0.6423)	
KJ	0.2617 ***	−0.0513 **	0.2104 ***	0.0010
	(2.9847)	(−2.1614)	(2.9407)	
GI	−0.0083	0.0016	−0.0066	−0.0002
	(−0.8289)	(0.7771)	(−0.8276)	
FDIGDP	0.0001	0.0001	0.0002	0.0001
	(−0.4266)	(0.4011)	(−0.4267)	
CONPC	0.0108	−0.0021	0.0087	0.0002
	(1.4097)	(−1.2702)	(1.4002)	

Note: *** significant at the 1% level; ** significant at the 5% level; * significant at the 10% level; T statistics in parentheses.

4.3. Robustness Check

4.3.1. Replacing the Spatial Weight Matrix

This paper adopts the spatial weight matrix of queen contiguity to verify the robustness of the estimation results. If there is a common boundary or a common contact point between two places, these two places are considered adjacent. Then we set the weight 1, otherwise 0 [57]. When performing the spatial measurement, this paper performs row normalization on the adjacent spatial weight matrix, that is, divides each element in the matrix (denoted as $\widetilde{w}_{ij}$) by the sum of the elements in its row to ensure the sum of the elements in each row is 1, that is, $w_{ij} = \widetilde{w}_{ij} / \sum_j \widetilde{w}_{ij}$. Row normalization yields the average of each region's neighbors. In order to verify the true spatial autocorrelation relationship, the GDP per capita of the explained variable is selected to calculate Moran's index. The significance is tested by the Monte Carlo simulation method. Moran's index is greater than 0, and the Z value is greater than 1.96. After performing up to 99,999 permutations, the p value is always close to 0, indicating the existence of spatial autocorrelation.

After the spatial autocorrelation relationship is verified, the adjacent spatial weight matrix is used to construct a spatial econometric model. In this paper, SDM, SAR, and POLS models are set as follows:

$$GDPPC_{it} = \rho \sum_{j=1}^{n} W'_{ij}GDPPC_{jt} + \beta''CORE_{it} + \theta \sum_{j=1}^{n} W'_{ij}CORE_{ijt} + \beta''CONT_{it} + \theta \sum_{j=1}^{n} W'_{ij}CONT_{ijt} + \mu_i + \xi_t + \varepsilon_{it} \quad (12)$$

$$GDPPC_{it} = \rho \sum_{j=1}^{n} W'_{ij}GDPPC_{jt} + \beta''CORE_{it} + \beta''CONT_{it} + \mu_i + \xi_t + \varepsilon_{it} \quad (13)$$

$$GDPPC_{it} = \beta''CORE_{it} + \beta''CONT_{it} + \mu_i + \xi_t + \varepsilon_{it} \quad (14)$$

where Equations (12)–(14) are the SDM, SAR, and POLS models, respectively. The explained variable GDP per capita, core explanatory variables, control variables, and other variables are consistent with the base model, but W'_{ij} is replaced by the queen contiguity weighting matrix and normalized. The model results are shown in Table 7.

Table 7. Results of robustness check.

	Dependent Variable: GDP Per Capita		
Variable	**SAR**	**SAR**	**SDM**
L.GDPPC	−0.0046 ***	−0.0043 ***	−0.0062 ***
	(−2.9053)	(−2.5937)	(−3.6498)
FT	−0.7828 *	−0.7665 *	−0.6810 *
	(−1.9554)	(−1.8866)	(−1.7427)
FT2	0.0991 ***	0.1016 ***	0.1013 ***
	(9.8301)	(9.7626)	(9.2241)
CSGDP	−0.0239 *	−0.0286 **	−0.0289 **
	(−1.7610)	(−2.0085)	(−1.9654)
ISO2GDP	0.0031 ***	0.0047	0.0036
	(0.3543)	(0.5082)	(0.4049)
WLI	−0.0016	−0.0011	−0.0029
	(−0.7919)	(−0.5279)	(−1.3508)
EUGDP	−0.0524 ***	−0.0526 ***	−0.0518 ***
	(−3.8967)	(−3.7191)	(−3.7152)
URB	0.0048	0.0054 *	0.0078 **
	(1.5648)	(1.7179)	(2.3585)
SECGDP	0.0001	0.0001	0.0001
	(−0.6400)	(−0.4969)	(−0.1923)
KJ	0.2312 ***	0.2232 ***	0.1446
	(2.9572)	(2.5833)	(1.6013)
GI	−0.0076	−0.0061	−0.0075
	(−0.8504)	(−0.6715)	(−0.8437)
FDIGDP	0.0001	0.0001	0.0001
	(−0.4924)	(−0.2024)	(0.2674)
CONPC	0.0095	0.0097	0.0110
	(1.3533)	(1.3396)	(1.5560)
R^2	0.5854	0.6175	0.6341
Obs	533	533	533

Note: *** significant at the 1% level; ** significant at the 5% level; * significant at the 10% level; T statistics in parentheses.

4.3.2. Adjusted Sample Period

In 2018, President Xi Jinping announced the integrated development of the YRD region and raised it as a national strategy. From a regional perspective, the integrated development of the YRD solves the problem of unbalanced regional development, and the integration of the YRD can play a guiding and exemplary role. Shanghai, Jiangsu, and Zhejiang are all developed regions, while Anhui is underdeveloped. The integration of the YRD needs to drive the neighboring regions with poor conditions through the regions with good conditions so that the elements flow and gather in Anhui. The YRD region should strengthen cooperation in infrastructure construction, technological innovation, industrial development, and ecological environment construction to achieve collaborative innovation and development. Therefore, the article adopts the method of reducing the sample period to avoid the impact of the policy, that is, to exclude data from 2018 and 2019. The model sample period focuses on 2007–2017, and other explained variables, explanatory variables, and control variables remain unchanged. The article also reports the results of the SAR and SDM models to reduce the bias of the model results. The results are shown in Table 8.

Table 8. Results of robustness checks from 2007–2017.

| | Dependent Variable: GDP Per Capita | | | |
| | Spatial Model | | | Non-Spatial Model |
Variable	SAR	SAR	SDM	OLS
L.GDPPC	−0.0043 ***	−0.0039 **	−0.0017	−0.0042 **
	(−2.6701)	(−2.3535)	(−0.7280)	(−2.4758)
FT	−0.9632 **	−0.9393 **	−0.9477 **	−0.9698 **
	(−2.2326)	(−2.0669)	(−1.7653)	(−2.0963)
FT2	0.0852 ***	0.0876 ***	0.0643 ***	0.0958 ***
	(6.8397)	(6.6826)	(3.2157)	(7.3667)
CSGDP	−0.0197	−0.0231	0.0068	−0.0276 *
	(−1.3633)	(−1.5413)	(0.3264)	(−1.8157)
ISO2GDP	0.0023	0.0034	0.0078	0.0037
	(0.2929)	(0.4047)	(0.6696)	(0.4296)
WLI	−0.0021	−0.0016	0.0025	−0.0018
	(−1.1307)	(−0.8612)	(0.9360)	(−0.9445)
EUGDP	−0.0432 ***	−0.0429 ***	−0.0367 **	−0.0447 ***
	(−3.2563)	(−3.0546)	(−2.0033)	(−3.1328)
URB	0.0014	0.0027	0.0001	0.0037
	(0.4392)	(0.7919)	(0.0020)	(1.0781)
SECGDP	0.0001	0.0001	0.0001	0.0001
	(−0.7393)	(−0.7308)	(−2.5473)	(−0.5204)
KJ	0.2253 ***	0.2198 ***	0.0832	0.2460 ***
	(2.9960)	(2.7051)	(0.7572)	(2.9855)
GI	−0.0114	−0.0127	−0.0264**	−0.0137
	(−1.2976)	(−1.3367)	(−2.0907)	(−1.4166)
FDIGDP	0.0001	0.0001	0.0001	0.0001
	(−0.8932)	(−0.6854)	(1.5175)	(−0.8663)
CONPC	0.0074	0.0082	0.0162 *	0.0078
	(1.0692)	(1.1208)	(1.7552)	(1.0480)
R^2	0.5613	0.5930	0.5618	0.4636
Obs	451	451	451	451

Note: *** significant at the 1% level; ** significant at the 5% level; * significant at the 10% level; T statistics in parentheses.

4.3.3. Replace the Dependent Variable

In 2017, President Xi Jinping proposed five development goals of "innovation, coordination, greenness, openness, and contribution" while considering economic development, achieving resource conservation, and environmental protection. This paper adopts the method of replacing the explained variables to verify further the impact of forest resource abundance on economic development. This section replaces GDP per capita with green total factor productivity (GTFP), which is essentially one of the ways to measure green economic growth [58]. Meanwhile, to examine the linear and non-linear effects of forest resource abundance on green total factor productivity, the core explanatory variable CORE first considers the forest coverage rate FT alone. It then simultaneously considers the forest coverage rate FT and its square. In order to stabilize the results and alleviate the endogeneity problem, the static and dynamic SAR and SDM are constructed as follows. The results are shown in Table 8.

$$GTFP_{it} = \rho \sum_{j=1}^{n} W'_{ij} GTFP_{jt} + \beta'' CORE_{it} + \beta'' CONT_{it} + \theta \sum_{j=1}^{n} W'_{ij} CONT_{ijt} + \mu_i + \xi_t + \varepsilon_{it} \tag{15}$$

$$GTFP_{it} = \rho \sum_{j=1}^{n} W'_{ij} GTFP_{jt} + \beta'' CORE_{it} + \theta \sum_{j=1}^{n} W'_{ij} CORE_{ijt} + \beta'' CONT_{it} + \theta \sum_{j=1}^{n} W'_{ij} CONT_{ijt} + \mu_i + \xi_t + \varepsilon_{it} \tag{16}$$

$$GTFP_{it} = \tau GTFP_{i,t-1} + \rho' \sum_{j=1}^{n} W_{ij} GTFP_{j,t-1} + \rho \sum_{j=1}^{n} W'_{ij} GTFP_{jt} + \beta'' CORE_{it} + \theta \sum_{j=1}^{n} W'_{ij} CORE_{ijt}$$
$$+ \beta'' CONT_{it} + \theta \sum_{j=1}^{n} W'_{ij} CONT_{ijt} + \mu_i + \xi_t + \varepsilon_{it} \tag{17}$$

where Equations (15)–(17) represent the static SAR, the static SDM, and the dynamic SDM, respectively. The explained variable GTFP is green total factor productivity. The core explanatory variables, control variables, other variables, and symbols are consistent with the benchmark model, and the spatial weight matrix adopts the Gaussian kernel function distance weight matrix.

As shown in Tables 7 and 8, the signs of all variables are basically unchanged. The phenomenon of economic convergence exists at the city level in the YRD region, and there is a lot of room for improvement in areas with relatively backward economies. The impact of forest resource abundance on the level of economic development has always maintained a U-shaped trend. Forest resources at the urban level in the YRD will play different roles in different stages of economic development, especially under the goal of low-carbon economic development. Resources play a pivotal role in carbon sequestration. This proves that the results of the model argument are robust. After the optimization and upgrading of the industrial structure, the energy utilization rate is high, and the investment in urbanization and science and technology education can effectively promote the economic level of the YRD region.

Table 9 shows the results of the influence of forest resource abundance on the green total factor productivity. SAR (1) and SAR (2) examine the linear and non-linear relationship between the abundance of forest resources and the level of green economic development. SDM (1) and SDM (2) represent the static and dynamic models, respectively. The results of the models show inertia in developing a green economy in the YRD region. The U-shaped characteristics of the impact of forest resource abundance on green economic development have not been verified. However, the impact of forest resource abundance on the development of the green economy is always positive. It shows that increasing the abundance of forest resources in the YRD will help improve its green economic development level and contribute to sustainable economic development. Greening is conducive to solving the trade-off dilemma of "economic growth, environmental friendliness, and resource conservation" in economic development. This is also in line with China's current "14th Five-Year Plan" and the strategic need for sustainable economic development. All results prove that the influence of forest resource abundance on economic development in the YRD region presents a U-shaped feature, and forest resource abundance is conducive to improving green total factor productivity.

Table 9. Estimation results of effects of forest resource abundance on GTFP.

	Dependent Variable: GTFP			
Variable	SAR (1)	SAR (2)	SDM (3)	SDM (4)
L.GTFP	0.4277 ***	0.4261 ***	0.4515 ***	0.5420 ***
	(0.0318)	(0.0318)	(0.0391)	(0.0441)
FT	0.3878 *	0.8019 **	0.7673 *	0.9644 *
	(0.0381)	(0.3924)	(0.2998)	(0.5759)
FT2		−0.6917		−0.9728
		(0.5206)		(0.7812)
Control variables	Yes	Yes	Yes	Yes
R^2	0.3527	0.3539	0.3697	0.4279
Obs	533	533	533	492

Note: *** significant at the 1% level; ** significant at the 5% level; * significant at the 10% level; standard errors are in parentheses.

4.4. Reverse Causation

In order to verify the impact of economic development on forest resource abundance, this paper constructs the dynamic SDM model, where the dependent variable is the Forest Resource Abundance (FT), and the core explanatory variables are the economic development, namely GDP per capita (GDPPC) and its square to adopt the environmental Kuznets curve. Other control variables are kept consistent with the baseline model. The spatial weight matrix adopts the queen adjacent weight matrix. Furthermore, this paper also examines the linear relationship between economic development and forest resource abundance. All variables are standardized. The estimation results of the SAR model are also shown to check the robustness of the results.

As shown in Table 10, although there is a U-shaped curve between the level of economic development and the abundance of forest resources, the results are not stable. However, the level of economic development inhibits the abundance of forest resources. There are two main reasons. First, the cost of land-use is high in YRD. This region contributes a lot to the national economy and is one of the important urban agglomerations in China. The total economic volume accounts for 25% of the country's total. In addition, the two most powerful harbors in the world, Shanghai and Zhoushan, are located here. Among the top 20 cities in terms of GDP, cities in the YRD region account for one-third at least. However, the area of such an economically developed region is only 358,000 km^2, accounting for about 4% of the country's total. Second, forest resources are limited by land and depend largely on territorial planning, which is difficult to change in the short term. The growth of forest resources has always been a key concern of forestry. However, from the perspective of economic development and forest land use, large, continuous forests have the potential to be transformed into smaller, isolated fragmentation processes, which exacerbate the degree of forest fragmentation [59].

Table 10. Estimation results of effects of economic development on forest resource abundance.

	Dependent Variable: Forest Cover Rate			
Variable	SDM (1)	SAR (2)	SDM (3)	SAR (4)
GDPPC	−0.0680 **	−0.0578 ***	−0.1268 *	−0.0987
	(0.0277)	(0.0279)	(0.0749)	(0.0733)
Square of GDPPC			0.0388	0.0271
			(0.0468)	(0.0459)
Control variables	Yes	Yes	Yes	Yes
R^2	0.9906	0.9905	0.9906	0.9904
Obs	492	492	492	492

Note: *** significant at the 1% level; ** significant at the 5% level; * significant at the 10% level; standard errors are in parentheses.

5. Discussions

This paper mainly uses the method of spatial econometrics, based on the data of 41 cities in the YRD region from 2007 to 2019, to analyze the influence and spatial effects of the abundance of forest resources in the YRD region on the level of economic development. The robustness check of the results is carried out by changing the spatial weight matrix, adjusting the sample period, and changing the explained variables. The results of the model estimates are proven to be robust. The main results about our hypotheses are as follows.

Hypothesis 1 was verified. The influence of the abundance of forest resources on the economic development level in the YRD has a U-shaped non-linear characteristic [60]. At the city level in the YRD, forest resources will play different roles in different stages of economic development. The curse of forest resources will gradually evolve into welfare with economic development. In the initial stage of economic development, the abundance of forest resources has a certain inhibitory effect on the level of economic development. When the economy develops to a certain level, abundant forest resources can promote the level of economic development [15].

Hypothesis 2 was verified. Spatial spillover effects of forest resources exist [61]. Spatial factors play an important role in economic growth and convergence. The spatial autocorrelation relationship of economic development level exists. Ignoring spatial factors may lead to biased estimation results. The spatial effects of forest resource abundance on local and surrounding areas are non-linear. However, when we implement the policy of increasing forests and reducing carbon emissions, it must be within a reasonable range. Excessive forest resources may have social costs [56]. It may eventually crowd out the production and living resources of the region and affect economic development.

Hypothesis 3 was verified. The abundance of forest resources can also promote the development of green total factor productivity. This aligns with the great vision of carbon peaking and carbon neutrality. In the long run, the abundance of forest resources will help improve the quantity and quality of economic development, which is consistent with the strategic goals of China's "14th Five-Year Plan" [62].

Hypothesis 4 was not verified. The impact of economic development on the abundance of forest resources does not show any evidence of an EKC curve in the YRD region. Furthermore, the level of economic development inhibits the forest growth. This is mainly due to the expensive land cost in the YRD region and the government's land-use planning. Generally, competing land-use values can lead to changes in land use, which in turn affect increases or decreases in forest cover [63]. In addition, economic and population growth will increase the demand for forest products and may reduce forest resources [44].

6. Conclusions

Based on the theoretical basis of the resource curse and economic growth, this article attempts to verify the impact of urban forest resource abundance on economic development in the YRD region from the perspective of renewable resources. The research results of this paper provide a reference for the coordinated development of resources and the economy.

The influence of forest resource abundance on economic development in YRD region presents a U-shaped non-linear relationship, and the phenomenon of the resource curse will evolve into resource welfare with the development of society. In addition, the abundance of forest resources can directly promote the improvement of green total factor productivity. However, economic development may inhibit the forest growth, mainly because of the high cost of land in the YRD region. Therefore, there is only a one-way causal relationship between forest resources and economic performance, while economic performance has no feedback effect on forest resources [64].

The impact of forest resource abundance on economic development is affected by many factors, including the initial level of the economy, carbon sequestration potential, energy consumption, urbanization level, science and technology, education level, and other factors. All act together on economic growth.

Although forest resources play a very important role in the critical period of economic transformation and development, we cannot achieve green growth and carbon reduction at the expense of economic development. Economic development and urbanization can go hand in hand with forest development but requires sound management models and additional measures [65].

The YRD region should further adjust and optimize their industrial structure. Each region should implement the industrial dislocation development strategy according to its advantages, give full play to the agglomeration effect and scale effect, and realize the integrated development of modern service and advanced manufacturing industries. Much more attention should be paid to the agglomeration of high-quality talents, capital, and resources. Urbanization is still regarded as a driving force for the upgrading and transformation of the economic structure. Urbanization should gradually realize coordinated development with the economy and gradually transform the extensive development mode into the intensive economic growth mode. In addition, technological innovation is very important to realize the economic development of the YRD region. Especially in the era of the digital economy, increasing investment in scientific research can effectively improve the level of economic development [66]. Meanwhile, the government should moderately intervene in economic development.

However, the article does have certain flaws. For example, the measure of forest resource abundance is relatively simple on how to manage forest resources efficiently. This is mainly due to the lack of statistical data and methods of forest resources and technology. In future research, forest resources should be refined and classified to analyze better the impact of forest resource abundance on the economy and specific paths and then provide useful suggestions for policymakers.

Author Contributions: Conceptualization, Q.Z.; methodology, Q.Z.; formal analysis, Q.Z.; investigation, D.T.; data curation, Q.Z.; writing—original draft preparation, Q.Z.; writing—review and editing, V.B. and D.T.; visualization, Q.Z.; supervision, D.T.; All authors have read and agreed to the published version of the manuscript.

Funding: This study was funded by the National Social Science Foundation of China (Grant No. 21BJY085) and the Soft Science Project of Jiangsu Provincial Department of Science and Technology (Grant No. BR2021003).

Data Availability Statement: The data presented in this study are available on request from the corresponding author. The data are not publicly available due to other unpublished articles.

Conflicts of Interest: The authors declare no conflict of interest.

References

1. Wang, J.; Feng, L.; Palmer, P.I.; Liu, Y.; Fang, S.; Bösch, H.; O'Dell, C.W.; Tang, X.; Yang, D.; Liu, L.; et al. Large Chinese land carbon sink estimated from atmospheric carbon dioxide data. *Nature* **2020**, *586*, 720–723. [CrossRef]
2. Wu, X.; Tian, Z.; Kuai, Y.; Song, S.; Marson, S.M. Study on spatial correlation of air pollution and control effect of development plan for the city cluster in the Yangtze River Delta. *Socioecon. Plann. Sci.* **2021**, *83*, 101213. [CrossRef]
3. Xia, C.; Zheng, H.; Meng, J.; Li, S.; Du, P.; Shan, Y. The evolution of carbon footprint in the yangtze river delta city cluster during economic transition 2012-2015. *Resour. Conserv. Recycl.* **2022**, *181*, 106266.
4. Zhang, Y.; Cao, W.; Zhang, K. Mapping Urban Networks through Inter-Firm Linkages: The Case of Listed Companies in Yangtze River Delta, China. *J. Geosci. Environ. Prot.* **2020**, *08*, 23–36. [CrossRef]
5. Angrist, J.D.; Kugler, A.D. Rural windfall or a new resource curse? coca, income, and civil conflict in Colombia. *Rev. Econ. Stat.* **2008**, *90*, 191–215. [CrossRef]
6. Dauvin, M.; Guerreiro, D. The Paradox of Plenty: A Meta-Analysis. *World Dev.* **2017**, *94*, 212–231. [CrossRef]
7. Mather, A.S.; Needle, C.L.; Fairbairn, J. Environmental Kuznets Curves and Forest Trends. *Geography* **1999**, *84*, 55–65.
8. Lantz, V. Is there an Environmental Kuznets Curve for clearcutting in Canadian forests? *J. For. Econ.* **2002**, *8*, 199–212. [CrossRef]
9. Mbatu, R.S.; Otiso, K.M. Chinese economic expansionism in Africa: A theoretical analysis of the environmental Kuznets Curve Hypothesis in the Forest Sector in Cameroon. *African Geogr. Rev.* **2012**, *31*, 142–162. [CrossRef]
10. Viet, V.Q. Gross regional product (GRP): An introduction. *United Nations Stat. Div. Retrieved August* **2010**, 15–17. Available online: https://unstats.un.org/unsd/economic_stat/China/background_paper_on_GRP.pdf (accessed on 30 August 2022).
11. Naidoo, R. Economic growth and liquidation of natural capital: The case of forest clearance. *Land Econ.* **2004**, *80*, 194–208. [CrossRef]
12. Pak, M.; Türker, M.F.; Öztürk, A. Total economic value of forest resources in Turkey. *African J. Agric. Res.* **2010**, *5*, 1908–1916. [CrossRef]
13. Damette, O.; Delacote, P. The Environmental Resource Curse Hypothesis: The Forest Case. *INRA Lab. d'Economie For.* **2009**, *4*, hal-01189378. [CrossRef]
14. Li, Y.; Mei, B.; Linhares-Juvenal, T. The economic contribution of the world's forest sector. *For. Policy Econ.* **2019**, *100*, 236–253. [CrossRef]
15. Huang, Y.; Raza, S.M.F.; Hanif, I.; Alharthi, M.; Abbas, Q.; Zain-ul-Abidin, S. The role of forest resources, mineral resources, and oil extraction in economic progress of developing Asian economies. *Resour. Policy* **2020**, *69*, 101878. [CrossRef]
16. Hogarth, N.J.; Belcher, B.; Campbell, B.; Stacey, N. The Role of Forest-Related Income in Household Economies and Rural Livelihoods in the Border-Region of Southern China. *World Dev.* **2013**, *43*, 111–123. [CrossRef]
17. Lu, J.; Li, M. Research on the Development of Forest Recreation Industry Based on SWOT and AHP Model-Take the Forest Recreation Industry of Xiying Street in Jinan City as an Example. *IOP Conf. Ser. Earth Environ. Sci.* **2022**, *966*. [CrossRef]
18. Song, M.; Xie, Q.; Tan, K.H.; Wang, J. A fair distribution and transfer mechanism of forest tourism benefits in China. *China Econ. Rev.* **2020**, *63*, 101542. [CrossRef]
19. van Kooten, G.C.; Nijnik, M.; Bradford, K. Can carbon accounting promote economic development in forest-dependent, indigenous communities? *For. Policy Econ.* **2019**, *100*, 68–74. [CrossRef]

20. Zhang, Y.; Zhang, T.; Zeng, Y.; Cheng, B.; Li, H. Designating National Forest Cities in China: Does the policy improve the urban living environment? *For. Policy Econ.* **2021**, *125*, 102400. [CrossRef]
21. Swamy, L.; Drazen, E.; Johnson, W.R.; Bukoski, J.J. The future of tropical forests under the United Nations Sustainable Development Goals. *J. Sustain. For.* **2018**, *37*, 221–256. [CrossRef]
22. Zhang, Y.; Chen, J.; Han, Y.; Qian, M.; Guo, X.; Chen, R.; Xu, D.; Chen, Y. The contribution of Fintech to sustainable development in the digital age: Ant forest and land restoration in China. *Land use policy* **2021**, *103*, 1–9. [CrossRef]
23. Miller, D.C.; Rana, P.; Nakamura, K.; Irwin, S.; Cheng, S.H.; Ahlroth, S.; Perge, E. A global review of the impact of forest property rights interventions on poverty. *Glob. Environ. Chang.* **2021**, *66*, 102218. [CrossRef]
24. Cheng, S.H.; MacLeod, K.; Ahlroth, S.; Onder, S.; Perge, E.; Shyamsundar, P.; Rana, P.; Garside, R.; Kristjanson, P.; McKinnon, M.C.; et al. A systematic map of evidence on the contribution of forests to poverty alleviation. *Environ. Evid.* **2019**, *8*, 1–22. [CrossRef]
25. Zhang, Q.; Brouwer, R. Is China Affected by the Resource Curse? A Critical Review of the Chinese Literature. *J. Policy Model.* **2020**, *42*, 133–152. [CrossRef]
26. Liu, Z.; Yao, S.; Liu, Y. Forest resource curse based on spatial panel modeling. *Resour. Sci.* **2015**, *37*, 379–390.
27. Gibson, J. Forest Loss and Economic Inequality in the Solomon Islands: Using Small-Area Estimation to Link Environmental Change to Welfare Outcomes. *Ecol. Econ.* **2018**, *148*, 66–76. [CrossRef]
28. Aslan, A.; Altinoz, B. The impact of natural resources and gross capital formation on economic growth in the context of globalization: Evidence from developing countries on the continent of Europe, Asia, Africa, and America. *Environ. Sci. Pollut. Res.* **2021**, *28*, 33794–33805. [CrossRef]
29. Caravaggio, N. Economic growth and the forest development path: A theoretical re-assessment of the environmental Kuznets curve for deforestation. *For. Policy Econ.* **2020**, *118*, 102259. [CrossRef]
30. Hao, Y.; Xu, Y.; Zhang, J.; Hu, X. Relationship between forest resources and economic growth: Empirical evidence from China. *J. Clean. Prod.* **2019**, *214*, 848–859.
31. Ahmed, Z.; Asghar, M.M.; Malik, M.N.; Nawaz, K. Moving towards a sustainable environment: The dynamic linkage between natural resources, human capital, urbanization, economic growth, and ecological footprint in China. *Resour. Policy* **2020**, *67*, 101677. [CrossRef]
32. Li, T.; Wang, Y.; Zhao, D. Environmental Kuznets Curve in China: New evidence from dynamic panel analysis. *Energy Policy* **2016**, *91*, 138–147. [CrossRef]
33. Zhao, H.; Uchida, E.; Deng, X.; Rozelle, S. Do Trees Grow with the Economy? A Spatial Analysis of the Determinants of Forest Cover Change in Sichuan, China. *Environ. Resour. Econ.* **2011**, *50*, 61–82. [CrossRef]
34. García, V.R.; Caravaggio, N.; Gaspart, F.; Meyfroidt, P. Long-and short-run forest dynamics: An empirical assessment of forest transition, environmental kuznets curve and ecologically unequal exchange theories. *Forests* **2021**, *12*, 431. [CrossRef]
35. Culas, R.J. Deforestation and the environmental Kuznets curve: An institutional perspective. *Ecol. Econ.* **2007**, *61*, 429–437. [CrossRef]
36. Culas, R.J. REDD and forest transition: Tunneling through the environmental Kuznets curve. *Ecol. Econ.* **2012**, *79*, 44–51. [CrossRef]
37. Choumert, J.; Combes Motel, P.; Dakpo, H.K. Is the Environmental Kuznets Curve for deforestation a threatened theory? A meta-analysis of the literature. *Ecol. Econ.* **2013**, *90*, 19–28. [CrossRef]
38. Ajanaku, B.A.; Collins, A.R. Economic growth and deforestation in African countries: Is the environmental Kuznets curve hypothesis applicable? *For. Policy Econ.* **2021**, *129*, 102488. [CrossRef]
39. Koop, G.; Tole, L. Is there an environmental Kuznets curve for deforestation? *J. Dev. Econ.* **1999**, *58*, 231–244. [CrossRef]
40. Zhang, X.; Huang, G.; Liu, L.; Zhai, M.; Li, J. Ecological and economic analyses of the forest metabolism system: A case study of Guangdong Province, China. *Ecol. Indic.* **2018**, *95*, 131–140. [CrossRef]
41. Foster, A.D.; Rosenzweig, M.R. Economic growth and the rise of forests. *Q. J. Econ.* **2003**, *118*, 601–637. [CrossRef]
42. Benedek, Z.; Fertő, I. Does economic growth influence forestry trends? An environmental Kuznets curve approach based on a composite Forest Recovery Index. *Ecol. Indic.* **2020**, *112*, 106067. [CrossRef]
43. Aguilar, F.X.; Wen, Y. Socio-economic and ecological impacts of China's forest sector policies. *For. Policy Econ.* **2021**, *127*, 102454. [CrossRef]
44. Zhang, K.; Song, C.; Zhang, Y.; Zhang, Q. Natural disasters and economic development drive forest dynamics and transition in China. *For. Policy Econ.* **2017**, *76*, 56–64. [CrossRef]
45. Shi, L.; Zhao, S.; Tang, Z.; Fang, J. The changes in China's forests: An analysis using the forest identity. *PLoS ONE* **2011**, *6*, e20778. [CrossRef]
46. Tang, X.; Guan, X.; Lu, S.; Qin, F.; Liu, X.; Zhang, D. Examining the spatiotemporal change of forest resource carrying capacity of the Yangtze River Economic Belt in China. *Environ. Sci. Pollut. Res.* **2020**, *27*, 21213–21230. [CrossRef] [PubMed]
47. Cheng, S.; Xu, Z.; Su, Y.; Zhen, L. Spatial and temporal flows of China's forest resources: Development of a framework for evaluating resource efficiency. *Ecol. Econ.* **2010**, *69*, 1405–1415. [CrossRef]
48. Cropper, M.; Griffiths, C. The interaction of population growth and environmental quality. *Am. Econ. Rev.* **1994**, *84*, 250–254.
49. Wang, J.; Xin, L.; Wang, Y. Economic growth, government policies, and forest transition in China. *Reg. Environ. Chang.* **2019**, *19*, 1023–1033. [CrossRef]

50. Kim, J.S.; Lee, T.J.; Hyun, S.S. Estimating the economic value of urban forest parks: Focusing on restorative experiences and environmental concerns. *J. Destin. Mark. Manag.* **2021**, *20*, 100603. [CrossRef]

51. Meng, L.; Huang, B. Shaping the Relationship Between Economic Development and Carbon Dioxide Emissions at the Local Level: Evidence from Spatial Econometric Models. *Environ. Resour. Econ.* **2018**, *71*, 127–156. [CrossRef]

52. Anselin, L. *Spatial Econometrics: Methods and Models*; Springer Science & Business Media: Dordrecht, The Netherlands, 1988; ISBN 978-90-481-8311-1.

53. Elhorst, J.P. *Spatial Econometrics From Cross-Sectional Data to Spatial Panels*; Springer: New York, NY, USA, 2014; ISBN 978-3-642-40339-2.

54. Kang, D.; Dall'erba, S. Exploring the spatially varying innovation capacity of the US counties in the framework of Griliches' knowledge production function: A mixed GWR approach. *J. Geogr. Syst.* **2016**, *18*, 125–157. [CrossRef]

55. Bernard, A.B.; Durlauf, S.N. Interpreting tests of the convergence hypothesis. *J. Econom.* **1996**, *71*, 161–173. [CrossRef]

56. Castro-Magnani, M.; Sanchez-Azofeifa, A.; Metternicht, G.; Laakso, K. Integration of remote-sensing based metrics and econometric models to assess the socio-economic contributions of carbon sequestration in unmanaged tropical dry forests. *Environ. Sustain. Indic.* **2021**, *9*, 100100. [CrossRef]

57. LeSage, J. *The Theory and Practice of spatial Econometrics*; Department of Economics, University of Toledo: Toledo, OH, USA, 1999.

58. Li, J.; Xu, B. Curse or Blessing: How Does Natural Resource Abundance Affect Green Economic Growth in China? *Econ. Res. J.* **2018**, *139*, 151–167.

59. Li, L.; Deng, D.; Zhang, D.; Yang, W. Analysis on Socio-economic Determinants of Forest Fragmentation in Beijing-Tianjin-Hebei Region. *For. Econ.* **2021**, *04*, 5–16. [CrossRef]

60. Wang, S.; Liu, C.; Wilson, B. Is China in a later stage of a U-shaped forest resource curve? A re-examination of empirical evidence. *For. Policy Econ.* **2007**, *10*, 1–6. [CrossRef]

61. Gong, Z.; Gu, L.; Yao, S.; Deng, Y. Effects of bio-physical, economic and ecological policy on forest transition for sustainability of resource and socioeconomics development. *J. Clean. Prod.* **2020**, *243*, 118571. [CrossRef]

62. Lin, B.; Ge, J. Carbon sinks and output of China's forestry sector: An ecological economic development perspective. *Sci. Total Environ.* **2019**, *655*, 1169–1180. [CrossRef] [PubMed]

63. Barbier, E.B.; Delacote, P.; Wolfersberger, J. The economic analysis of the forest transition: A review. *J. For. Econ.* **2017**, *27*, 10–17. [CrossRef]

64. Xie, M.; Irfan, M.; Razzaq, A.; Dagar, V. Forest and mineral volatility and economic performance: Evidence from frequency domain causality approach for global data. *Resour. Policy* **2022**, *76*, 102685. [CrossRef]

65. Choi, Y.; Lim, C.H.; Chung, H.I.; Kim, Y.; Cho, H.J.; Hwang, J.; Kraxner, F.; Biging, G.S.; Lee, W.K.; Chon, J.; et al. Forest management can mitigate negative impacts of climate and land-use change on plant biodiversity: Insights from the Republic of Korea. *J. Environ. Manage.* **2021**, *288*, 112400. [CrossRef]

66. Li, L.; Hao, T.; Chi, T. Evaluation on China's forestry resources efficiency based on big data. *J. Clean. Prod.* **2017**, *142*, 513–523. [CrossRef]

Article

Exploring the Measurement of Regional Forestry Eco-Efficiency and Influencing Factors in China Based on the Super-Efficient DEA-Tobit Two Stage Model

Junlan Tan [1], Xiang Su [1] and Rong Wang [2,*]

[1] School of Economics and Management, Jiangsu University of Science and Technology, Zhenjiang 212100, China
[2] Business School, Nanjing Xiaozhuang University, Nanjing 211171, China
*   Correspondence: rongwang@njxzc.edu.cn

Abstract: This paper adopts the super-efficient DEA (data envelopment analysis) model to measure the forestry eco-efficiency (FECO) of 30 Chinese provinces and cities from 2008 to 2021, and then introduces the Tobit model to explore the influencing factors of FECO to better understand the sustainable development level of forestry. It draws the following conclusions: (1) The average value of FECO in China is 0.504, which is still at a low level, and the FECO of each region has significant regional heterogeneity; the provinces with higher FECO are mainly concentrated in the eastern region, while the FECO of the central and western regions is lower; (2) In terms of the main factors affecting FECO in China, the regression coefficients of market-based environmental regulations are significantly positive in the national, eastern and central regions, while they are significantly negative in the western region. The coefficient of impact of scientific research funding investment on forestry industry eco-efficiency is negative and shows a significant promotion effect in the eastern region, but the elasticity coefficient in the central and western regions is negative but not significant. Economic development has a positive but insignificant effect on FECO, with the eastern region showing a positive correlation, while the central and western regions are insignificant. Industrial structure has a significant negative effect on FECO in the national, eastern and central regions, but the effect of industrial structure on FECO in the western region is not significant. The effect of foreign direct investment on FECO was negative for the national, central and western regions, but the central region did not pass the significance test, while the eastern region reflected a significant promotion effect.

Keywords: forestry; eco-efficiency; sustainable development; forestry resources

Citation: Tan, J.; Su, X.; Wang, R. Exploring the Measurement of Regional Forestry Eco-Efficiency and Influencing Factors in China Based on the Super-Efficient DEA-Tobit Two Stage Model. *Forests* **2023**, *14*, 300. https://doi.org/10.3390/f14020300

Academic Editors: Noriko Sato and Tetsuhiko Yoshimura

Received: 19 January 2023
Revised: 1 February 2023
Accepted: 1 February 2023
Published: 3 February 2023

1. Introduction

From a worldwide perspective, energy, resource and ecological emergencies have become serious challenges. Economist Daly [1] pointed out that man-made capital has become relatively abundant and the constraints to human development have been transformed into scarce natural capital. Zeb et al. [2] argued that the depletion of natural resources and the adverse effects of environmental degradation, including desertification, drought, land degradation, lack of freshwater resources and loss of biodiversity are increasing and worsening the various challenges facing humanity. Forestry has multiple benefits and can provide practical solutions to address these issues. The green development of forestry and the improvement of forestry eco-efficiency have become the focus of attention of the international community [3,4].

As an important basic industry of the national economy, forestry plays a pivotal role in human societal development. For a long time, China's forestry economic growth mode has been characterized by a large number of factor inputs; this is a crude economic growth mode relying on the massive consumption of resources to promote [5] it. Under the current conditions of resource scarcity in China, this type of crude forestry economic growth will encounter a "limit"; when growth approaches its "limit" or people are aware of this "limit", economic growth needs to improve the use efficiency of input factors of production to break through this "limit", that is, to increase the share of the total factor productivity contribution to forestry economic growth. According to relevant economic theories, the growth of output can depend on factor inputs as well as technological progress [6,7]. The development of China's forestry industry is so rapid that it is obvious to measure it numerically, but the question remains: what is the real quality of this forestry development? In today's increasingly prominent resource and environmental problems, ensuring rapid economic development while reducing environmental pollution and resource waste is not only a sizable challenge for social development, but also one of the urgent tasks facing the future development of forestry. Therefore, a scientific and objective discussion of the quality of forestry development is conducive to a better understanding of what will be the sustainable development level of forestry in the future [8–10]. Schaltegger and Sturm [11] proposed an eco-efficiency concept, which has been widely recognized by scholars in various fields. Specifically in the field of forestry, eco-efficiency is a measure of the level of sustainable development of forestry, and its objective is to measure whether forestry can minimize environmental pollution and resource consumption while achieving multifaceted value in a comprehensive manner, with the premise of ensuring the quality of forest products [12]. Focusing on FECO improvement is conducive to advocating for modern forestry development through the concept of coordinated and sustainable development of society, economy and ecology [13,14].

The following are this paper's main contributions: (1) Existing studies mainly study productivity and explore its efficiency from the perspective of production, while the issue of forest efficiency from an ecological perspective has rarely been addressed; this paper considers the inputs and undesired outputs and measures regional forest eco-efficiency in China, which is a useful supplement to the existing studies; (2) Existing studies mainly focus on specific regions and lack comparative studies between regions; this paper provides theoretical support by comparing the east, central and west regions; (3) Exploring forest eco-efficiency in the context of sustainable development and proposing corresponding policies can provide a reference for regional green development. Therefore, this paper adopts the super-efficient DEA model to measure the forestry eco-efficiency of 30 Chinese provinces and cities (except Hong Kong, Macao, Taiwan and Tibet) for 14 years from 2008 to 2021, and then introduces the Tobit model to analyze the influencing factors for forestry eco-efficiency in order to better understand the sustainable development level of forestry and provide some theoretical support for the sustainable development of modern forestry in the future.

2. Literature Review

2.1. Connotation of Eco-Efficiency in Forestry

Eco-efficiency is the ratio of the value of economic and social development (GDP) to the physical amount of resource and environmental consumption, which creatively connects three indicators: resources, economy and environment. It emphasizes the unity of environmental and economic benefits, establishes a connection between the best economic and environmental goals and provides a link between regions; it provides an important evaluation tool for the sustainable development of regions and industries, and becomes an important reference for policy makers [15,16]. This concept is generally accepted and widely used in the evaluation and research of the forest industry by domestic and foreign scholars, such as Wu and Zhang [17]. They considered that the eco-efficiency of the forest industry refers to the direct impact on the forest's ecological environment caused by the inputs of technological improvement and environmental management, such as air quality improvement and wastewater treatment, in order to ameliorate the ecological impact brought by the forestry industry, i.e., industrial efficiency in the forestry industry. According to Chen et al. [18], FECO is a measure of the sustainable development of forestry, and its objective is to measure whether forestry can minimize environmental pollution and resource consumption while ensuring the quantity and quality of forest products and achieving its multifaceted value in a comprehensive manner. Hong et al. [19] pointed out that for forestry to be sustainable, coordination between input and output factors should be ensured. Zheng and Yin [20] argued that FECO facilitates a win-win input and output relationship between economic and social benefits.

2.2. Measurement of FECO

Academics have analyzed FECO from different perspectives, and the differences are mainly reflected in the research methods and regional selection. In the process of measuring FECO, the stochastic frontier approach (SFA), the ratio method and the DEA evaluation model are mainly used in China. For example, Can et al. [21] used the SFA method to measure the ecological efficiency of plain forestry and analyzed the degree of ecological contribution of farmland forest networks and small forests to total agricultural output value, plantation output value and livestock output value. Weerawat et al. [22] used the ratio method to analyze several rubber plantation areas in Thailand and made recommendations. Most studies using the DEA evaluation, such as Li et al. [23], measured the forestry input-output efficiency using the DEA model, but the output indicators they selected did not reflect the social benefits of forestry; in addition, the only reflected the situation in 2006. Tian and Xu [24] measured the forestry input-output efficiency from 1993 to 2010 in China. At the local level, Lai and Zhang [25] used the super-efficient DEA model to measure the forestry input-output efficiency of 21 cities in Guangdong province and ranked and classified them; Zang et al. [26] measured the technical efficiency of forestry production and its influencing factors from the perspective of forestry production in Chongqing. Zhang and Kang [27] selected the super-efficient DEA-Tobit model to measure forestry production efficiency in 30 Chinese provinces from 2000 to 2014, and pointed out that its efficiency is lower, with significant spatial divergence. Luo et al. [28] also used the DEA model to measure the forestry efficiency of each province and further analyzed the spatial differences using the Gini coefficient and Moran index and concluded that forestry efficiency increased year by year but was still low in general. Tian et al. [29] used the C2R-DEA model (DEA model without considering returns to scale) and the SE-DEA model to measure forestry production efficiency in China. In 2012, Tian et al. [29] analyzed and measured the input-output efficiency of forestry using the super-efficient DEA model.

2.3. Influencing Factors of FECO

In recent years, many scholars have used DEA measurements to evaluate regional FECO and constructed an index system to study the key factors affecting FECO in the region. For example, Zheng et al. [30] examined the impact of industrial agglomeration on FECO through an econometric model and found that the level of industrial agglomeration and eco-efficiency in provinces with high levels of forestry industry development in China showed different degrees of increase during the study period. Chen and Geng [31] pointed out that the economy of scale efficiency of the forestry industry would be affected by property rights factors, local government intervention behavior, market concentration, market entry barriers and externalities. Jiang et al. [32] argued that the industrialization of forestry is an objective requirement, an inevitable result of the market economy and an effective way to improve efficiency in modern forestry. Hou [33] believed that the economic productivity or environmental productivity of forestry can be improved through the division of labor and specialization. Li and Tang [34] considered that the rationalization of the industrial structure is an effective means to improve the efficiency of forestry. Tian and Xu [24] measured the input-output efficiency of forestry in China from 1993 to 2010 and analyzed its influencing factors in depth. Zang et al. [26] measured the technical efficiency of forestry production and its influencing factors from the perspective of forestry production of Chongqing foresters. Zhang and Xiong [35] concluded that forestry ecological construction and protection, forest ecological compensation and forestry prevention and control inputs have negative and significant effects on the comprehensive FECO.

The studies that have been conducted lack an analysis of forestry eco-efficiency, and in the field of forestry in China, a unified definition of forestry eco-efficiency has not been clearly established. This is mainly because, although the development of forestry has the dual attributes of economic value and ecological value, there are problems such as long investment recovery cycles and economic externalities, which, together with the emergence of natural and man-made disasters, do not guarantee that forestry development can meet people's expectations. In the process of pursuing the economic value of forestry, people gradually ignore the ecological benefits. Therefore, this paper measures the forestry eco-efficiency of 30 provinces in China from the provincial panel data of 30 provinces from 2008–2021 (due to limited data sources, Hong Kong, Macao, Taiwan and Tibetan areas are not considered for the time being), and further explores the regional differences and spatial and temporal characteristics of forestry eco-efficiency. On this basis, the factors affecting forestry eco-efficiency are explored to better protect the diversity of forestry ecosystems. While the forestry ecological economy is developing continuously, we should also pay attention to the protection of forestry ecology, so as to realize people's expectations for a better ecological environment in the near future.

3. Methods

The main methods currently used for eco-efficiency evaluation are the single ratio method, the indicator system method and the model analysis (DEA) method. Although the single ratio method is relatively simple, it cannot distinguish the impact of different environments, give the optimal set of ratios or give decision makers flexibility in their choices. The indicator system method can reflect the level of development and coordination of social, economic and natural subsystems, but it requires artificial weighting, so there is often subjective interference. The use of the model analysis method can better compensate for the above shortcomings, so the data envelopment analysis method is widely used in the empirical study of eco-efficiency.

3.1. FECO Measurement Method

3.1.1. Super-Efficient DEA

Data envelopment analysis is a typical class of nonparametric analysis, referred to as DEA analysis, which was proposed by Charnes et al. [36]. They studied the optimization of resource allocation in the production process by evaluating the relative efficiency ratio of inputs and outputs within each decision unit. In the production process, the ratio between the input quantity of resource consumption and the output quantity of the product determines the production efficiency value within the decision unit, and the weighting of the input and output values can be used to analyze multiple input and output problems. For the study of green issues, in the process of constructing the DEA model, pollution factor indicators and negative ecological indicators can be classified as non-desired outputs, and the DEA model will be based on the "asymmetric" curve measurement of each type of output to accurately estimate the eco-economic efficiency value. The DEA model will accurately estimate the eco-economic efficiency values based on "asymmetric" curve measures for each type of output, and through projection analysis, ensure sufficient output and appropriate inputs while strictly controlling the amount of undesired output [37]. The estimation of desired and undesired outputs in the production process is achieved by means of the radial measure of the curve and the inverse of the curve measure, respectively. The DEA is the curve radial measure, without function expressions and without hypothesis testing [38].

1. Introduction of CCR-DEA model

The CCR-DEA model was proposed by Charnes et al. [39]; it is an input-oriented DEA model. The CCR model is the most basic DEA model with n DMU ($i = 1, 2, \ldots , n$), which satisfies the assumption of homogeneity and are all comparable. Each DMU has the same t inputs, and the input vector is:

$$x_i = (x_{1i}, x_{2i}, \ldots, x_{ti})^T, i = 1, 2, \ldots, n . \tag{1}$$

Each DMU has the same s outputs, then there is an output vector of

$$y_i = (y_{1i}, y_{2i}, \ldots, y_{si})^T, i = 1, 2, \ldots, n, \tag{2}$$

i.e., each DMU has the same t inputs and s output types. Where x_{ji} denotes the input quantity of the i-th DMU to the j-th DMU, and y_{ji} denotes the output quantity of the i-th DMU to the j-th DMU. In order to integrate all the DMUs in a uniform way, each input and output needs to be assigned a value, so that the weight vectors of the input and output are

$$v = (v_1, v_2, \ldots, v_j)^T, \tag{3}$$

$$u = (u_1, u_2, \ldots, u_r)^T, \tag{4}$$

where v_j denotes the j-th type of input weight and u_r denotes the rth type of output weight. At this point, the combined value of the i-th decision unit input is $\sum_{j=1}^{t} v_j x_{ji}$, and the combined value of the output is $\sum_{r=1}^{s} u_r y_{ri}$, so the efficiency evaluation index of each DMUi is defined as

$$h_i = \frac{\sum_{r=1}^{s} u_r y_{ri}}{\sum_{j=1}^{t} v_j x_{ji}}, \tag{5}$$

$$\begin{cases} maxh_{i0} = \frac{\sum_{r=1}^{s} u_r y_{ri0}}{\sum_{j=1}^{t} v_j x_{ji0}} \\ \frac{\sum_{r=1}^{s} u_r y_{ri}}{\sum_{j=1}^{t} v_j x_{ji}} \le 1, i = 1, 2, \ldots, n \\ v = (v_1, v_2, \ldots, v_j)^T \ge 0 \\ u = (u_1, u_2, \ldots, u_r)^T \ge 0 \end{cases}. \tag{6}$$

2. Introduction to the BCC-DEA model

The CCR-DEA model based on the premise of constant returns to scale is slightly different from reality; the returns to scale are not constant in actual economic production activities, so Banker et al. [39] proposed an extension of the DEA analysis with the variable returns to scale model, namely the BBC-DEA model analysis method.

The assumptions in the BBC-DEA model are variable payoffs of scale, and also a decomposition of technical efficiency into two components. The BBC-DEA model incorporates convexity constraints, and its input-oriented model is:

$$\begin{cases} max\left(u^T y_0 + \mu_0\right) \\ s.t. \omega^T x_i - \mu^T y_i \ge 0 \\ \omega^T x_0 = 0 \\ \omega \ge 0, \mu \ge 0, i = 1, 2, \ldots n \end{cases}, \tag{7}$$

where μ_0 denotes the payoff of scale; ω is the portfolio ratio of effective decision units

3. Introduction of the super-efficient DEA analysis method

In the analysis results of the traditional DEA model, there will be multiple effective decision units at the same time, i.e., there are multiple efficiency values of 1, and thus the individual decision units cannot be ranked according to their high efficiency values. For this situation, then some scholars proposed Super-efficient DEA [40,41]. The Super-Efficiency DEA analysis method allows the simultaneous efficient decision units to be further analyzed and all DMUs reordered. With variable payoffs to scale, the super-efficient DEA is as follows:

$$\begin{cases} min\left[\theta - \varepsilon\left(\hat{e}^T S^- + e^T S^+\right)\right] \\ s.t. \sum_{i=1}^{n} \lambda_i x_i + S^- = \theta x_0 \\ \sum_{i=1}^{n} \lambda_i y_i + S^+ = y_0 \\ S^- \ge 0, S^+ \ge 0, \lambda_i \ge 0, i = 1, 2, \ldots n \end{cases}, \tag{8}$$

where λ represents the slack variable, and the next step introduces the slack variable S^+ and the residual variable S^-; θ is the efficiency value required in the paper.

3.1.2. Variable Selection and Data Sources

1. Input variables

One of the most critical aspects of the super-efficient DEA evaluation model is the selection of input-output indicators and samples because they have a great impact on the final evaluation results. According to Pedraja—Chaparro and Salinas-Jimenez [42], for ensuring the credibility of the model's results, the inputs and outputs must be highly correlated. Forestry inputs should be the various factors of production that are invested to promote forestry development. In this paper, we study the ecological efficiency of forestry, therefore, the eco-forestry input and output index system should not only include the resource consumption factor component, but also the environmental pollution factor.

Forestry labor input: Labor input is the first influencing factor. As in other means of production, human capital is also a of means of production and plays an important role in production activities. In his study of economics, Quaker pointed out that people are the primary factor in the process of wealth creation. Labor input affects technical efficiency through both quantity and quality. In this paper, labor input refers only to the quantity of labor input, using the year-end number of forestry system employees to measure forestry labor input.

Forestry capital input: This paper uses the amount of forestry fixed asset investment in the current year as the capital input variable. Investment in forestry fixed assets refers to the monetary sum of man-hours or costs required for the construction and acquisition of forestry fixed assets in forestry production. Technical efficiency improvement of forestry production by forestry fixed assets is continuous and long-term and has an important role in forestry production. The impact of forestry fixed investment on eco-efficiency is not only expressed in the scale of investment, but also in the stability and continuity of the sources of forestry investment which will also improve FECO. For this reason, forestry fixed asset investment can well reflect funds and the smoothness of funding channels impact on FECO.

Forestry ecological input: It is expressed by the forestry ecological construction input.
Forestry livelihood inputs: It is expressed by the forestry infrastructure inputs.

2. Output variables

Desired output: This is the total output value of the forestry industry in each province, converted to constant prices in 2008 according to the CPI index, in order to exclude the influence of price changes

Non-desired output: Considering that it is difficult to characterize the environmental pressure by a single indicator, this paper uses "three waste" emissions from each region to represent pollution emissions: wastewater emissions from the secondary forestry industry in each province for wastewater, gas emissions from the secondary forestry industry in each province for gas emissions, and solid waste emissions from the secondary forestry industry in each province for solid waste. Specific variables and descriptive statistics are shown as follows in Table 1.

3.2. FECO Impact Factor Analysis Method

3.2.1. DEA Two-Stage Method and Tobit Model

The DEA two-stage method is an advanced model derived to further explore the influencing factors and their degree of influence on efficiency values. In the first stage, the efficiency value of each DMU is calculated using the DEA method; in the second stage, the efficiency value calculated in the first stage is used as the dependent variable, and the factors influencing efficiency are used as the independent variables for regression analysis. The efficiency values measured by the DEA model are between 0 and 1, and the direct use of ordinary least squares (OLS) would cause bias and inconsistency problems. Therefore, the Tobit model is used in this paper and the maximum likelihood estimation method is applied for regression analysis [43].

Table 1. Results of input-output indicator selection and descriptive statistics.

Indicators	Category	Indicator Measure	Unit	Mean	Standard Deviation
Input Indicators	Forestry Labor Input	Number of employees in forestry system at the end of the year	people	125,626.98	143,349.54
	Forestry Capital Inputs	Forestry fixed asset investment	million yuan	67,432.332	74,093.23
	Forestry Livelihood Inputs	Forestry infrastructure investment	million yuan	19,225.982	21,923.442
	Forestry Ecological Inputs	Forestry ecological construction investment	million yuan	11,228.453	12,664.021
Output Indicators	Desired Output	Total output value of forestry industry by province	million yuan	48,958.332	51,027.816
	Non-desired Output	Forestry secondary industry wastewater emissions	million tons	44.985	29.384
		Forestry secondary industry solid waste emissions	million tons	499,683.331	987,350.119
		Forestry secondary industry waste gas emissions	million tons	998.672	884.013

The sample data used for the analysis are obtained from the China Forestry Statistical Yearbook and the provincial Statistical Yearbooks from 2008–2021.

The Tobit model is suitable for regressions where the dependent variable is restricted. The standard form of Tobit model is:

$$Y_i = \begin{cases} \beta_0 + \sum_{t=1}^{n} \beta_t x_t + \mu_t, \ if \ \beta_0 + \sum_{t=1}^{n} \beta_t x_t + \mu_t > 0 \\ 0, \ if \ \beta_0 + \sum_{t=1}^{n} \beta_t x_t + \mu_t \leq 0 \end{cases}, \tag{9}$$

where Y_i denotes the actual dependent variable, i.e., the efficiency value of the ith DMU; x_t denotes the independent variable; β_0 denotes the constant term; β_t denotes the regression coefficient of the independent variable; μ_t denotes the independent error disturbance term and obeys a normal distribution of $N(0, \sigma^2)$.

3.2.2. Variable Selection and Data Sources

1. Explained variables

FECO. In this paper, the forestry eco-efficiency values of different regions in China will be selected as the explanatory variables, the statistical data of 30 provincial administrative regions in China will be selected as the support, and the values measured using super-efficiency DEA will be analyzed from the national sample and different regions, respectively.

2. Explanatory variables

Industrial structure (IS). With the total ban on commercial logging in state-owned forest areas, forest areas that once had tree harvesting and wood product processing as their main economic development model must adjust their industrial structure in order to steadily develop their economy while ensuring sustainable forest development. At present, different forestry bureaus in forest areas have different dominant industries, and therefore their main mode of operation is not the same. The primary forestry industry's main mode of operation is the above-mentioned planting and breeding of forest products; the secondary forestry industry's main mode of operation is the processing of wood products; and the tertiary forestry industry's main mode of operation is the vigorous development of forest tourism and the service industry in the process. In the process of development and operation of these industries, there is bound to be spatial spillover and spatial benefit of forestry economic development, therefore the industrial structure is also one of the influencing factors of FECO. To make this study dynamic, this paper uses the proportion of forest industry output value to total forestry output value to express [44].

Economic development (PGDP). This paper uses GDP per capita to reflect the sum of the value of products in a country or region in that year, which to a certain extent shows the degree of economic development of a region. Economic growth can promote scientific and technological progress, and a high level of economic development will lead to a higher demand for technology by the inhabitants of that place, which will lead to a higher eco-efficiency of local production through a demand-induced effect [45]. In economically developed areas, urbanization is bound to develop rapidly, which leads to the rapid development of the real estate and decoration markets, which provide good opportunities for the development of the forest products market. Meanwhile, people's demand for a good ecological environment is increasing. Urban gardening, greening, forest tourism, tourism forestry and other industries will develop rapidly. In summary, a good economic environment can promote forestry industry development as well as the optimization and upgrading of the forestry structure.

Foreign direct investment (FDI). Based on the analysis of FDI technology spillover channels, from the perspective of the competition effect, foreign enterprises usually enter China's forestry field by virtue of their capital, technology, scale and other advantages. On the one hand, they will introduce advanced technology, which is conducive to improving forest enterprises' ability to introduce, absorb and apply new technologies and promote technological progress; these can be understood as improvements brought by competition, which are positive spillover effects. On the other hand, forestry FDI generally does not choose to invest in greenfield sites, but is more involved in competition for existing forestry resources and markets, leading to a reduction in the market share of local enterprises, and coupled with China's preferential policies for foreign investors, it is easy to squeeze out domestic capital, thus inhibiting the improvement of TFP, which is a negative spillover effect. Whether the positive or negative effect is larger or smaller has not been determined, but to some extent, it can explain the negative effect of FDI on total factor productivity in Chinese forestry [46]. In this paper, the ratio of forestry FDI to total forestry output is used to measure FDI intensity.

Investment in scientific research (TI). Forestry is one of the most special industries in the national economic system, with very strong benefit spillovers. In addition to providing a large number of products and services for people's lives and social production, there are also generally sizable ecological benefits. In addition, forestry has an important role in ensuring the basic livelihood of forest farmers in forest areas and in rural revitalization. The most prominent feature of forestry is the long cycle, which determines a longer scientific research cycle, so the adequacy and stability of scientific research funding is particularly important [47]. In the paper, the ratio of research funding to GDP is chosen to represent the intensity of research funding.

Market-based environmental regulation (ER). Environmental regulation includes command-and-control and market-based. The former mainly includes setting environmental standards, pollutant emission standards and technical standards; the latter mainly includes establishing an emission charging or taxation system and an emission rights trading system [48]. In this paper, we use the proportion of the total emission fee levied by the forestry industry to the total regional forestry output value.

Due to the constraints of the eco-efficiency index calculation formula, the eco-efficiency values measured by the DEA method range between 0 and 2. In this case, if the traditional ordinary least squares (OLS) method is used to analyze the actual effect of each influencing factor on eco-efficiency, the results will be biased and inconsistent. In order to avoid the bias, the Tobit model was selected to analyze the factors influencing forestry eco-efficiency. Based on the above variable selection, the Tobit model was constructed as follows:

$$\text{FECO}_{it} = c + \beta_1 IS_{it} + \beta_2 PGDP_{it} + \beta_3 FDI_{it} + \beta_4 TI_{it} + \beta_5 ER_{it} + \varepsilon_{it}, \tag{10}$$

where β is the elasticity coefficient of variables; FECO_{it} is the explanatory variable of FECO; IS_{it}, $PGDP_{it}$, FDI_{it}, TI_{it} and ER_{it} are the industrial structure, economic development, foreign direct investment, investment in scientific research and environmental regulation, respectively. The data are mainly obtained from the statistical yearbooks of each province.

4. Results

4.1. Results of Regional FECO Measurement in China

This paper is based on the input-output panel data from 2008 to 2021 (Tibet is not included in the analysis because of incomplete data), and the FECO of each province is measured based on EMS software. Considering the existence of regional heterogeneity, this paper divides the 30 provinces (regions and municipalities) into three major regions: east, central and west, with 11 provinces (municipalities) in the east, 8 provinces in the central and 11 provinces in the west. The measurement results are shown in Table 2.

Table 2. Results of forestry eco-efficiency by province in China from 2008 to 2021.

	Region	2008	2009	2010	2011	2012	2013	2014	2015	2016	2017	2018	2019	2020	2021	Mean
Eastern	Beijing	0.546	0.557	0.588	0.589	0.593	0.603	0.611	0.614	0.628	0.637	0.655	0.672	0.679	0.699	0.619
	Tianjin	0.446	0.458	0.468	0.473	0.487	0.499	0.513	0.522	0.546	0.578	0.611	0.624	0.633	0.664	0.537
	Hebei	0.412	0.435	0.458	0.478	0.495	0.513	0.543	0.577	0.593	0.623	0.658	0.711	0.733	0.766	0.571
	Liaoning	0.233	0.245	0.257	0.267	0.278	0.288	0.291	0.294	0.301	0.311	0.323	0.326	0.335	0.345	0.292
	Shanghai	0.712	0.733	0.756	0.788	0.804	0.825	0.866	0.889	0.934	1.114	1.246	1.289	1.355	1.467	0.984
	Jiangsu	0.678	0.698	0.716	0.775	0.812	0.866	0.894	0.911	0.928	0.945	0.968	0.977	1.112	1.213	0.892
	Zhejiang	0.544	0.576	0.593	0.627	0.655	0.689	0.705	0.727	0.756	0.789	0.822	0.856	0.894	0.912	0.725
	Fujian	0.711	0.722	0.746	0.768	0.789	0.844	0.868	0.898	0.843	0.889	0.945	0.978	1.117	1.232	0.882
	Shandong	0.678	0.698	0.735	0.787	0.856	0.893	0.944	0.969	1.014	1.115	1.213	1.236	1.278	1.301	0.980
	Guangdong	0.588	0.598	0.611	0.624	0.645	0.698	0.723	0.759	0.798	0.834	0.876	0.912	0.944	0.976	0.756
	Hainan	0.445	0.457	0.466	0.483	0.496	0.523	0.546	0.569	0.597	0.611	0.635	0.657	0.689	0.788	0.569
Central	Shanxi	0.344	0.367	0.379	0.389	0.428	0.458	0.481	0.544	0.566	0.589	0.595	0.604	0.621	0.677	0.503
	Jilin	0.214	0.223	0.236	0.247	0.255	0.267	0.273	0.288	0.301	0.311	0.314	0.318	0.334	0.358	0.281
	Heilongjiang	0.211	0.215	0.226	0.236	0.245	0.266	0.274	0.279	0.286	0.293	0.301	0.311	0.315	0.325	0.270
	Anhui	0.364	0.375	0.398	0.423	0.447	0.459	0.476	0.498	0.533	0.576	0.589	0.627	0.655	0.725	0.510
	Jiangxi	0.675	0.688	0.698	0.734	0.779	0.823	0.839	0.856	0.873	0.895	0.988	1.118	1.128	1.216	0.879
	Henan	0.445	0.457	0.476	0.489	0.511	0.537	0.566	0.599	0.613	0.633	0.644	0.658	0.687	0.712	0.573
	Hubei	0.388	0.398	0.421	0.433	0.445	0.476	0.498	0.511	0.533	0.556	0.588	0.615	0.633	0.697	0.514
	Hunan	0.387	0.391	0.398	0.412	0.425	0.452	0.487	0.513	0.539	0.546	0.568	0.59	0.644	0.662	0.501
Western	Neimenggu	0.113	0.116	0.145	0.152	0.167	0.183	0.196	0.207	0.218	0.234	0.258	0.276	0.283	0.311	0.204
	Guangxi	0.168	0.178	0.199	0.216	0.236	0.256	0.289	0.318	0.334	0.357	0.387	0.399	0.416	0.498	0.304
	Chongqing	0.435	0.447	0.457	0.487	0.498	0.512	0.523	0.535	0.546	0.567	0.572	0.583	0.591	0.657	0.529
	Sichuan	0.301	0.311	0.319	0.326	0.334	0.347	0.358	0.366	0.378	0.399	0.411	0.417	0.424	0.446	0.367
	Guizhou	0.244	0.259	0.266	0.279	0.299	0.311	0.326	0.338	0.348	0.362	0.379	0.398	0.411	0.423	0.332
	Yunnan	0.211	0.223	0.236	0.247	0.268	0.287	0.295	0.311	0.325	0.346	0.357	0.377	0.398	0.443	0.309
	Shanxi	0.339	0.347	0.354	0.367	0.377	0.379	0.382	0.393	0.402	0.411	0.422	0.431	0.438	0.447	0.392
	Gansu	0.099	0.102	0.116	0.136	0.142	0.148	0.152	0.166	0.173	0.179	0.188	0.193	0.225	0.258	0.163
	Qinghai	0.111	0.114	0.121	0.125	0.132	0.137	0.144	0.165	0.179	0.189	0.193	0.216	0.226	0.268	0.166
	Ningxia	0.278	0.299	0.311	0.314	0.325	0.329	0.334	0.356	0.366	0.371	0.374	0.388	0.389	0.394	0.345
	Xinjiang	0.087	0.094	0.104	0.113	0.124	0.135	0.147	0.159	0.172	0.188	0.197	0.214	0.223	0.257	0.158
National mean		0.380	0.393	0.408	0.426	0.445	0.467	0.485	0.504	0.521	0.548	0.576	0.599	0.627	0.671	0.504

From the above table, it can be seen that from 2008–2021, the integrated efficiency of Shandong and Shanghai exceeds or approaches 1.0, and the integrated efficiency of Jiangsu, Jiangxi and Fujian approaches 0.9. Shandong, Shanghai and Jiangsu show an upward trend, indicating that the above regions have been effective in adjusting the balance between forestry production output and environmental pollution. The comprehensive efficiency of provinces such as Inner Mongolia, Heilongjiang, Jilin and Liaoning is below 0.3. Although the forestry resource stock and forestry output values of these provinces are high, the input consumption in forestry production is too large, which makes their comprehensive efficiency hover at a low level. The low ecological overall efficiency of forestry in Beijing and Tianjin is due to the innate condition of their limited forestry resource stocks. The comprehensive ecological efficiency of forestry in western provinces such as Xinjiang, Gansu and Qinghai is less than 0.2. These regions are constrained by topography, resources and other factors that do not release the scale benefit; the desertification of land is serious, which affects forestry production and leads to low comprehensive efficiency.

The average FECO values of the three major economic regions in the country in the 14-year period are, in descending order, the eastern, central and western regions. The mean value FECO in the eastern region is significantly higher than the national average and the rest of the provinces and cities, except Hebei, have relatively high efficiency. Hebei province is a special case and in the preferred position to undertake the transfer of high pollution industries from Beijing and Tianjin, resulting in its low FECO. Beijing, because of its special location and position as a political and economic center, has a low level of FECO due to its inherent condition of limited forestry resource stock; generally speaking, the forestry input and output of the eastern provinces and cities are more reasonable. The FECO in the central region is basically above 0.5, except for Jilin and Heilongjiang. Shanxi Province, with a forest cover of 20.5% in 2016, has relatively few forestry resources and mainly focuses on coal energy production, and pays less attention to the development and utilization of forestry resources, resulting in a backward production technology level; this leads to a low FECO. Hunan Province, with abundant forestry resources, has a low FECO level in the early stage due to the model of exchanging resources for economic development, and then in the process of undertaking industrial transformation, actively adjusts the strategic structure for industry. In the process of undertaking industrial transformation, Hunan Province actively adjusted its strategic structure for industrial upgrading and transformation, and FECO reached above 0.5 after 2015. The overall FECO value in the western region is low, but Chongqing's FECO is much higher than other western regions due to its special geographical location and economic level in the west.

4.2. Analysis of Factors Influencing FECO

4.2.1. Multicollinearity Test

Before establishing the model, it is necessary to judge whether there is multicollinearity among variables. If there is multicollinearity, the partial least squares model is the best choice, hence it is necessary to conduct the diagnosis of covariance. The higher the variance inflation factor is, the more serious the multicollinearity is, and there is a positive correlation between them. Usually, the largest variance inflation factor among all independent variables is used as an indicator of multicollinearity. If the value is greater than 10, it means that the corresponding independent variables are approximately linearly combined with other variables in the linear regression analysis. This will affect the least squares estimation and the accuracy and reliability of the linear regression analysis cannot be guaranteed, i.e., the multicollinearity is serious and not suitable for linear regression analysis. Based on this, SPSS was used to verify the results, and the results can be seen in Table 3.

Table 3. Results of multicollinearity test.

Method	IS	PGDP	ER	TI	FDI
VIF	3.44	4.56	1.47	3.55	2.47
1/VIF	0.291	0.219	0.680	0.282	0.405

Analyzing the multicollinearity among the influencing factors of FECO, the variance inflation factor among the independent variables has a maximum of 4.56, thus it can be determined that there is no significant correlation among the respective variables; hence, there is no multicollinearity problem. If the problem is serious, the least squares model cannot be chosen, and the construction of the Tobit model is a better choice.

4.2.2. Unit Root Test

Panel data may have variable data that are not smooth, thus affecting the authenticity and credibility of the final estimation results and creating the problem of pseudo-regression. In order to obtain more robust regression results, the unit root test is conducted on the panel data before parameter estimation. The more commonly used testing methods are the Im-Pesaran-Shin (IPS) test, the Augmented Dickey-Fuller (ADF) test and the Levin-Lin-Chu (LLC) test, whose original hypothesis is the existence of a unit root. To avoid the problem of inaccurate test results brought by a single test method, this paper integrates the above three test methods to verify whether there is a unit root in the panel data. If the above three methods are passed, this indicates that there is no unit root in the sample data. The results are shown in Table 4.

Table 4. Unit root test results.

Variables	LLC Test	IPS Test	ADF Test	Unit Root
IS	−94.1390 ***	0.9986	11.4808 ***	exist
PGDP	−4.2216 ***	3.1436	2.7853 ***	exist
ER	−5.4437 ***	−1.2346	5.7784 ***	exist
TI	18.2231	8.5521	−0.7622	exist
FDI	6.3242	9.0053	0.9843	exist
D.IS	−66.1453 ***	−2.3348 ***	9.0622 ***	not exist
D.PGDP	−33.4571 ***	−4.6782 ***	7.7768 ***	not exist
D.ER	−39.8953 ***	−5.12398 ***	7.0954 ***	not exist
D.TI	−9.6473 ***	−4.3346 ***	3.5562 ****	not exist
D.FDI	3.3456 ***	3.4516 ***	1.5674 ***	not exist

Note: ***, **** means $p < 0.05$ and $p < 0.01$, respectively.

As can be seen from Table 4, all variables fail to reject the original hypothesis, indicating the existence of a unit root for each variable, while the first-order difference terms of each variable can reject the original hypothesis at the 1% significance level. This indicates that there is no unit root in the first-order difference series of each variable, which in turn indicates that the panel data are first-order single integer and can be subjected to the next step of regression analysis.

4.2.3. Model Selection Test

There are two methods for model regression, i.e., the fixed-effects model and the random-effects model; Hausman's test is required to determine the choice of model. The original hypothesis model was selected as the random effects model, and the Hausman test results (in Table 5) were obtained through Eview 6.0.

Table 5. Hausman test results.

Test Summary	Chi-Sq.Statistic	Chi-Sq.d.f	Prob.
Cross-section random	77.1389	6	0.0000

From the analysis results, the p value is 0, so the original hypothesis is rejected, i.e., the original random effects model hypothesis is rejected, and the fixed effects model is adopted

4.2.4. Model Regression Results

According to the regression model and test method, the regressions were conducted in China as a whole, and the east, central and west to explore the effects of the same influencing factors on different regions; this is because regions face various differences in their development environment, their geography and their development level which are heterogeneous. The results can be seen in Table 6.

Table 6. Regression results.

Explanatory Variables	National	East	Centra	West
IS	−0.1499 **	−0.1543 ***	−0.0998 ***	−0.1987
PGDP	0.2102	0.1125 ***	0.0761	0.1024
ER	0.0495 **	0.1203 ***	0.0833 *	−0.0293 ***
TI	−0.0428 ***	0.0908 ***	−0.0765	−0.1239
FDI	−0.0765 ***	0.0345 ***	−0.1235 ***	−0.1897 ***

Note: *, **, *** means $p < 0.1$, $p < 0.05$ and $p < 0.01$, respectively.

The regression coefficients of market-based environmental regulation are significantly positive in the country, in the eastern and central regions and significantly negative in the west. Due to the significant differences in the marketization process between regions, some developed provinces in the east and central regions, most of which have entered the post-industrialization period, have more sound market mechanisms and a higher degree of marketization; they rely mainly on market incentives to solve environmental problems. The western region has relatively low environmental costs, and in order to promote regional economic development, it takes over some of the low-end forestry industries transferred from home and abroad. Some forestry enterprises take the payment of sewage charges as an excuse to continue polluting, and the expenditure on pollution control follows the marginal decreasing cost: the sewage charges paid are far from enough to cover the costs of environmental treatment and ecological restoration.

As a whole, investment in scientific research will promote FECO, but from the empirical perspective, the current impact coefficient of research funding input intensity on the eco-efficiency of the forestry industry in China is negative and the correlation is significant. This indicates that current investment in scientific research in China fails to meet the needs of the forestry industry to improve eco-efficiency. To a certain extent this hinders the development of forestry eco-safety, as well as the subsequent need to further improve the technological content and technological level of the forestry industry to make the development of modern forestry go in the direction of the circular economy and eco-efficiency improvement. The intensity of scientific research investment in the eastern region shows a significant promotion effect, which indicates that forestry industry development in the eastern region tends to be intensive; the eastern economy is developed, and sufficient investment in scientific research can meet the needs of regional forestry development, thus showing a promotion effect. Forestry industry development in the central and western regions is still based on the scale of rough growth, rather than high-tech orientation, which in the long run is important for forestry industry development and the improvement of FECO. However, the elasticity coefficient of central and western China is negative but not significant, indicating that the negative impact is not significant.

Economic development has a positive but insignificant effect on FECO. Forestry plantations and the forest industry itself are a more complete industrial chain. The higher the level of economic development, the more likely it is to make use of industrial policies, extend the industrial chain, acquire advanced resources such as technology and increase productivity. Forestry cannot be separated from forests, which have long regeneration cycles and extremely uneven distribution, and are also constrained by specific factors such as land. The higher the level of economic development, the stronger the demand for the multifunctional use of forests, i.e., the more wood, carbon sink and tourism use. The more prominent the multifunctional problems are, the more obvious the problem of wood resource supply constraints faced by forestry development. Therefore, even though the degree of economic development is high, it is difficult to obtain an effective allocation of capital and technology in the case of shortage of timber resources, which in turn restricts the improvement of total factor productivity. However, there are large differences among regions: the eastern region shows a positive correlation, indicating that the eastern region is more mature in the allocation of resources and technology and is qualitatively better than the central and western regions; hence, in the east it is significant to enhance eco-efficiency, while it is not significant in the central and western regions.

The industrial structure inhibits FECO. This means that the FECO of the regions with a high proportion of secondary forestry industry in China is relatively low at present, i.e., the ecological and environmental costs of increased output value in the forestry industry are large. Specifically, the negative effect of the eastern region is more significant, which may be due to the fact that the paper industry, as the main component of the forest industry, is mostly concentrated in the eastern region. The central region is the main concentration of China's wood processing industry, probably because it is a lightly polluting industry, so the impact of the industrial structure is less than in the east; the impact of industrial structure on FECO in the western region is not significant. However, with further optimization of the three industrial structures, it is expected that FECO will be gradually improved in the future as the proportion of tertiary industries in forestry increases.

The impact of external openness on regional FECOs varies greatly among regions. The elasticity coefficients of foreign direct investment in the eastern, central and western regions were 0.0345, −0.1235 and −0.1897, respectively, and all regions passed the significance level test except for the central region, which did not pass the significance test. It can be seen that the environmental access threshold of the forestry industry in the eastern region is higher than other regions, and the quality requirements of foreign businessmen are higher, thus reflecting a positive effect. Foreign enterprises have a "pollution transfer effect" on the forestry industry in the central and western regions, i.e., the pollution-intensive forestry industry undertaken by these regions aggravates the regional environmental pollution problem. However, the negative correlation at the national level indicates that foreign direct investment is still dominated by the pollution transfer effect, but it does not pass the significance test, indicating that this negative effect is tending to be insignificant, and that FECO can be improved by improving the quality of foreign investment introduced.

4.3. Discussion

This paper adopts the super-efficient DEA model to measure the forestry eco-efficiency of 30 Chinese provinces and cities (except Hong Kong, Macao, Taiwan and Tibet) for 14 years from 2008 to 2021, and then introduces the Tobit model to analyze the influencing factors of forestry eco-efficiency in order to better understand the sustainable development level of forestry. Existing studies mainly focus on studying forestry production efficiency. For example, Xu et al. [49], Wei [50] and Tan et al. [51] all measured the regional forestry production efficiency in China based on the Malmquist-DEA model without considering the impact of forestry industry development on the environment, thus this paper incorporates environmental factors to study forestry eco-efficiency; this is complementary to existing studies. Different regions showed significant heterogeneity with large regional differences in this study, which is consistent with the conclusions reached by most scholars. For example, Zheng and Yin [20] concluded that the eastern region has significantly higher eco-efficiency values than other regions and has been at a high efficiency level of about 0.9, while the western and central regions have slowly increased eco-efficiency values to approximately the 0.6 level. Wu and Zhang [52] also found the same trend for forestry eco-efficiency.

However, there are some scholars who came to different conclusions. Chen et al. [18] and Hong et al. [19] concluded that the western region is higher than the central region, and the opposite conclusion exists with this paper. This is probably because the above scholars used the traditional DEA model in evaluating eco-efficiency and came to the result that the maximum is one, and the ones greater than the data are all one. This paper uses the super-efficient DEA model to break through the efficiency boundary of one and can be greater than one, which more accurately reflects the actual value of the results and provides a more accurate depiction of the problem. Meanwhile, this paper uses the Tobit model to verify the influence of some economic variables on forestry eco-efficiency; it analyzes the influence on forestry eco-efficiency from environmental regulation, marketization, research funding, industrial structure and openness to the outside world, respectively, which are more variables than previous scholars had considered. The data used are also up-to-date, which can more accurately illustrate the influence relationship between variables. However, compared with other scholars, the research in this paper is dominated by linear relationship research, and does not explore the nonlinear relationship between variables; for example, Jiang et al. [53] verified the study of the threshold effect of environmental regulation on forestry eco-efficiency, which is the direction of future research for this paper, i.e., exploring nonlinear relationships between variables.

5. Conclusions and Implications

5.1. Conclusions

In this study, the super-efficient DEA model is used to measure the FECO of 30 provinces from 2008 to 2021, and then the Tobit model is introduced to explore the influencing factors of FECO. The conclusions are as follows.

1. The average value of FECO in the three major economic regions of the country over 14 years is 0.5586, which is still at a low level; this may be related to the rough development model of Chinese forestry. There is still more room for improvement when compared with developed countries. The FECO of each region has significant regional heterogeneity, and the provinces with higher FECO are mainly concentrated in the eastern region, while the FECO in the central and western regions is lower. However, FECO in the central region is higher than that in the western region.

2. From the viewpoint of the main factors affecting FECO in China, the regression coefficients of market-based environmental regulations are significantly positive in the national, eastern and central regions, while they are significantly negative in the western region mainly because the environmental costs in the western region are relatively low. There is no environmental regulation system in place in order to promote regional economic development and undertake the transfer of some low-end forestry industries at home and abroad; the intensity of environmental regulations is low, which means that the ecological efficiency of forestry in this type of regions has a negative impact. The impact coefficient of scientific research funding investment on forestry industry eco-efficiency is negative and shows a significant promotion effect in the eastern region; the elasticity coefficient in the central and western regions is negative but not significant, indicating that the negative impact is not significant. The level of economic development has a positive but insignificant effect on the eco-efficiency of China's forestry industry. However, it varies significantly across regions, with the eastern region showing a positive correlation, while the central and western regions are not significant. Industrial structure has a significant negative effect on forestry industry eco-efficiency in the national, eastern and central regions, but the effect of industrial structure on FECO in the western region is not significant. Foreign direct investment has a negative effect on FECO in the national, central and western regions, but the central region does not pass the significance test because the environmental access threshold of the forestry industry in the eastern region is higher than other regions, and the quality requirements of foreign investors are higher, thus reflecting a positive effect.

5.2. Recommendations

Rational utilization of foreign investment means taking advantage of forest areas. The use of foreign capital in forestry significantly affects regional ecological efficiency, and foreign capital is a major thrust for forestry industry development in terms of capital. The reasonable and effective use of foreign capital is conducive to improving regional forestry ecological efficiency. To this end, we should continue to develop the advantages of regional and forest resources and make reasonable use of foreign capital. Specific suggestions are as follows: (1) Broaden the channels of attracting foreign capital and diversify the sources of foreign capital; (2) Rationalize the use and distribution of foreign capital and optimize the utilization structure of foreign capital; (3) Create and establish local forestry characteristic brands, study and learn from foreign advanced forestry experience, and strengthen international exchange and cooperation in forestry; (4) Develop and implement realistic industrial environmental protection standards and guide enterprises in technological innovation; (5) Increase the research, development and application of industrial environmental protection technologies; (6) Promote advanced and mature production technologies to improve resource utilization efficiency.

Local governments should focus on the construction and improvement of environmental protection infrastructure and increase investment in the development of environmental protection technologies; in addition, they should strengthen environmental supervision and regulate the awareness and behavior of forestry enterprises in environmental protection with legal systems to promote FECO. At the same time, the ecological audit system should be implemented in the internal management process of forestry enterprises to incorporate ecological and environmental management into the daily production process of enterprises and to realize the supervision and evaluation of forestry production processes. A reduction in production costs and environmental pollution, and ultimately the enhancement of the comprehensive competitiveness of forest products in domestic and international markets, can be achieved through continuous technological innovation and management innovation of enterprises to improve the quality and grade of forest products.

There must be rational use of forestry resources and a change of forestry production methods. Whether between regions or provinces, the quantity of forestry resources invested, and the level of eco-efficiency are not absolutely the same. The average value of FECO is east > central > west. The rational use of resources and appropriate transformation of forestry production methods to enhance FECO are beneficial. Sending also fully illustrates the importance of rational use of resources and development of forestry according to local conditions. The economically-developed eastern major forestry provinces such as the provinces of Laoning and Fujian can vigorously develop the local forestry economy. The central provinces such as Jiangxi and Anhui, which are rich in forestry resources and have a low level of economic development, can make full use of their abundant forestry resources to vigorously develop low-carbon industries such as ecological forestry. The provinces and cities with scattered and low-scale forestry production methods in each region should change their active forestry development methods. Only in this way can the FECO be improved and forestry be placed on the road of sustainable development.

Each region should design differentiated environmental regulation tools for the forestry industry according to its ecological carrying capacity and the current situation of the FECO. With the aim of improving the FECO, targeted regulatory objectives and policy measures should be formulated according to the actual situation of the region. For most of the central and western regions with low FECO and backward development levels, more attention should be paid to the strength and manner of environmental regulation in future development, especially to prevent the ecological destruction and environmental damage caused by the westward migration of polluting industries from the developed eastern regions. The developed eastern regions with strong economic, financial and technological strength can consider further increasing the intensity of environmental regulations to stimulate enterprises to strengthen the research, development and application of new technologies in the field of ecology and environment. When formulating environmental regulation policies and measures, all regions should consider including the requirement to improve the eco-efficiency of forestry in their environmental policies.

There are several guidelines for reducing the "pollution paradise" effect. The "pollution paradise" effect has caused affected forestry enterprises in provinces and municipalities that have reformed their sewage charges to move their production lines to provinces and municipalities that have not reformed their sewage charges and to "export" their pollution to other provinces. This makes the effectiveness of the policy implementation greatly reduced and further places the pollution pressure on the forestry industry in the provinces that have not been reformed. Therefore, in order to reduce the "pollution paradise" effect, the relevant departments should introduce relevant policies to guide the expectations of local governments, so that the provinces that have not reformed the sewage levy standards will not give unconditional acceptance to high-pollution high-emission industries in economic development considerations. This will also enable the levy standard to be improved comprehensively, and the decline of eco-efficiency will be mitigated to a certain extent as the transfer of high-pollution and high-emission industries receives restrictions. On the other hand, the relevant departments should also guide the expectations of enterprises in the provinces where the reform is carried out so that they will not take any chances and transfer high-pollution and high-emission enterprises out of the country recklessly; this would make the implementation of the reform of the emission levy standard achieve the reduction in relevant pollutant emissions in the provinces where it is carried out.

5.3. Limitations and Prospects

1. There is no uniform definition in existing research for forestry eco-efficiency; hence, this paper provides a definition based on previous scholarly research and the actual situation of China's forestry development. This may be subjective and needs to be improved in the future;
2. The study of forestry eco-efficiency is relatively new and the literature of related studies is relatively small; thus, in terms of the selection of influencing variables, they can only be selected exploratively in combination with previous scholars' studies. Verifying the influence of more variables on forestry eco-efficiency in future studies is worthwhile;
3. Economic variables have spatial correlation, for example, forestry eco-efficiency improvement in this region may have an impact on forestry eco-efficiency in neighboring regions, but this paper does not consider spatial effects which may lead to bias in the research results; thus; the spatial effects of variables can be further explored in future studies.

Author Contributions: Conceptualization, J.T.; methodology, J.T.; software, J.T.; validation, X.S., formal analysis, X.S.; investigation, X.S.; resources, X.S.; data curation, J.T.; writing—original draft preparation, R.W.; writing—review and editing, R.W.; visualization, R.W.; supervision, R.W.; project administration, R.W.; funding acquisition, R.W., All authors have read and agreed to the published version of the manuscript.

Funding: The paper is supported by Social Science Foundation of Jiangsu Province [21GLD010].

Institutional Review Board Statement: Not applicable.

Informed Consent Statement: Not applicable.

Data Availability Statement: The data used to support the findings of this study are available from the corresponding author upon request.

Conflicts of Interest: The authors declare no conflict of interest.

References

1. Daly, H.E. Economics in a full world. *Sci. Am.* **2005**, *293*, 100–107. [CrossRef]
2. Zeb, R.; Salar, L.; Awan, U.; Zaman, K.; Shahbaz, M. Causal links between renewable energy, environmental degradation and economic growth in selected SAARC countries: Progress towards green economy. *Renew. Energy* **2014**, *71*, 123–132. [CrossRef]
3. Lundmark, R.; Lundgren, T.; Olofsson, E.; Zhou, W. Meeting challenges in forestry: Improving performance and competitiveness. *Forests* **2021**, *12*, 208. [CrossRef]
4. Banaś, J.; Utnik-Banaś, K.; Zięba, S.; Janeczko, K. Assessing the technical efficiency of timber production during the transition from a production-oriented management model to a multifunctional one: A case from Poland 1990–2019. *Forests* **2021**, *12*, 1287. [CrossRef]
5. Näyhä, A. Transition in the Finnish forest-based sector: Company perspectives on the bioeconomy, circular economy and sustainability. *J. Clean. Prod.* **2018**, *209*, 1294–1306. [CrossRef]
6. Baumgartner, R.J. Sustainable Development Goals and the Forest Sector—A Complex Relationship. *Forests* **2019**, *10*, 152. [CrossRef]

7. Nitoslawski, S.A.; Galle, N.J.; Van Den Bosch, C.K.; Steenberg, J.W. Smarter ecosystems for smarter cities? A review of trends, technologies, and turning points for smart urban forestry. *Sustain. Cities Soc.* **2019**, *51*, 101770. [CrossRef]

8. Verkerk, P.J.; Costanza, R.; Hetemäki, L.; Kubiszewski, I.; Leskinen, P.; Nabuurs, G.J.; Palahí, M. Climate-smart forestry: The missing link. *For. Policy Econ.* **2020**, *115*, 102164. [CrossRef]

9. Rehman, A.; Ma, H.; Ahmad, M.; Irfan, M.; Traore, O.; Chandio, A.A. Towards environmental Sustainability: Devolving the influence of carbon dioxide emission to population growth, climate change, Forestry, livestock and crops production in Pakistan. *Ecol. Indic.* **2021**, *125*, 107460. [CrossRef]

10. Neykov, N.; Krišťáková, S.; Hajdúchová, I.; Sedliačiková, M.; Antov, P.; Giertliová, B. Economic Efficiency of Forest Enterprises—Empirical Study Based on Data Envelopment Analysis. *Forests* **2021**, *12*, 462. [CrossRef]

11. Schaltegger, S.; Sturm, A. Ecological rationality: Approaches to design of ecology-oriented management instruments. *Die Unternehm* **1990**, *4*, 273–290.

12. Koskela, M.; Vehmas, J. Defining Eco-efficiency: A Case Study on the Finnish Forest Industry. *Bus. Strat. Environ.* **2012**, *21*, 546–566. [CrossRef]

13. Koskela, M. Measuring eco-efficiency in the Finnish forest industry using public data. *J. Clean. Prod.* **2015**, *98*, 316–327. [CrossRef]

14. Costa, M.P.; Schoeneboom, J.C.; Oliveira, S.A.; Vinas, R.S.; de Medeiros, G.A. A socio-eco-efficiency analysis of integrated and non-integrated crop-livestock-forestry systems in the Brazilian Cerrado based on LCA. *J. Clean. Prod.* **2018**, *171*, 1460–1471. [CrossRef]

15. Puertas, R.; Guaita-Martinez, J.M.; Carracedo, P.; Ribeiro-Soriano, D. Analysis of European environmental policies: Improving decision making through eco-efficiency. *Technol. Soc.* **2022**, *70*, 102053. [CrossRef]

16. Grassauer, F.; Herndl, M.; Nemecek, T.; Fritz, C.; Guggenberger, T.; Steinwidder, A.; Zollitsch, W. Assessing and improving eco-efficiency of multifunctional dairy farming: The need to address farms' diversity. *J. Clean. Prod.* **2022**, *338*, 130627. [CrossRef]

17. Wu, Y.Z.; Zhang, Z.G. The DEA-Tobit Model Analysis of Forestry Ecological Safety Efficiency and Its Influencing Factors—Based on the Symbiosis between Ecology and Industry. Resour. Environ. *Yangtze River Basin* **2021**, *30*, 76–86.

18. Chen, K.; Li, X.T.; Pu, H.L. Measurement of FECO and eco-productivity in China. *For. Econ. Issues* **2016**, *36*, 115–120.

19. Hong, M.Y.; Long, J.; Lou, L. FECO: Spatial and temporal characteristics and influencing factors—An empirical analysis based on data from 31 provinces in China. *Ecol. Econ.* **2022**, *38*, 91–97.

20. Zheng, Y.M.; Yin, S.H. An empirical study on eco-efficiency of forestry industry–analysis of panel data based on 15 provinces. *Collection* **2016**, *11*, 36–40.

21. Liu, C.; Yu, F.J.; Ren, H.C. Measurement and analysis of eco-efficiency of plains forestry: Huai'an City, Jiangsu Province as an example. *China Rural. Econ.* **2004**, *6*, 47–53.

22. Weerawat, O.; Thunwadee, T.S.; Kitikorn, C. Selection of the sustainable area for rubber plantation of Thailand by eco-efficiency. *Procedia Soc. Behav. Sci.* **2012**, *40*, 58–64.

23. Li, C.H.; Li, N.; Luo, H.Y. Analysis of forestry production efficiency and optimization path based on DEA method in China. *China Agron.* **2011**, *27*, 55–59.

24. Tian, S.Y.; Xu, W.L. Evaluation of China's forestry input-output efficiency based on DEA modeling. *Resour. Sci.* **2012**, *34*, 1944–1950.

25. Zhang, Z.H.; Luo, H. Path analysis of input-output efficiency optimization of forestry in Guangdong based on DEA method. *Guangdong Sci. Technol.* **2008**, *20*, 10–12.

26. Zang, L.Z.; Zhi, L.; Qi, X.M. Factors influencing the technical efficiency of forestry production of farmers in the Tianbao Project area: The case of Wulong, Chongqing. *J. Beijing For. Univ. Soc. Sci. Ed.* **2009**, *12*, 59–64.

27. Zhang, H.L.; Kang, X. Research on the spatial and temporal variation of forestry productivity in China and the influencing factors. *J. Northwest For. Coll.* **2017**, *32*, 301–305.

28. Luo, S.F.; Li, Z.L.; Li, R.R. Study on the spatial and temporal differences of forestry productivity and its influencing factors in China. *Arid. Zone Resour. Environ.* **2017**, *31*, 95–100.

29. Tian, S.Y.; Li, Y.; Dong, W. The evolution of forestry ecological economy, development path and practice model in China. *For. Econ.* **2017**, *39*, 71–76.

30. Zheng, Y.; Gao, C.; Lei, G. Empirical analyses of forestry industry agglomeration and eco-efficiency: Based on panel data test on 15 Provinces in China. *Econ. Geogr.* **2017**, *37*, 136–142.

31. Chen, X.H.; Geng, Y.D. Analysis of scale efficiency and influencing factors of forestry industry in Northeast state-owned forest area. *Economist* **2014**, *8*, 22–26.

32. Jiang, H.; Su, Z.Y.; Xian, F. Theoretical analysis of forestry industrialization. *World For. Res.* **1998**, *11*, 64–69.

33. Hou, Y.Z. The economic basis of forestry division of labor theory. *World For. Res.* **1998**, *11*, 1–8.

34. Li, H.; Tang, S.Z. A review of forest industry structure research. *World For. Res.* **2000**, *13*, 41–46.

35. Zhang, Y.; Xiong, X. An empirical analysis of eco-efficiency evaluation and influencing factors of Chinese forestry in the context of green development–based on DEA analysis. *J. Cent. South Univ. For. Sci. Technol.* **2020**, *40*, 149–158.

36. Charnes, A.; Cooper, W.; Lewin, A.Y.; Seiford, L.M. Data envelopment analysis theory, methodology and applications. *J. Oper. Res. Soc.* **1997**, *48*, 332–333. [CrossRef]

37. Rosano-Peña, C.; Pensado-Leglise, M.D.R.; Marques Serrano, A.L. Agricultural eco-efficiency and climate determinants: Application of dea with bootstrap methods in the tropical montane cloud forests of Puebla, Mexico. *Sustain. Environ.* **2022**, *8*, 2138852. [CrossRef]
38. Rosano-Peña, C.; Teixeira, J.R.; Kimura, H. Eco-efficiency in Brazilian Amazonian agriculture: Opportunity costs of degradation and protection of the environment. *Environ. Sci. Pollut. Res.* **2021**, *28*, 62378–62389. [CrossRef]
39. Charnes, A.; Cooper, W.W.; Rhodes, E. Measuring the efficiency of decision making units. *Eur. J. Oper. Res.* **1978**, *2*, 429–444. [CrossRef]
40. Banker, R.D.; Thrall, R.M. Estimation of returns to scale using data envelopment analysis. *J. Oper. Res.* **1992**, *62*, 74–84. [CrossRef]
41. Andersen, P.; Petersen, N.C. A Procedure for Ranking Efficient Units in Data Envelopment Analysis. *Manag. Sci.* **1993**, *39*, 1261–1264. [CrossRef]
42. Pedraja-Chaparro, F.; Salinas-Jimenez, J. An assessment of the efficiency of Spanish Courts using DEA. *Appl. Econ.* **1996**, *28*, 1391–1403. [CrossRef]
43. Tobin, J. Estimation of relationships for limited dependent variables. *Econom. J. Econom. Soc.* **1958**, *26*, 24–36. [CrossRef]
44. Su, L.; Ji, X. Spatial-temporal differences and evolution of eco-efficiency in China's forest park. *Urban For. Urban Green.* **2020**, *57*, 126894. [CrossRef]
45. Li, S.; Ren, T.; Jia, B.; Zhong, Y. The Spatial Pattern and Spillover Effect of the Eco-Efficiency of Regional Tourism from the Perspective of Green Development: An Empirical Study in China. *Forests* **2022**, *13*, 1324. [CrossRef]
46. Zhou, Y.; Kong, Y.; Wang, H.; Luo, F. The impact of population urbanization lag on eco-efficiency: A panel quantile approach. *J. Clean. Prod.* **2019**, *244*, 118664. [CrossRef]
47. Ning, Y.; Liu, Z.; Ning, Z.; Zhang, H. Measuring Eco-Efficiency of State-Owned Forestry Enterprises in Northeast China. *Forests* **2018**, *9*, 455. [CrossRef]
48. Pan, D. Effects of command-and-control and market-incentive environmental regulations on afforestation area: Quasi-natural experimental evidence from the county level in China. *Resour. Sci.* **2021**, *43*, 2026–2041.
49. Xu, W.; Feng, Y.; Bao, Q.F. Forestry productivity measurement and analysis of regional differences in China-inter-provincial panel data based on Malmquist-DEA model. *For. Econ.* **2015**, *37*, 85–88.
50. Wei, Y.N. Analysis of forestry production efficiency measurement in China based on DEA model. *Market Week* **2016**, *7*, 43–44.
51. Tan, Q.; Dou, Y.Q.; Li, Y. Study on the spatial and temporal evolution of forestry productivity in China. *Anhui Agron. Bull.* **2021**, *27*, 59–61+69.
52. Wu, Y.Z.; Zhang, Z.G. SBM—Malmquist measure and spatio-temporal characteristics analysis of ecological safety efficiency of forestry industry. *Sci. Technol. Manag. Res.* **2019**, *24*, 259–267.
53. Jiang, W.; Liu, J.C.; Hu, H. Study on the spatial and temporal evolution of forestry eco-efficiency and the threshold effect of environmental regulation in China. *J. Cent. South Univ. For. Technol.* **2020**, *40*, 166–174.

 *forests*

Article

Dynamic Evaluation of Coupling and Coordinating Development of Environments and Economic Development in Key State-Owned Forests in Heilongjiang Province, China

Xiangyue Liu [1], Muhammad Arif [2,*], Zhifang Wan [1,*] and Zhenfeng Zhu [1]

[1] College of Economics and Management, Northeast Forestry University, Harbin 150040, China
[2] Biological Science Research Center, Academy for Advanced Interdisciplinary Studies, Southwest University, Chongqing 400715, China
* Correspondence: muhammadarif@swu.edu.cn (M.A.); zhifang_wan@126.com (Z.W.)

Abstract: This study examines state-owned forest areas in Heilongjiang Province, China, and uses statistical data from 2011 to 2019 to evaluate the dynamic coupling and coordination relationship between the forest environment and economic development. The study aims to provide guidelines for the sustainable development of forest areas. The study concludes that: (1) There is a significant interaction between the environment and economic development, which manifests in coercion and restriction effects during the ecological construction and economic development processes. (2) The forest area environment in 2011–2019, within the coupling and coordination relationship with economic development, was generally of a high quality. (3) Forest environment construction achieved remarkable results in 2011–2019 and benefitted from China's new position on ecological restoration in key state-owned forest areas. (4) The economic development of forest areas after 2015 showed a lag, which restricted the level and progress in the coordinated development of the environment and the economies of the forest areas. (5) During the 14th Five-Year Plan period (2021–2025), the key state-owned forest areas still fully incorporated the strategic positioning of ecological protection and economic development coordination. This study provides countermeasures and suggestions to further improve the ecological and economic development of key state-owned forest areas.

Keywords: key state-owned forest area; system coupling; ecological construction; economic transformation; grey relational analysis (GRA)

Citation: Liu, X.; Arif, M.; Wan, Z.; Zhu, Z. Dynamic Evaluation of Coupling and Coordinating Development of Environments and Economic Development in Key State-Owned Forests in Heilongjiang Province, China. *Forests* **2022**, *13*, 2069. https://doi.org/10.3390/f13122069

Academic Editors: Noriko Sato and Tetsuhiko Yoshimura

Received: 6 November 2022
Accepted: 2 December 2022
Published: 4 December 2022

Publisher's Note: MDPI stays neutral with regard to jurisdictional claims in published maps and institutional affiliations.

1. Introduction

The international community still faces a structural shortage of resources when dealing with climate change and environment protection, despite the positive progress that has been made [1,2]. Some countries and regions, faced with the challenges of alleviating poverty, economic development, and ecological restoration, and resource protection, have to make difficult choices to achieve an effective balance between ecology and sustainable economic development [3,4]. It is a major topic that countries and regions, including China, continue to explore and study [5]. The 19th National Congress of the Communist Party of China called for modernization in which man and nature coexist in harmony along with greater efforts to protect ecosystems. Major projects need to be carried out to protect and restore important ecosystems, improve ecological security barriers, build ecological corridors and biodiversity protection networks, and improve the quality and stability of ecosystems. A good balance between the environment and economic development is an important part of promoting economic and social progress, as well as an important path for building a moderately prosperous society and realizing the goals of "five-in-one" socialism with Chinese characteristics. While China's economic development continues to improve and public welfare programs continue to make progress, the country still faces a series of ecological challenges [6].

Key forestry areas have supported the accumulation of original national capital and resources for social and economic development for a long time. These key state-owned forest areas have experienced a shift from timber production to ecological construction after entering the 21st century. Exploring the coupling and coordination relationship between the environment and economic development in key state-owned forest areas in the current period provides important content for the evaluation of the reform and transformation of state-owned forest areas. Through the clear judgment and positioning of the coupling and coordination relationship, this study summarizes the experiences gained from the long-term development and provide guidelines for the forest areas to achieve win–win cooperation in improving the environment and developing the forestry economy.

1.1. Research Literature

The dynamic relationship between ecological protection and economic development has attracted increasing attention from experts and scholars in China and globally [7–9]. Foreign scholars have studied the relationship between ecology and economy since 1936 and have revealed the internal law of coordination between environment and economy through different models [10–13]. Using mathematical and simulation models, the literature shows that the relationship between ecology and economy is a coupling relationship that coordinates, influences, and promotes each other [14,15].

Since 1980, domestic researchers have actively discussed the theory and practice of eco-cities from an ecological perspective and identified three systematic theories: social, economic, and natural [16]. Ren Jizhou [17] first applied the concept of system coupling to the field of ecological economy, followed by Liu et al. [18], Zhu and He [19], and Gui-lin Lei [20]. Chinese scholars have subsequently also applied system coupling to the field of agro-ecology. Studies have also used input–output models to investigate the environment and the level of economic development. Researchers examined the interaction between the coupling coordination model and the environment and coordinated economic development based on a time and space evolution analysis. Scholars used the grey relational analysis (GRA) model and system dynamics model to forecast the environment and analyze coordinated economic development in three aspects [21–23]. The relevant literature provides sufficient references and inspiration for exploring the coupling and coordinated development relationship between the environment and economic development in key state-owned forest areas in terms of providing technical tools and research paradigms. Wenjun Wang [24] explored ways to promote the sustainable development of rural economies during ecological construction. For forestry economic development and environmental protection, scholars such as Li Jieying [25], Sheng Wang [26], and Su Lizhuo [27], have suggested that forestry economic development plays an important role in forestry ecological protection and sustainable development in China.

This study analyzes the current status of forestry environment protection and economic development, and discusses China's forestry environment protection within its forestry economic development strategy to promote the sustainable development of China's forestry economy.

The Kuznets curve was created in the 1950s by economist Kuznets to demonstrate that income disparity increases with economic development. Panayotou [28] constructed a model with cross-sectional data to verify that forest resource consumption shows an inverted U-shape with per capita income growth. Cao et al. [29] argued that reducing the arc of the environmental Kuznets curve is of great significance for state forest conservation. Song et al. [30] claimed that the development process of state-owned forest areas goes through four stages of forest deficit sequentially: expansion, maintenance, reduction, and surplus, i.e., the process of gradual decoupling of forest resources from timber production. Li et al. [31] reasoned that measure such as industrial restructuring and investment can be taken to reduce the arc of the Kuznets curve and achieve the recovery of forest resources.

Many studies have explored ways for state-owned forest areas to stop the commercial harvesting of natural forests in terms of economic development [32,33]. Using the Yichun forest district in the Lesser Khingan Mountains as an example, studies have highlighted the necessity of strengthening policy support for state-owned forest areas after logging activities come to an end [34,35]. The complete cessation of commercial logging in natural forests presents various challenges that seriously affect the sustainable development of forest areas. Counter measures and suggestions have been put forward, such as increasing policy support at a national level and adjusting tax policies [36]. Once commercial logging in natural forests has stopped in key state-owned forest areas, the state should adjust the logging tax policy and use fiscal and tax levers to adjust and support industrial development with ecological construction as the main focus. Commercial logging of natural forests has stopped in key state-owned forest areas in Heilongjiang Province (KSOFAsHP) and maintaining ecological security and providing ecological products has become the primary strategic task [37].

Scholars have explored the topic by combining state-owned forest areas and the ecological economic system [38]. Under the guidance of sustainable development theory and system science theory, this paper analyzed the social and economic structure of state-owned forest areas in Heilongjiang Province and the causal feedback relationship among various factors, and then established a system dynamics model of social and economic development of state-owned forest areas in Heilongjiang Province. Geng and Zhang [39] analyzed the composition of forestry industries and ecosystems in the state-owned forest areas in northeast China. Some studies examined the coordination of forest eco-economic development based on system theory [25–27]. An evaluation model for the coordinated development of forest eco-economic systems has been developed based on the evaluation index system. The coordinated development of forest eco-economic systems in Heilongjiang Province has been empirically studied based on the evaluation index system [39].

On the coupling of state-owned forests and the eco-economic system, Zheng et al. [40] and Dong et al. [41] combined the slope correlation degree and grey theory (where white refers to complete certainty of information, black refers to complete uncertainty of information, and gray refers to uncertainty of information, i.e., grey theory). The principle is to seek the degree of similarity or dissimilarity of the trends between factors by using sample characteristics data and making full use of the small amount of data to propose a coupling degree model for the forestry industry and forest ecosystem. Zhang et al. [42] used a coupling coordination degree model to evaluate the coupling coordination level between economic development and environment in Yuhuan City. Tang et al. [43] quantitatively analyzed economic development and the environment in Shaanxi Province from 2008 to 2018 to explore the coupling and coordination relationship between the two and the main influencing factors of coupling and coordinated development to provide a reference for the coordinated and sustainable development of the economic environment in Shaanxi Province. Xie et al. [44] constructed a development level index for the economic development and environment subsystems of Qinghai Province using the entropy weight method. The coupling coordination degree model was used to quantitatively analyze the coupling coordination degree between economic development subsystem and environment subsystem and its dynamic changes, and evaluate the ecological restoration level of Qinghai Province.

1.2. Purpose of the Research

This study's research objectives have two aspects. First, we scientifically evaluate the coupling and coordination relationship between environment construction and economic development in key state-owned forest areas, and summarize the development rules of the forest eco-economic system in the dynamic coupling and coordination relationship. Second, based on a dynamic analysis of the coupling and coordination relationship, we identify the policy factors affecting the coordinated and healthy development of different forest systems to lay the foundation for inspiring forest areas to explore innovative development paths.

2. Research Area and Research Logic

2.1. Study Area

The study area refers to the state-owned forest areas in Heilongjiang Province in northeast China (Figure 1).

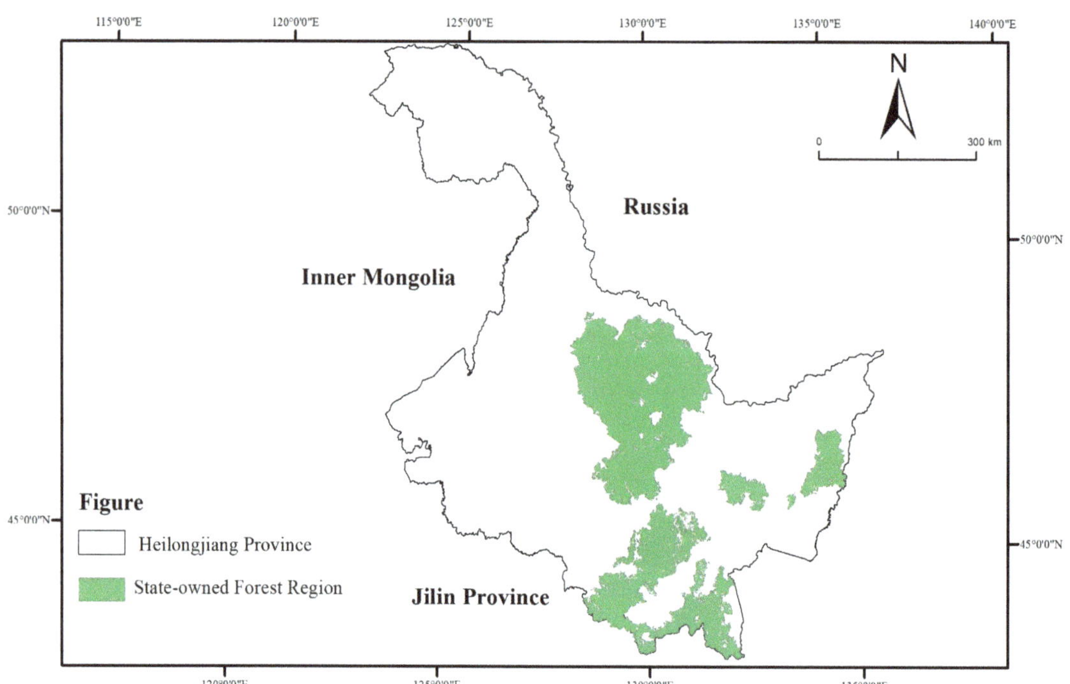

Figure 1. Regional map of state-owned forest areas in Heilongjiang Province, China.

Heilongjiang has a key state-owned forest region in northeast China. In 1998, it began to reposition itself from primarily timber production to a greater emphasis on ecological development. This new phase represented a functional change to environmental construction and economic and social development. In 2014, Heilongjiang Province, the largest key state-owned forest area in China, took the lead in a nationwide pilot project for the implementation of a new policy to completely stop the commercial logging of natural forest resources, following the International Union for Conservation of Nature policies [45]. This policy, which is China's strictest management and control policy of natural forest resources in history, posed an unprecedented challenge for the coordination between ecological protection and economic and social development in key state-owned forest areas.

It also provided an opportunity for theoretical research into the deep coupling and coordinated development of environments and economy in forest areas. After years of practice and exploration, the KSOFAsHP have made some progress in the process of coupling and coordination between environment and economic development. On the one hand, relying on the natural forest protection policy, the construction of national ecological function zones and sustained long-term investment in environment construction has achieved positive results. On the other hand, despite the closure of logging activities and the loss of subsistence industries, enforced forest economic transformation and the vigorous development of non-wood replacement industries has led a gradual recovery in economic growth. In this context, the forest environment and the interactive relationship between economic development and coupling coordination level on empirical research, focus on the collaboration between environment and economic development, the relationship between bound and limit characteristics for identification and dynamic evaluation, so as to other resources dependent region cooperation between environment and economic sustainable development to provide inspiration.

2.2. Research Logic

Using typical representatives of KSOFAsHP as the research object, this study conducts a quantitative identification and dynamic evaluation of the interaction and coordination relationship between forest areas' environment and economic development since 2011.

This study is based on the GRA, which identifies and examines the interactive and cooperative relationship between the forest environment and economic development. As viewed from a system perspective, the characteristics of the relationship between the environment and economic development are explored. Next, the degree of coupling coordination model was used to measure the level of coordinated development between the environment and economic development in forest areas, and the degree of coordination between them was obtained, and a dynamic analysis of its change trend was carried out. Last, combined with changes in the degree of coordination, the environment evaluation index and the economic development evaluation index in the coupled coordination degree model were used to deconstruct and analyze the factors or obstacles that drive the coupling coordination relationship between the environment and economic development in forest areas. The results are used to provide inspiration and ideas for the continuous optimization of the relationship.

3. Materials and Methods

3.1. Grey Relational Analysis Model

The basic idea of GRA is to judge whether the relationship between sequence curves is close according to the degree of similarity of their geometric shapes. The closer the curve, the greater the degree of correlation between corresponding sequences and in the case of the opposite, the smaller the degree of correlation [46]. This paper aims to examine the basic relationship of interaction and cooperation between environment and economic development in key state-owned forest areas using the GRA model. The specific steps of the analysis steps are:

1. Index forward;
2. Determine the analysis sequence, including the parent sequence (similar to the dependent variable Y, denoted as X here$_0$) and sub-series (similar to independent variable X, denoted as (X) here$_1$, X_2, ... , X_m));
3. Variable preprocessing (removing dimensional influence and reducing variable range to simplify calculation);
4. Define and calculate the GRA correlation coefficient;
5. Obtain the GRA correlation degree matrix.

Among them, the first three steps have been widely used in academic circles, and will not be repeated. The GRA correlation coefficients between different indicators are as follows:

$$y(x_0(k), x_i(k)) = \frac{a + \rho b}{|x_0(k) + x_i(k)| + \rho b} (i = 1, 2, \ldots, m; \ k = 1, 2, \ldots, n) \tag{1}$$

$$a = min_i min_k |x_0(k) - x_i(k)| \tag{2}$$

$$b = max_i max_k |x_0(k) - x_i(k)| \tag{3}$$

In the above equations, a is the minimum difference between two poles, b is the maximum difference between two poles; p is the resolution coefficient, $0 < p < 1$. If p is smaller, the difference between the correlation coefficients is larger, and the discrimination ability is stronger. Usually, p is 0.5. The GRA correlation degree matrix is finally obtained in Equation (1).

3.2. Coupling Coordination Degree Model

The basic idea of the coupling coordination degree model is to determine the relationship between the environment and economic development based on the coupling degree and coordination degree, and obtain the interaction relationship and coupling coordination development level between them [47]. The coupling degree reflects the degree of interaction between the environment and economic development, regardless of the advantages and disadvantages. The degree of coordination refers to the degree of benign coupling in the interaction, which reflects the quality of the coordination, and can determine whether the system functions of key state-owned forest areas promote each other at a high level or restrict each other at a low level.

The model of coupling coordination degree between environment and economic development can be presented as follows:

$$D = \sqrt{C \cdot T} \tag{4}$$

$$C = \left\{ \frac{f(x) \times g(y)}{\left[\frac{f(x)+g(y)}{2} \right]^2} \right\}^{1/2} \tag{5}$$

$$T = \alpha f(x) + \beta g(y) = \alpha \sum_{i=1}^{m} a_i x_i' + \beta \sum_{i=1}^{n} b_i y_i' \tag{6}$$

In the above equation, D is the coupling coordination degree between the environment and economic development in key state-owned forest areas, C is the coupling degree, and T is the comprehensive evaluation index of the coupling coordination development level between them. Where, $f(x)$ and $g(y)$ are the environment evaluation index and economic development evaluation index of key state-owned forest areas, respectively. In addition, in view of the objective reality of the functional orientation of key state-owned forest areas at the present stage, the coefficients α and β in Equation (6) are set as 0.6 and 0.4, respectively.

To effectively distinguish the coupling coordination level between the environment and economic development, the concept of coordination level is introduced. According to the quantitative relationship between $f(x)$ and $g(y)$, if $f(x) > g(y)$, it is an economic lag type T1. If $f(x) = g(y)$, it is T of the synchronous type of environment and lag type 2. If $f(x) < g(y)$, it is the environment lag type T3. The specific classification standards and types of coupling coordination degree are summarized in Table 1.

Table 1. Classification standards and types of coupling coordination degrees.

Coordination Layer	Coordination Index	Coordination Level	Coordination Type
	0.00–0.09	Extremely dysregulated recession class I_1	I_1T_1, I_1T_2, I_1T_3
Coordination recession class I	0.10–0.19	Severe dysregulation recession class I_2	I_2T_1, I_2T_2, I_2T_3
	0.20–0.29	Moderate dysregulation recession class I_3	I_3T_1, I_3T_2, I_3T_3
	0.30–0.39	Mild dysregulation recession class I_4	I_4T_1, I_4T_2, I_4T_3
Excessive class II	0.40–0.49	On the verge of dysregulation decline class II_1	II_1T_1, II_1T_2, II_1T_3
	0.50–0.59	Barely coordinated development class II_2	II_2T_1, II_2T_2, II_2T_3
	0.60–0.69	Junior coordination class III_1	III_1T_1, III_1T_2, III_1T_3
Coordination of class III	0.70–0.79	Intermediate coordination class III_2	III_2T_1, III_2T_2, III_2T_3
	0.80–0.89	Good coordination class III_3	III_3T_1, III_3T_2, III_3T_3
	0.90–1.00	Good coordination class III_4	III_4T_1, III_4T_2, III_4T_3

3.3. Evaluation Index System of the Relationship between Environment and Economic Development

This study examines the state-owned forests in Heilongjiang Province and its specific ecological protection practices and economic development results. In order to develop a comprehensive evaluation index system, this study follows scientific, systematic, typicality, operability, and comprehensive basic principles. For instance, it examines how the environment is related to economic development in Heilongjiang Province and develops a comprehensive evaluation index system based on this information [48]. The results are summarized in Table 2. Among them, the first-level indicators of the different systems are environment and economic development. The first-level indicators are further observed by the second- and third-level indicators. The environment construction of key state-owned forest areas is reflected by ecological construction and environmental protection, while economic development is reflected by economic scale and economic structure. Table 2 provides the basic index system for the subsequent empirical study of the GRA correlation degree and coupling coordination degree model. The weight of each index in the table is obtained according to the entropy weight method (in view of the extensive application of the entropy weight method, the process will not be described in detail).

Table 2. Comprehensive evaluation index system of the relationship between environment and economic development.

Quasi-Lateral Layer Index	Secondary Indicators	Level 3 Indicators	Index Description	Index of the Unit
Eco-environment X	Ecological construction X_1	Forest construction X_{11}	Forestation area of barren hills and sandy land in the forest area Barren mountains and sandy land afforestation area	ha
		Closing hillsides to facilitate afforestation X_{12}	In mountainous areas with suitable natural resources, the area of forest vegetation restored after the implementation of measures such as regular closure of mountains for reforestation and prohibition of land reclamation and grazing.	ha
		Forest administration and protection X_{13}	The area for which workers are paid to enter into a responsible system of forest protection	ha
		Silvicultural input X_{14}	Cumulative forestation and afforestation investment	10,000 yuan
		Forest management and protection input X_{15}	Actual investment in forest management and conservation	10,000 yuan
		Forest tending investment X_{16}	Actual investment in forest nurturing	10,000 yuan
	Environmental protection X_2	Environmental protection input X_{21}	Investment in environmental protection as a share of GDP	%
Economic development Y	Economic scale Y_1	Economic aggregate Y_{11}	Total output value of forest industry	10,000 yuan
		Economic growth Y_{12}	Growth rate of total output value of forest industry	%
		Fixed asset investment Y_{13}	Fixed asset investment in forestry as a share of GDP	%
		Labor compensation Y_{14}	Average wage of employees	yuan
	Economic structure Y_2	The share of secondary industry in GDP Y_{21}	The share of forest industry in GDP in the study area	%
		The Share of tertiary industry in GDP Y_{22}	The share of forest services in GDP in the study area	%
		Non-forest specific gravity Y_{23}	The share of non-forest and non-wood industries in GDP in the study area	%

3.4. Data Sources

The study's main data sources included:

1. China Forestry Statistical Yearbook (2012–2018), State Forestry Administration, China Forestry Press, 2012–2018.
2. State Forestry and Grassland Administration, China Forestry and Grassland Statistical Yearbook (2019–2020), China Forestry Press, 2019–2020.
3. National Bureau of Statistics, Heilongjiang Statistical Yearbook (2012–2020), China Statistics Press, 2012–2020.
4. Statistics published on the website of the State Forestry and Grassland Administration (www.forestry.gov.cn/) and others accessed on 15 July 2022.
5. Data from the Ninth National Forest Resources Inventory of China (China Forestry Network http//:www.forestry.gov.cn/) accessed on 21 July 2022.

Data translated with www.DeepL.com/Translator (free version) accessed on 1 August 2022 are from the China Forestry Statistical Yearbook issued by China Forestry Publishing House from 2012 to 2020 (China Forestry and Grassland Statistical Yearbook after 2018); the Heilongjiang Statistical Yearbook issued by China Statistics Press from 2012 to 2020; statistics published on the website of the National Forestry and Grassland Administration; and data from China's ninth National Forest Resources Inventory.

4. Results and Discussion

4.1. Analysis of the Stress and Limitation between the Environment and Economic Development

The correlation degree matrix of the interaction between environment and economic development indicators at various levels in the key state-owned forest areas of Heilongjiang Province from 2011 to 2019 was calculated using the GRA model (Table 3). Table 3 shows that the degree of correlation between the environment and the indicators of all levels of economic development was higher than 0.5, which is a medium correlation. First, the results confirm the basic hypothesis of an effective correlation between the environment and economic development in key state-owned forest areas. Further, by ranking the average value of the correlation degree among indicators, we obtained the interaction stress and limiting factors of the corresponding indicators.

Table 3. Grey relational analysis correlation degree matrix between the environment and economic development.

Indicators at Each Level		Ecological Construction (0.6158)						Environmental Protection (0.6551)	Average
		X_{11}	X_{12}	X_{13}	X_{14}	X_{15}	X_{16}	X_{21}	
Size of economy (0.6353)	Y_{11}	0.8736	0.4984	0.7264	0.6361	0.6308	0.6929	0.6807	0.6770
	Y_{12}	0.6201	0.5977	0.5794	0.5536	0.5600	0.5711	0.5877	0.5814
	Y_{13}	0.5590	0.6445	0.6962	0.5641	0.6398	0.5757	0.5589	0.6055
	Y_{14}	0.8960	0.4755	0.6848	0.6288	0.6091	0.7068	0.7383	0.6771
Economic structure (0.6029)	Y_{21}	0.5657	0.5642	0.5390	0.5740	0.5187	0.5071	0.6110	0.5543
	Y_{22}	0.8411	0.4618	0.6553	0.6279	0.6292	0.7160	0.7257	0.6653
	Y_{23}	0.5970	0.5279	0.5280	0.5726	0.5427	0.6731	0.6831	0.5892
Average	-	0.7075	0.5386	0.6299	0.5939	0.5901	0.6347	0.6551	-

4.1.1. Analysis of the Stress and Limitation of the Environment on Economic Development

According to the second-level indicators under the criterion level, the average correlation degree between the two indicators—ecological construction and environmental protection, which constitute the environment system, and the economic development system of the KSOFAsHP—was higher than 0.6. The results show that environmental protection has a more prominent restriction effect on economic development.

Specifically, the second phase of the Natural Forest Protection Project (2010–2020) was officially implemented in key state-owned forests in 2011. The first phase of the project (2000–2010) aimed to reduce timber harvesting by 50%. The volume of timber harvested in the forest area decreased from 4.025 million cubic meters in 2010 to 774,000 cubic meters in 2013. Logging of natural forest resources stopped after 2014.

The strictly enforced policy of prohibiting the logging of natural forest resources has had a direct impact on traditional forest industries in the forest areas. Forest industries of different sizes have either shrunk or closed down, and economic development in the forest areas has suffered from the stress effect of environment construction. In addition, based on furthering the ecological construction of forest areas, forest management, financial input in engineering, forest to ecological construction and protection, state investment reached 9.585 billion yuan in 2018, with an additional 60.36 million yuan invested in the development of forestry industries. The huge difference in ecological protection capital investment has become an important factor that restricts the economic development of forest industry logging.

Among the three levels of indicators, the forest construction index X in ecological construction11 (0.7075) has the most prominent stress effect on the economic development of the KSOFAsHP. Combining the original forests and artificial afforestation, forest management plays an important role in the ecological construction of the whole system and involves a large amount of capital and labor input. The core role in the woodland multifunctional use of the forest region is to develop alternative industries and speed up the upgrading of traditional abilities and the scale of the logging industry objectively and confine the formation stress effect. That is, the investment is mainly in the construction of forestry, rather than in the harvesting and processing industry. Therefore, the input index of environmental protection X_{21} (0.6551) has an obvious limiting effect on the economic development of the forest areas, mainly due to an unbalanced allocation of capital, policy, labor, and other key factors in forest areas, which produces an insufficient driving force and support for the economic development of forest areas, and further weakens the economic development ability and efficiency of the forest areas.

4.1.2. Analysis of the Stress and Restriction of Economic Development on the Environment

From the perspective of the second-level indicators under the criterion level, the average degree of correlation between the economic scale and economic structure of the economic development system and the ecological construction system of the KSOFAsHP is 0.6353 and 0.6029, respectively. Both are lower than the average correlation level of 0.6355 between the forest ecological construction system and economic development (namely, the average of the two indicators of ecological construction and environmental protection), indicating that economic development has a more significant stress restriction effect on forest ecological construction.

In other words, economic scale has played a more important role than economic structure in two ways: (1) Since 2011, the economic scale of the KSOFAsHP has shown a basic trend of total increase but slow and weak growth. The value of the total economic output of the forest area increased from 36.316 billion yuan in 2011 to 60.729 billion yuan in 2018, with a growth rate of 67.23%, but the average annual growth rate decreased from 12.95% in 2011 to 10.12% in 2018, and declined further to −3.86% in 2014. Forest undergrowth enterprises took a long time to speed up the development of a variety of alternative industries. In addition to the development of forest tourism, food, and other industries, cottage industries, such as understory plant breeding, represent small-scale, decentralized operations that are unable to support regional economic development. In the forest community, the burden of obtaining employment is becoming increasingly serious due to the huge loss of talent. This has a limiting effect on accelerating the promotion of new breakthroughs in ecological construction in the forest areas [49]. (2) From the perspective of economic structure, the prominent labor-intensive characteristics of the primary industry, the severe contraction of forest product processing industries, and the structural imbalance of the three industries cannot provide an effective economic basis and value feedback for forest environment construction.

From the three-level indicators under the criterion level, labor remuneration in economic development is Y_{14} (0.6771) and the economic aggregate is Y_{11} (0.6770). The two indexes have a high degree of correlation with the environment construction of the KSOFAsHP, and have relatively obvious and almost indistinct restrictive and limiting effects on the ecological construction and environmental protection of the forest areas. Specifically, there are two main effects: (1) In terms of labor remuneration, in 2019, logging enterprises in the key state-owned forest region in Heilongjiang Province had more than 200,000 employees. When logging activities stopped, these employees were impacted by layoffs, but the development of alternative industries had not yet reached scale benefits and therefore not able to provide jobs and higher pay [50]. (2) In terms of economy development, the current economic benefits of the key state-owned forest region from primary industries and the increase in tertiary industries related to forest tourism and ecological services are still unable to form effective decoupling. Forest resources rather than the wood of the forestry industry are still heavily dependent on capital and resources. There is still a large gap in realizing the expected goal of the economic development of non-forest and non-wood industries feeding back into environment construction.

4.2. Dynamic Analysis of the Coupling Coordination Relationship between the Environment and Economic Development

Equations (4)–(6) and the basic data, show the changes in the coupling coordination relationship between the environment and economic development in the key state-owned forest areas of Heilongjiang Province from 2011 to 2019, and are summarized in Table 4.

Table 4. The coupling coordination relationship between environment and economic development in key state-owned forest areas of Heilongjiang Province from 2011 to 2019.

Year	Coupling Coordination	Stage of Development	Type of Coupled Coordinated Development
2011	0.5351	Barely coordinated development class	II_2T_3, eco-lagged type
2012	0.6749	Junior coordination class	III_1T_3, eco-lagged type
2013	0.3747	Mild dysregulation recession class	I_4T_3, eco-environmental lag type
2014	0.4361	On the verge of dysregulation decline class	II_1T_3, eco-lagged type
2015	0.4506	On the verge of dysregulation decline class	II_1T_3, eco-lagged type
2016	0.9384	Good coordination class	III_4T_1, economic development lag type
2017	0.9842	Good coordination class	III_4T_1, economic development lag type
2018	0.9015	Quality coordination class	III_4T_1, economic development lag type
2019	0.9112	Quality coordination class	III_4T_1, economic development lag type

4.2.1. The Coupling and Coordination Relationship between the Environment and Economic Development Tends to Be Good

Table 4 shows the level of coupling coordination between the environment and economic development in key state-owned forest areas of Heilongjiang Province since 2011; the stage and type of coupling coordination relationship between them are dynamically identified. The relationship between them tends to be generally excellent.

First, in terms of development stage, the relationship between the environment and economic development in the forest area has experienced a relatively complex transition process. The primary coordination between the two was realized from 2011 to 2012, which means that the key state-owned forest areas steadily cut back on logging activities and reduced production levels. By actively developing multiple understory management initiatives during the first phase of the national forest protection project, they gradually adapted to find a balance between ecology and economy and win–win development solutions [51]. However, in 2012, the large-scale reduction in timber output of the leading traditional forest industry had a huge impact. The emerging alternative industries could not effectively support the forest regions' economic growth, and the fragile balance between ecology and economy was broken. Before 2015, both were on the verge of dysregulation and recession. Ecological prioritization of time for space did not begin to change until 2016 [52,53]. After 2016, the environment construction and economic development of the forest area adjusted and re-adapted over nearly four years, getting rid of the previous dysregulation and recession dilemma, and started to enter a phase of high-quality coordinated development. The reasons are directly related to the acceleration of the adaptive development of the understory economy, an increase in the government's capital investment in forest areas, and the continuous improvement in the environment. In spite of the positive development of the economic and ecological sectors since 2016, the relationship between them has not remained stable. In order to ensure stable development and effectively promote the high-quality development of the regional forest and grass industries during the period of change, it became imperative that the department responsible for the key state-owned forest region pay more attention by the end of 2019 to ensure stability in development.

4.2.2. Improvement in the Economic Development of Forest Areas under the Orientation of Ecological Construction

Table 4 shows the key state-owned forest regions in Heilongjiang Province from 2011 to 2019 during the coordinated development of the environment and changes in economic development type (as shown in Figure 2), specifically the changing trends in the evaluation index of the environment and economic development, as well as the impact of the coupled coordination of the forest environment and economic development.

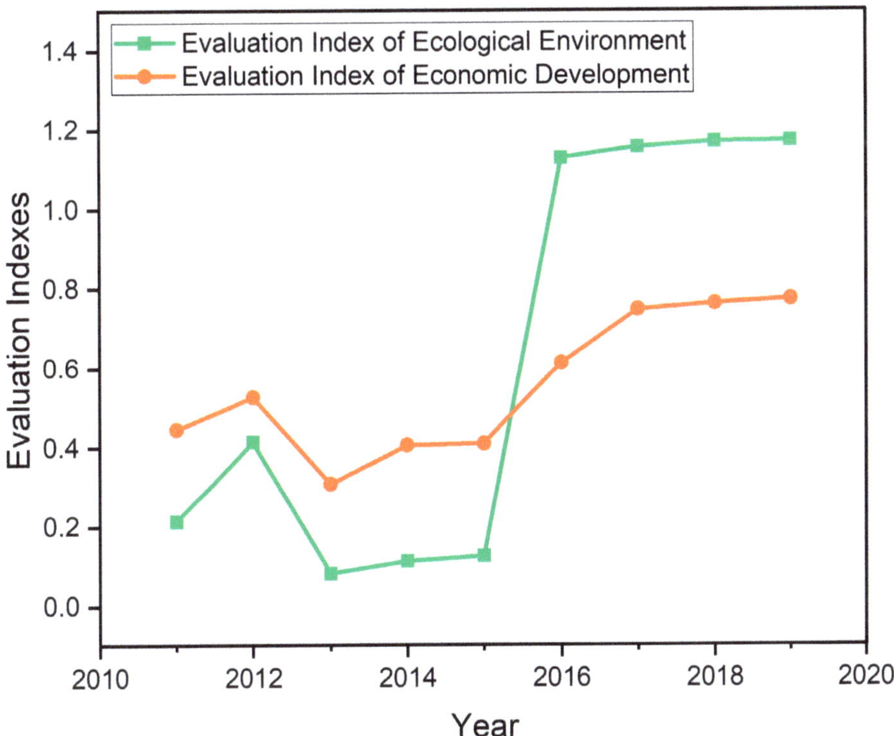

Figure 2. Change trends in the evaluation index of the environment and economic development from 2011 to 2019.

Figure 2 shows that the evaluation indexes of environment and economic development in the key state-owned forest areas of Heilongjiang Province maintained a highly consistent upward trend from 2011 to 2019, indicating that the forest areas had achieved positive results in the fields of ecological construction and economic development in recent years. This is related to the long-term implementation of effective ecological engineering policies in the forest areas and the reversal of understory economic transformation and development. Especially after 2015, both development indexes showed significant growth as the result of the effective adaptation to the no-logging policy decision in 2014, the stimulus provided by the comprehensive launch of the state-owned forest resource management system in 2015, and the restructuring and reform policy of forest industrial enterprises in the state-owned forest areas [54].

From a development perspective, the environment development evaluation index began to exceed the economic development evaluation index after 2015. The forest areas also officially changed from the environment lag period to the economic development lag period as the 100-year history of commercial logging of natural forests in the key state-owned forest areas of Heilongjiang Province came to an end. The forest areas entered a new phase in which the mountains were closed for afforestation and recuperation. The ecological priority strategy of exchanging time for space makes the environment construction in forest areas the core task for the future [55,56]. At the same time, the reform of the system mechanism became the focus of forest management after 2015. Despite the government's fiscal reform of the forest areas and the provision of special funds, economic development has been challenged by insufficient talent, capital, industry, and multiple factors such as the supply of resources. Development results are far less than environment construction [57].

The reform of the management system of the key state-owned forest region in Heilongjiang Province in 2020 still needs to strengthen all kinds of industry development and support to achieve the protection and restoration of the natural forests against the background of environment construction. In particular, there is still room for the non-wood forestry economy to develop and improve.

4.3. Results and Discussion

Key state-owned forest areas have undertaken the task to protect natural forest resources, practice green economic development, and support national ecological progress. The quantitative evaluation and analysis of the coupling and coordination between environment and economic development of forest areas have deepened the understanding of sustainable development of forest areas. However, there are still some deep-seated problems worthy of further discussion and exploration.

Firstly, based on the construction concept of ecological priority, it is necessary to explore policies and measures to stimulate economic development from the perspective of resource allocation [25]. Weak economic development in key state-owned forest areas has limited environment construction. To achieve the goal of ecological protection and economic transformation in forest areas, the central government issued relevant economic support policies and investment stimulus policies. These policies need to be refined further. On the premise of adhering to the concept of ecological priority construction, it is necessary to optimize the factor supply structure and scale for the cultivation of replacement industries and the development of enterprises in forest areas, especially in traditional factors such as capital and labor, which are moderately inclined to the economic transformation.

Secondly, a new policy design for forest areas is very critical to promote its environment and economic development. The central government has issued a new ecological policy for key state-owned forest areas to protect and restore natural forests. It is necessary to speed up the development of the key state-owned forest areas' forest ecosystem protection and restoration measures [58]. On the basis of fully using the natural forest protection repair system plan and the implementation scheme of protection and restoration of natural forests in Heilongjiang Province, the key ecological forests, wetlands, lakes, river basin areas related to the natural ecological system, wild animals, and plant resources and habitats require comprehensive repair management and protection to achieve the goal of ecological protection [59].

Thirdly, it is necessary to expand the research horizon and strengthen the integrated development of key state-owned forest areas and their surrounding areas. The reformed management system of forest resources in key state-owned forest areas has been devolved from the central government to local governments, and cooperation should be carried out with local regions for ecological restoration, resource protection, and environmental pollution prevention and control, rather than independent decision making and management of the past, to obtain maximum ecological protection benefits. At the same time, in terms of economic development, we should actively promote the integration of the forestry industry and local agricultural resources and industries [57], accelerate the flow of production factors across regions, explore a broader spatial scope of economic and social interaction and integration development model [60], and improve the factors and market environment of forest economic transformation and development [61–63].

5. Conclusions

Based on the empirical evaluation of the long-term coupling and coordination relationship between the environment and economic development in the key state-owned forest areas of Heilongjiang Province, the following conclusions are drawn:

(1) From 2011 to 2019, the key state-owned forest areas achieved positive results in ecological construction and economic development, especially in ecological construction. The ecological construction evaluation index increased by 448.99%, and the economic development evaluation index increased by 73.56%.

(2) There was a significant interaction between environment and economic development in key state-owned forest areas from 2011 to 2019. The average correlation degree between ecological construction and environmental protection and economic development of forest areas was 0.6158 and 0.6551, respectively. The average correlation degree between economic scale and economic results and forest environment was 0.6353 and 0.6029, respectively.

(3) From 2011 to 2019, the interaction between the environment and economic development in key state-owned forest areas showed a gradual high-quality trend. The coupling coordination degree increased from 0.5351 in 2011 to 0.9112 in 2019, and the coupling coordination relationship shifted from a barely coordinated development stage to a high-quality coordinated development stage.

(4) The average correlation degree between economic development and environment in key state-owned forest areas was 0.6191, which was lower than between the environment and economic development (0.6355), indicating that economic development in forest areas had certain stress and restriction effects on environment construction.

(5) Key state-owned forest areas began to enter the stage of economic development lag from the stage of eco-environmental development lag in 2015. It is an important task for forest areas to continuously accelerate economic transformation and development in the future.

Author Contributions: Conceptualization, X.L. and M.A.; methodology, X.L., M.A. and Z.W.; software, X.L., M.A. and Z.W.; validation, X.L., M.A. and Z.W.; formal analysis, X.L., M.A. and Z.W.; investigation, X.L., M.A. and Z.W.; resources, X.L., M.A., Z.W. and Z.Z.; data curation, X.L., M.A. and Z.W.; writing—original draft preparation, X.L.; writing—review and editing, X.L., M.A., Z.W. and Z.Z.; visualization, X.L., M.A. and Z.W.; supervision, M.A. and Z.W.; project administration, M.A. and Z.W.; funding acquisition, X.L., M.A., Z.W. and Z.Z. All authors have read and agreed to the published version of the manuscript.

Funding: This research was funded by the International Young Talents Program (No. QN2022168001L), China and the Heilongjiang Provincial Key Foundation for Philosophy and Social Sciences (No. 21G-LA449), China.

Institutional Review Board Statement: Not applicable.

Informed Consent Statement: Not applicable.

Data Availability Statement: The data supporting the findings of this study are available from the corresponding author upon reasonable request.

Conflicts of Interest: The authors declare no conflict of interest.

References

1. Geng, Q.; Muhammad, A.; Yuan, Z.; Zheng, J.; He, X.; Ding, D.; Yin, F.; Li, C. Plant species composition and diversity along successional gradients in arid and semi-arid regions of China. *For. Ecol. Manag.* **2022**, *524*, 120542. [CrossRef]
2. Zheng, J.; Arif, M.; He, X.; Ding, D.; Zhang, S.; Ni, X.; Li, C. Plant community assembly is jointly shaped by environmental and dispersal filtering along elevation gradients in a semiarid area, China. *Front. Plant Sci.* **2022**, *13*, 1–14. [CrossRef]
3. Muhammad, A.; Li, C. Impacts of environmental literacy on ecological networks in the Three Gorges Reservoir, China. *Ecol. Indic.* **2022**, *145*, 109571. [CrossRef]
4. Arif, M.; Li, J.; Tahir, M.; Zheng, J.; Li, C. Environmental literacy scenarios lead to land degradation and changes in riparian zones: Implications for policy in China. *Land Degrad. Dev.* **2022**, *4450*, 1–17. [CrossRef]

5. Wu, J.G.; He, C.Y.; Zhang, Q.Y.; Yu, D.Y.; Huang, G.L.; Huang, Q.X. Coupling model of global change and regional sustainable development and regulation countermeasures. *Adv. Earth Sci.* **2014**, *29*, 1315–1324.

6. Zhou, Y.; Pu, Y.; Chen, S.; Fang, F. Government support and development of new industries: A case study of new energy. *J. Econ. Res.* **2015**, *50*, 147–161.

7. Couto, E.V.; Oliveira, P.B.; Vieira, L.M.; Schmitz, M.H.; Ferreira, J.H.D. Integrating environmental, geographical and social data to assess sustainability in hydrographic basins: The ESI approach. *Sustainability* **2020**, *12*, 3057. [CrossRef]

8. Conceição, E.O.; Garcia, J.M.; Alves, G.H.Z.; Delanira-Santos, D.; Corbetta, D.d.F.; Betiol, T.C.C.; Pacifico, R.; Romagnolo, M.B.; Batista-Silva, V.F.; Bailly, D.; et al. The impact of downsizing protected areas: How a misguided policy may enhance landscape fragmentation and biodiversity loss. *Land Use Policy* **2022**, *112*, 105835. [CrossRef]

9. Ferreira, I.J.M.; Bragion, G.d.R.; Ferreira, J.H.D.; Benedito, E.; Couto, E.V.d. Landscape pattern changes over 25 years across a hotspot zone in southern Brazil. *South. For. J. For. Sci.* **2019**, *81*, 175–184. [CrossRef]

10. Fan, P.; Liang, L.; Li, Y.; Duan, L.; Wang, N.; Chen, C. Evaluation of coordinated urbanization development in Beijing-Tianjin-Hebei region from the perspective of system coupling. *Resour. Sci.* **2016**, *38*, 2361–2374.

11. Jin, F. The coordinated promotion strategy of ecological protection and high-quality development in the Yellow River Basin. *Reform* **2019**, *11*, 33–39.

12. Brown, L.R. Building a sustainable society. *Society* **1982**, *19*, 75–85. [CrossRef]

13. Allan, G.N.; Hanley, P.; McGregor, K. The impact of increased efficiency in the industrial use of energy: A computable general equilibrium analysis for the United Kingdom. *Energy Econ.* **2007**, *27*, 89–95. [CrossRef]

14. Muhammad, S.; Ilhan, O.; Talat, A.; Amjad, A. Revisiting the environment Kuznets curve in a global economy. *Renew. Sust. Energy Rev.* **2013**, *25*, 494–502.

15. Bradley, B.W. Local management of mangrove forests in the Philippines: Successful conservation and efficient resource exploitation. *Hum. Ecol.* **2004**, *32*, 177–195.

16. Ma, S.; Wang, R.; Ma, S.J.; Wang, S.S. Socio-economic-natural complex ecosystem. *Acta Ecol. Sin.* **1984**, *4*, 3–11.

17. Ren, J.Z. *Grassland Agroecology*; China Agricultural Press: Beijing, China, 1995; pp. 139–152. ISBN 7-109-02963-8/S.1885.

18. Liu, A.M.; Liu, Y.; Gao, X. Regional differentiation and coupling of agriculture and animal husbandry in Hexi region. *J. Econ. Geogr.* **2002**, *22*, 49–53.

19. Zhu, H.; He, S. Coupling effect in the development of agricultural resources. *J. Nat. Resour.* **2003**, *5*, 583–588.

20. Lei, G. Coupling model and benefit of mountain and oasis ecosystem. *Pratacultural Sci.* **2003**, *5*, 64–66.

21. Liu, X.X.; Wang, R.; Chen, X.Z. Ecological environment analysis and green trade transition of China's foreign trade: An empirical study based on embodied carbon. *Resour. Sci.* **2015**, *37*, 280–290.

22. Shu, X.L.; Gao, Y.B.; Zhang, Y.X.; Yang, C.Y. Coupling relationship and coordinated development of tourism industry and ecological civilization city. *China Popul. Resour. Environ.* **2015**, *25*, 82–90.

23. Zhou, C.; Feng, X.; Tang, R. Analysis and prediction of the coupled and coordinated development of regional economy, eco-environment and tourism industry: A case study of provinces and cities along the Yangtze River economic belt. *J. Econ. Geogr.* **2016**, *36*, 186–193.

24. Wang, W.J. How ecological civilization construction promotes rural economic development. *China Collect. Econ.* **2021**, *23*, 22–23.

25. Li, Y.J. Analysis of forestry economic development under the protection of forestry ecological environment. *Flowers* **2018**, *22*, 264–265.

26. Sheng, W. Based on the angle of ecological environment protection of forestry economic development strategy. *J. South. Agric.* **2021**, *21*, 73–74.

27. Su, L. Under the forestry ecological environment protection of forestry economic development study. *J. Shanxi Agric. Econ.* **2022**, *2*, 118–120.

28. Panayotou, T. *Empirical Tests and Policy Analysis of Environmental Degradation at Different Stages of Economic Development*; ILO Working Papers 992927783402676; International Labour Organization: Geneva, Switzerland, 1993.

29. Cao, Y.K.; Tang, X.W.; Wang, Y. Reducing the arc of the environmental Kuznets curve to protect state forest resources. *For. Econ.* **2000**, *3*, 70–72.

30. Song, W.M.; Yang, C. Forestry industry structure, spatial layout and its evolution mechanism since 1949. *For. Econ.* **2020**, *42*, 3–17.

31. Li, L.C.; Li, X.F.; Cheng, B.D.; Liu, J.L. Analysis of forest transformation in China based on environmental Kuznets curve. *World For. Res.* **2016**, *29*, 56.

32. Ma, M.Q.; Wang, M.Y.; Huang, Q. Study on supporting policy of state forest after logging suspension: A case study of Yichun Forest in Lesser Hinggan Mountains. *China For. Econ.* **2015**, *6*, 4.

33. Arif, M.; Qi, Y.; Dong, Z.; Wei, H. Rapid retrieval of cadmium and lead content from urban greenbelt zones using hyperspectral characteristic bands. *J. Clean. Prod.* **2022**, *374*, 133922. [CrossRef]

34. Wang, Z.W.; Wang, S. Study on the adjustment of sustainable development path of state-owned forest area after logging cessation: A case study of state-owned forest area in greater khingan mountains. *Issues For. Econ.* **2014**, *4*, 309–313.

35. Yuan, S.; Shen, Q.; Tian, J.; Bai, X. The full stop forest commercial cuts of Heilongjiang forest study. *J. For. Econ.* **2015**, *2*, 47–51.

36. Yu, L.C.; Liu, Y.Z. Discussion on tax and fee policy after the complete stop of commercial logging of natural forests in key state-owned forest areas. *Green Account.* **2015**, *6*, 3–5.

37. Bai, X.H. Reflections on the construction of ecological civilization in key state-owned forest areas of Heilongjiang Province. *For. Explor. Des.* **2017**, *3*, 17–18. [CrossRef]
38. Song, C.P.; Fang, F. Study on simulation and development strategy of social and economic system in state-owned forest areas of Heilongjiang Province. *China For. Econ.* **2012**, *3*, 26–29.
39. Geng, Y.; Zhang, Z. Study on Mechanism of niche construction in the evolution of forestry industry in state-owned forest areas. *Issues For. Econ.* **2012**, *32*, 195–199.
40. Zheng, L.J.; Wan, Z.F.; Li, W. Heilongjiang forest ecological economy system coordination development evaluation. *J. For. Econ.* **2018**, *40*, 28–32.
41. Dong, P.W.; Zhang, X.Z. Study on coupling degree of forestry industry and forest ecosystem. *China Soft Sci.* **2013**, *11*, 178–184.
42. Zhang, J.Y.; Long, F.; Bi, F.F.; Ge, C.Z.; Duan, X.M. Study on the coupling coordination between economic development and ecological environment of coastal industrial city: A case study of Yuhuan City. *Environ. Pollut. Prev.* **2021**, *2*, 255–258.
43. Tang, X.L.; Feng, Y.R.; Du, L. Study on the coupling and coordinated development of economic development and ecological environment in Shaanxi Province. *Environ. Pollut. Prev.* **2021**, *4*, 16–526.
44. Xie, W.J.; Wang, S.X. Coupling coordination between economic development and ecological environment on the Qinghai-Tibet Plateau: A case study of ecological civilization construction in Qinghai Province. *Plateau Sci. Res.* **2020**, *4*, 36–45.
45. Behzad, H.M.; Arif, M.; Duan, S.; Kavousi, A.; Cao, M.; Liu, J.; Jiang, Y. Seasonal variations in water uptake and transpiration for plants in a karst critical zone in China. *Sci. Total Environ.* **2022**, 160424. [CrossRef] [PubMed]
46. Zhang, J.C.; Zhi, R. Analysis and optimization adjustment of forestry industrial structure in Inner Mongolia: Based on grey relational analysis correlation degree and grey theory development decision analysis. *Issues For. Econ.* **2018**, *38*, 78–85, 110.
47. Jiang, Y.; Yang, Y.L. Analysis on the coupling coordination relationship between ecological environment and social economy in key state-owned forest areas of Heilongjiang province. *Chin. Econ.* **2022**, *1*, 134–135.
48. Wang, Y.; Hao, X.; Yao, L. Spatial-temporal evolution of the coupled and coordinated development of regional economy and eco-environment: A case study of Heilongjiang province. *Resour. Ind.* **2021**, *23*, 21–30.
49. Zhang, I.P. Forestry ecology construction and forests coordinated economic development way analysis. *J. Agric. Sci. Technol. Inf. Technol.* **2021**, *10*, 81–83.
50. Zhu, H.G.; Zhang, S.P.; Hu, Q.X. Comprehensive Ting Fa policy impact on household income of state-owned forest. *J. For. Econ.* **2019**, *33*, 6:457–464.
51. Wang, Y.Z.; Song, Y.; Wu, G.C. Research on the implementation performance, problems and countermeasures of Tianbao project in Heilongjiang province. *For. Econ.* **2012**, *10*, 68–71.
52. Gao, Y.J.; Guan, C. Study on the development of Understory economy and industry in Heilongjiang Province during the second phase of the Tianbao project. *Issues For. Econ.* **2020**, *40*, 618–625.
53. Li, C.H.; Wei, W.; Liu, S.X. Performance evaluation of natural forest resources protection project in state-owned forest areas: A case study of key state-owned forest areas in Heilongjiang province. *Sci. Silvae Sin.* **2021**, *57*, 153–162.
54. Chen, H.; Peng, Y. Study on the relationship between understory economic development and financial support in Heilongjiang province. *China For. Econ.* **2018**, *3*, 9–12, 76.
55. Dai, X.L. Suggestions on deepening the reform of management system of state-owned key forest areas in Heilongjiang province. *For. Explor. Des.* **2017**, *2*, 11–12.
56. Muhammad, A.; Jiajia, L.; Dongdong, D.; Xinrui, H.; Qianwen, G.; Fan, Y.; Songlin, Z.; Changxiao, L. Effect of topographical features on hydrologically connected riparian landscapes across different land-use patterns in colossal dams and reservoirs. *Sci. Total Environ.* **2022**, *851*, 158131. [CrossRef]
57. Wang, R.Z. Analysis on forestry ecological construction and understory economic industry development. *New Agric.* **2021**, *24*, 10.
58. General Office of the State Council of the People's Republic of China. Natural forest protection and restoration system scheme. *Hebei For.* **2019**, *9*, 16–18.
59. Yuancai, Q.; Arif, M.; Dong, Z.; Ting, W.; Qin, Y.; Bo, P.; Peng, W.; Wei, H. The effect of hydrological regimes on the concentrations of nonstructural carbohydrates and organic acids in the roots of Salix matsudana in the Three Gorges Reservoir, China. *Ecol. Indic.* **2022**, *142*, 109176. [CrossRef]
60. Arif, M.; Hussain, A.; Shahzad, M.K.; Khan, W.R.; Na, N.; Zarif, N.; Liu, X.; Amin, H.; Yukun, C. One belt one road initiative: Pakistan's forest-based sector investment opportunities within the context of globalization. *Custos Agronegocio* **2019**, *15*, 165–194.
61. Hira, A.; Arif, M.; Zarif, N.; Gul, Z.; Liu, X.; Cao, Y. Impacts of stressors on riparian health indicators in the upper and lower indus river basins in Pakistan. *Int. J. Environ. Res. Public Healthy* **2022**, *19*, 13239. [CrossRef]
62. Muhammad, A.; Behzad, H.M.; Tahir, M.; Changxiao, L. Nature-based tourism influences ecosystem functioning along waterways: Implications for conservation and management. *Sci. Total Environ.* **2022**, *842*, 156935. [CrossRef]
63. Arif, M.; Behzad, H.M.; Tahir, M.; Li, C. The impact of ecotourism on ecosystem functioning along main rivers and tributaries: Implications for management and policy changes. *J. Environ. Manag.* **2022**, *320*, 115849. [CrossRef] [PubMed]

 *forests*

Article

What Corporate Social Responsibility (CSR) Disclosures Do Chinese Forestry Firms Make on Social Media? Evidence from WeChat

Ma Zhong [2], Feifei Lu [1,*], Yunfu Zhu [2] and Jingru Chen [3]

1 College of Economics and Management, Qingdao Agricultural University, No. 700 Changcheng Road, Chengyang District, Qingdao 266109, China
2 College of Economics and Management, Nanjing Forestry University, No. 159 Longpan Road, Xuanwu District, Nanjing 210037, China
3 School of Economics, University of Bristol, 8 Woodland Road, Bristol BS8 1TN, UK
* Correspondence: feifei.lyu@qau.edu.cn

Abstract: Corporate social responsibility (CSR) disclosure serves as a vital bridge for forestry firms to communicate with their stakeholders and obtain legitimacy support. Existing studies focus on forestry firms' CSR disclosures based on CSR reports but lack consideration of such disclosures on social media. In this study, based on WeChat, the most widely used social media platform in China, we obtained 3311 tweets from 36 WeChat Official Accounts (WOA) of 63 Chinese-listed forestry firms in 2018 and used content analysis to classify the CSR information involved in these tweets based on the stakeholder dimensions. The main analysis results show that the top three CSR dimensions disclosed by Chinese forestry firms in social media are the shareholder (28.21%), customer (26.20%), and employee (23.64%) dimensions, and there are also great differences in the subcontent of disclosure concerns in each stakeholder dimension, e.g., approximately 86% of CSR disclosures for customers are product and service information. Additionally, we conducted a content analysis on the CSR reports of forestry firms using WOA. The results show that firms express different concerns in CSR reports than on social media, and the most mentioned dimensions in their reports are the environment (23.69%), employees (20.91%), and shareholders (20.21%). This indicates that there is a significant difference between the stakeholders that Chinese forestry firms focus on in social media and those that they focus on in CSR reports. This paper is the first study to focus on the CSR disclosure of Chinese forestry firms in social media and provides a reference for scholars to understand the information activities of forestry firms in social media.

Keywords: ESG; sustainable management; corporate social responsibility (CSR); non-financial disclosure; emerging market

Citation: Zhong, M.; Lu, F.; Zhu, Y.; Chen, J. What Corporate Social Responsibility (CSR) Disclosures Do Chinese Forestry Firms Make on Social Media? Evidence from WeChat. *Forests* **2022**, *13*, 1842. https://doi.org/10.3390/f13111842

Academic Editors: Noriko Sato and Tetsuhiko Yoshimura

Received: 11 October 2022
Accepted: 2 November 2022
Published: 4 November 2022

Publisher's Note: MDPI stays neutral with regard to jurisdictional claims in published maps and institutional affiliations.

1. Introduction

With the launch of the UN 2030 Agenda for Sustainable Development, the United Nations proposed the Global Sustainable Development Goals (SDGs). The 17 SDGs aim to promote better environmental and social performance globally [1]. As the world's largest emerging country, China is actively responding to global development visions and goals, implementing major strategies for carbon peaking and carbon neutrality, and actively assuming international responsibilities. The role of corporate social responsibility (CSR) in business operations and strategic development is becoming increasingly important [2–4]. CSR serves as an important bridge for firms to communicate with stakeholders, help firms to maintain their legitimacy, and obtain key resources. Thus, CSR disclosures are considered by firms to be the most important non-financial information activities [5]. Forests have an irreplaceable integrated value for sustainable economic and ecological development [6–8]; although forested areas only account for approximately 1/3 of the total land area, the carbon sink storage in forested vegetation areas accounts for almost half of the total terrestrial carbon pool. Forestry enterprises are microeconomic subjects of forestry industry development and are typical resource-sensitive and environment-sensitive enterprises, with the triple attributes of economic, social, and ecological benefits, which are of great significance for sustainable development [9,10].

According to the China Internet Network Information Center (Source: https://m.th epaper.cn/baijiahao_16888548, accessed on 7 June 2022), in 2021, the number of internet users in China was 1.032 billion, of which 1.029 billion were cellular data users, and the internet penetration rate had reached 73.0%. With previously unheard-of access to information and networks, many companies are finding a new voice in their interactions with consumers and other stakeholders. Among them, social media networks such as WeChat and Weibo are developing rapidly, replacing paper media and traditional online media to a large degree. Social media networks are also being used via an increasing number of firms to communicate for CSR [11–14], including forestry firms, who also use them as an important tool for CSR communication. Therefore, this paper conducts a study on CSR information disclosure on social media, which can serve as a new reference for the study of the CSR disclosure of forestry enterprises in various countries.

Our study aims to clarify the CSR disclosure of forestry firms on social media and draw on key differences between traditional media and social media. Related studies only focus on forestry firms' CSR disclosures based on CSR reports but lack consideration of such disclosures on social media. This study is the first to present the viewpoint of the CSR disclosure of Chinese forestry firms in social media and provides a reference to understand the characteristics of the information activities of forestry firms in social media. Potential contributions of this study include: (1) applying a new optic angle to investigate CSR disclosure of forestry firms in China, and extending the point of potential associating and differences between CSR reports and social media; (2) updating and expanding previous study on CSR topic in China, meanwhile, gaining a broader investigation of CSR data from social media; and (3) as the largest emerging market, our study provides a reference understanding of CSR disclosure in other emerging economies for practitioners, managers, and researchers.

The remainder of this paper begins with the literature review and theoretical background about corporate disclosures on social media and CSR disclosure in the forestry industry. Then we turn to the research design, and based on the framework, we describe the variables. The content analysis framework also provides insight into forestry companies activities via a variety of scenarios regarding the level of CSR disclosure. We conclude with a summary of the results with practical implications, limitations, and questions for future research.

2. Literature Review and Theoretical Background

2.1. Corporate Disclosure on Social Media

Before the advent of social media, corporate disclosure on the internet was of interest to researchers [15–17]. However, corporate online information disclosure in the presocial media era still had some deficiencies, which are similar to those of traditional paper media; for example, information dissemination required review by authorities and third-party platforms and there was a lack of two-way communication with audiences.

With the birth of social media, internet information dissemination has changed dramatically [18]. (1) The "hierarchy" was broken; for example, while traditional media can be divided into "regional", "national", and "international" types, social media does not have such hierarchical properties. (2) The "elite" lost their control, namely, the major "information gatekeepers" (e.g., large newspaper company) lost their control on social media, while non-government organizations (NGOs), consumers, and other weak stakeholders more easily took over the initiative of information spread; (3) social media provides higher dynamism and timeliness, unlike the fixed nature of traditional media; (4) the cost of dissemination of information on social media is extremely low because the use of social media platforms is free or very inexpensive; (5) information users have a stronger sense of trust of social media because the information dissemination therein is more based on personal relationships and trust rather than commercial speech; (6) social media makes communication much more direct, specifically, it provides more direct dialogue between firms and stakeholders than traditional media; (7) there is a potential for higher rates of public response, namely, information dissemination on social media can obtain more public responses; (8) information on social media achieves wider ranges of diffusion, because social media is not limited to specific information-spreading channels but can spread throughout the social network; (9) information dissemination on social media is subject to less institutional control and involvement but wider public opinion scrutiny.

Social media thus has dramatically changed corporate information activities [19]. First, social media helps to promote the efficiency of corporate information activities and reduce the impact of negative information. For example, Blankespoor, et al. [20] and Prokofieva [14] suggested that social media helps the spread of earnings information and then helps to gain more attention from investors. Yang and Liu [21] found that firms can use social media to mitigate the impact of negative financial information in some ways. In addition, social media also helps stakeholders communicate with each other, which in turn reinforces the monitoring for firms. For example, Ang et al. [22] found that social media plays a complementary governance role in the Chinese market and that stakeholder discussions on social media can effectively curb inefficient corporate mergers and acquisitions.

Social media also plays an important role in CSR information activities [23]. First, social media strengthens CSR communication between firms and their stakeholders [24–26], significantly improves the perception and recognition of CSR among customers, employees, and other stakeholders [27–30], and helps transform CSR into intangible values such as improving corporate reputation [29,31]. It is worth noting that compared to traditional communication, CSR information activities on social media are more indicative, which facilitates impression management for firms. She and Michelon [12] found that firms selectively direct their stakeholders on social media to avoid the proliferation of negative news. Pizzi et al. [32] found that CSR disclosures on social media are selective and not always responsive to stakeholder claims. Russo et al. [33] argued that CSR-oriented firms use social media more as a tool to achieve a higher level of legitimacy than as a tool to manage their sustainability strategies and related performance.

2.2. CSR Disclosure in the Forestry Industry

Existing research on CSR disclosure in the forestry industry focuses on traditional reporting vehicles such as CSR reports. First, some studies focus on the information topics in traditional CSR reports for forestry firms. Vidal and Kozak [34] conducted a textual analysis of CSR reports of forestry firms worldwide and found that the proportion of environmental and social dimensions is increasing. Colaço and Simão [35] analyzed CSR disclosures for international forestry firms in the Congo Basin and found that the most frequently disclosed topics are organizational certification and the environment, and that the level of CSR disclosure is higher in firms from Western countries than in those from Asian and African countries. D'Amato et al. [36] analyzed CSR reporting on a global scale involving land use (including forestry firms' CSR reports) and found that its focus was on the sustainability concepts of the circular, green, and bio-economy. Wang and Juslin [37] studied the CSR reports of Chinese forestry enterprises and argued that the most mentioned topic in these reports is economic responsibility. The driving factors of the CSR disclosures of forestry enterprises have also received great interest from scholars. Hansen, et al. [38], Panwar, et al. [39], and Li, et al. [40] studied the drivers of CSR disclosure in cross-country forestry firms based on globalization and firm-level factors. Wang and Juslin [37] analyzed the driving role of managerial characteristics in CSR disclosures. Li, Toppinen, Tuppura, Puumalainen and Hujala [40], and Lu et al. [41] explored the factors influencing CSR disclosure for Chinese forestry firms, focusing on the firm-level factors that characterize forestry firms' CSR disclosure. Lu et al. [42] conducted a study on the impact of managers' risk awareness on CSR disclosure in Chinese forestry firms. In summary, the research on forestry CSR disclosure only focuses on traditional information vehicles such as CSR reports, and no one discusses their CSR disclosure in emerging social media platforms. Specifically, it is worthwhile studying which kind of topics are preferred by forestry firms in the CSR reporting of social media and the difference between CSR disclosure in social media and traditional reports.

3. Research Design

3.1. Data Sources

As the largest developing economy, China has a huge forestry industry. According to a report by the China National Forestry and Grassland Administration (Source: http://www.forestry.gov.cn/main/62/20171221/1086586.html, accessed on 10 June 2022), gross output value of forestry reached 6.49 trillion RMB in 2016 and is expected to reach 9 trillion RMB by 2025 (Source: https://news.cgtn.com/news/2022-02-16/China-s-forestry-industry-output-to-reach-9t-yuan-by-2025-17He4nAk6li/index.html, accessed on 10 June 2022). Thus, Chinese forestry firms have a strong representation.

In China, due to regulatory policies, Facebook and Twitter are not available, and similar mainstream social media platforms are WeChat and Weibo. After consideration, we selected the WeChat platform for study. The reasons are as follows:

First, WeChat allows a higher information capacity for users and allows them to provide more substantive information. Compared with Weibo, which requires users to limit their tweets to 140 characters, WeChat does not have such limitations.

Second, WeChat has the largest user base and extremely high user stickiness in China. Its daily logged-in users reached 900 million in 2017, and the monthly active official account (The WeChat platform provides two types of account services: first, a public account, which allows all users to view the tweets of the account at will; the second category, a personal account, is only used by individuals who only want to allow their friends to view the tweets of their account) reached 3.5 million, with nearly 800 million monthly active followers (As a comparison, Weibo had only 376 million monthly active users in 2017 according to the Q3 2017 Sina earnings report. Source: https://www.sohu.com/a/203437993_667510, accessed on 10 June 2022). Moreover, WeChat provides an electronic payment service that is the most widely used by Chinese people (According to WeChat user report in 2018 year, its payment services get 600 million active users. Source: https://www.sohu.com/a/278496021_506058x, accessed on 10 June 2022), and furthermore, creates a high level of dependency and a long-term habit for its users. As a result, the information shared by enterprises through the WeChat Official Account (WOA) will be received by users more reliably and frequently.

The original sample is 63 listed firms in the China forest industry in 2018, and we excluded 27 samples that did not use the WOA or who had used it for less than 1 year. Therefore, there are 36 sample firms included in the analysis. Appendix B presents a list of the sample firms involved in WeChat. Moreover, 11 of 36 forestry firms used in this study also issued CSR reports in that year, which provided us with an opportunity to compare their CSR disclosures in social media channels and traditional channels. We used Python to obtain the WOAs of forestry firms, totaling 3311 articles. CSR reports are sourced from Cninfo (http://www.cninfo.com.cn). Financial and other firm-level data were obtained from CSMAR.

3.2. Content Analysis

We used the content analysis system created by Lu, Kozak, Toppinen, D Amato, and Wen [41] (Lu, Kozak, Toppinen, D Amato and Wen [39] design a content analysis system for Chinese forestry enterprises based on the guidance of CASS 3.0 and CNFPIA 2.0, which are released by Chinese Academy of Social Sciences and Chinese National Forestry Products Industry Association), which is for Chinese forestry firms, to categorize the tweets and CSR reports.

The system of Lu, Kozak, Toppinen, D Amato, and Wen [41] contains seven level-1 dimensions based on stakeholder theory: environment, customer, employee, supplier, community, government, and shareholder. Each level-1 dimension contains five to thirteen level-2 dimensions; for example, the employee dimension contains a total of seven level-2 dimensions: "Abidance by rule and laws (EM1)", "Percent of contract signing (EM2)", "Coverage of social insurance (EM3)", "Equal employment institution (EM4)", "Staff development training (EM5)", "Occupational health and safe producing (EM6)", "Staff relation management (EM7)" and "Other employee-related (EM0)". We scored according to the level-2 dimensions of this system; specifically, when the qualitative information of this dimension is provided in the tweet or CSR report, we scored "1", and when the quantitative information is provided, the score was "2". Then, we aggregated the scores of all level-2 dimensions to form level-2 variables. Finally, the level-2 dimension variables were aggregated to level-1 dimensions to form level-1 variables. Based on the above steps, we finally made 16 level-1 variables (eight variables for analysis on tweets and eight variables for analysis on CSR reports, variable definitions are reported in Table 1) and 42 level-2 variables (Since our interest of this paper is not the content of the CSR report, we only use level-2 subdimension variables for CSR disclosure on social media, and no longer use level-2 variables for the CSR report). The scoring index system and examples are reported in Appendix A. The software used for the content analysis is Atlas.ti 8.0.

Table 1. Variable definitions.

Variable	Level Annotation	Definition
Content variables for CSR disclosure on social media		
WeChat_All	Summary of level-1 variable for social media	Total level of CSR disclosure on social media, equal to the sum of seven level-1 content variable for tweets.
ep_w	Level-1 variable for social media	The level of the social media environment dimension CSR disclosure, equal to the sum disclosure scores of all subdimensions of the environment on social media.
cu_w	Level-1 variable for social media	The level of the social media customer dimension CSR disclosure, equal to the sum disclosure scores of all subdimensions of the customer on social media.
em_w	Level-1 variable for social media	The level of the social media employee dimension CSR disclosure, equal to the sum disclosure scores of all subdimensions of the employee on social media.
su_w	Level-1 variable for social media	The level of the social media supplier dimension CSR disclosure, equal to the sum disclosure scores of all subdimensions of the supplier on social media.
co_w	Level-1 variable for social media	The level of the social media community dimension CSR disclosure, equal to the sum disclosure scores of all subdimensions of the community on social media.
go_w	Level-1 variable for social media	The level of the social media government dimension CSR disclosure, equal to the sum disclosure scores of all subdimensions of the government on social media.
sh_w	Level-1 variable for social media	The level of the social media shareholder dimension CSR disclosure, equal to the sum disclosure scores of all subdimensions of the shareholder on social media.

Table 1. *Cont.*

Variable	Level Annotation	Definition
ep1_w/ep2_w/ep4 _w/ep5_w/ep6_w/ep7 _w/ep8_w/ep9_w/ep10 _w/ep11_w/ep12_w/ep0_w	Level-2 variable for social media	The level of environment disclosure of the level-2 subdimension on social media, equal to the sum of the scores of the information of the level-2 subdimension of the environment. For example, the variable ep1_w is equal to the sum of the qualitative and quantitative scores of the level-2 subdimension indicator EP1 "Environment management system". The description of the level-2 subdimension indicators are shown in
cu1_w/cu2_w/cu4 _w/cu6_w/cu0_w	Level-2 variable for social media	The level of customer disclosure of the level-2 subdimension on social media, equal to the sum of the scores of the information of the level-2 subdimension of the customer. For example, the variable cu1_w is equal to the sum of the qualitative and quantitative scores of the level-2 subdimension indicator CU1 "Product quality management system". The description of level-2 subdimension indicator are shown in
em1_w/em3_w/ep4 _w/em5_w/ep6_w/ep7 _w/em0_w	Level-2 variable for social media	The level of employee disclosure of the level-2 subdimension on social media, equal to the sum of the scores of the information of the level-2 subdimension of the employee. For example, the variable em1_w is equal to the sum of the qualitative and quantitative scores of the level-2 subdimension indicator EM1 "Abidance by rule and laws". The description of level-2 subdimension indicators are shown in
su1_w/su4_w/su5 _w/su6_w/su0_w	Level-2 variable for social media	The level of supplier disclosure of the level-2 subdimension on social media, equal to the sum of the scores of the information of the level-2 subdimension of the supplier. For example, the variable su1_w is equal to the sum of the qualitative and quantitative scores of the level-2 subdimension indicator SU1 "Responsibility purchasing system". The description of level-2 subdimension indicators are shown in
co1_w/co4_w/co6 _w/co0_w	Level-2 variable for social media	The level of community disclosure of the level-2 subdimension on social media, equal to the sum of the scores of the information of the level-2 subdimension of the community. For example, the variable co1_w is equal to the sum of the qualitative and quantitative scores of the level-2 subdimension indicator CO1 "The effect of enterprise operation on community". The description of level-2 subdimension indicators are shown in
go1_w/go2_w/go4 _w/go0_w	Level-2 variable for social media	The level of government disclosure of the level-2 subdimension on social media, equal to the sum of the scores of the information of the level-2 subdimension of the government. For example, the variable go1_w is equal to the sum of the qualitative and quantitative scores of the level-2 subdimension indicator GO1 "Enterprise management abided by rule". The description of level-2 subdimension indicators are shown in
sh1_w/sh2_w/sh3 _w/sh4_w/sh0_w	Level-2 variable for social media	The level of shareholder disclosure of the level-2 subdimension on social media, equal to the sum of the scores of the information of the level-2 subdimension of the shareholder. For example, the variable sh1_w is equal to the sum of the qualitative and quantitative scores of the level-2 subdimension indicator SH1 "Investor relation management". The description of level-2 subdimension indicators are shown in

Table 1. *Cont.*

Variable	Level Annotation	Definition
Content variables for CSR disclosure on CSR report		
Report_All	Summary of level-1 variable for the CSR report	Total level of CSR disclosure on the CSR report, equal to the sum of the seven level-1 content variables for the CSR report.
ep_r	Level-1 variable for CSR report	The level of CSR report environment dimension CSR disclosure, equal to the sum disclosure scores of all subdimensions of the environment on social media.
cu_r	Level-1 variable for CSR report	The level of the CSR report customer dimension CSR disclosure, equal to the sum disclosure scores of all subdimensions of the customer on social media.
em_r	Level-1 variable for CSR report	The level of the CSR report employee dimension CSR disclosure, equal to the sum disclosure scores of all subdimensions of the employee on social media.
su_r	Level-1 variable for CSR report	The level of the CSR report supplier dimension CSR disclosure, equal to the sum disclosure scores of all subdimensions of the supplier on social media.
co_r	Level-1 variable for CSR report	The level of the CSR report community dimension CSR disclosure, equal to the sum disclosure scores of all subdimensions of the community on social media.
go_r	Level-1 variable for CSR report	The level of the CSR report government dimension CSR disclosure, equal to the sum disclosure scores of all subdimensions of the government on social media.
sh_r	Level-1 variable for CSR report	The level of the CSR report shareholder dimension CSR disclosure, equal to the sum disclosure scores of all subdimensions of the shareholder on social media.

3.3. Descriptions of Variables

Table 1 reports the variables used in the analysis of this paper:

(1) Content variables for CSR disclosure on social media. These include the total social media disclosure variable WeChat_All, social media environment dimension disclosure variable ep_w, social media customer dimension disclosure variable cu_w, social media employee dimension disclosure variable em_w, social media supplier dimension disclosure variable su_w, social media community dimension disclosure variable co_w, social media government dimension disclosure variable go_w, and social media shareholder dimension disclosure variable sh_w. Moreover, we defined forty-two level-2 subdimension variables, e.g., variable ep1_w stands for the reporting level of the social media environment EP1 subdimension (Some of the level-2 subdimensions indicators of [41], 10 dimensions in all, among them the forest biodiversity conservation dimension EP3 and dispute resolution mechanism dimension CU3, are not addressed in the social media disclosure, so we do not create variables for these dimensions).

(2) Content variables for CSR disclosures in CSR reports. These also include eight variables: the total CSR report disclosure variable Report_All, CSR report environment dimension disclosure variable ep_r, CSR report customer dimension disclosure variable cu_r, CSR report employee dimension disclosure variable em_r, CSR report supplier dimension disclosure variable su_r, CSR report community dimension disclosure variable co_r, CSR report government dimension disclosure variable go_r, and CSR report shareholder dimension disclosure variable sh_r.

4. Results

4.1. Content Analysis of CSR Disclosures on Social Media

4.1.1. Total and Level-1 Dimension Analysis on Social Media

Table 2 reports the descriptive statistics of the overall and level-1 dimension variables of social media CSR disclosure for 36 forestry firms. The mean value of the overall disclosure variable WeChat_All is 103.3, and the median value is 72.5. Rows (2) to (8) of Table 2 show the statistics of the environment dimension variable ep_w, customer dimension variable cu_w, employee dimension variable em_w, supplier dimension variable su_w, community dimension variable co_w, government dimension variable go_w, and shareholder dimension variable sh_w, where the top three highest disclosure dimensions are shareholder, customer, and employee, respectively. The mean (median) values of the shareholder dimension variable sh_w, customer dimension variable cu_w, and employee dimension variable em_w are 29.14 (18.5), 27.06 (8.5), and 24.42 (5), respectively. The last four variables are the government dimension variable go_w, environment dimension variable ep_w, community dimension variable co_w, and supplier dimension variable su_w, with mean (median) values of 8.78 (3), 8.14 (2), 5.17 (3), and 0.64 (0), respectively. Column (4) shows the proportion of subdimension disclosure to total disclosure. Among them, the top three dimensions with the highest proportion of disclosure are shareholders, customers and employees, with 28.21%, 26.20%, and 23.64% (The information proportion of level–1 subdimension is equal to the mean value of the subdimension indicator divided by the mean value of the total rating indicator (WeChat_All), for example, the mean value of the environment dimension variable ep_w is 8.14, divided by the mean value of the total rating indicator WeChat_All, 103.3, which equals 7.88%), respectively, which cumulatively account for approximately 80% of the total CSR disclosure on social media. The above results show that the stakeholders that forestry firms care most about in their social media CSR disclosures are shareholders, customers, and employees, and these three stakeholders account for approximately 80% of the disclosures.

Table 2. Results of total and level-1 dimension analysis on social media.

	Variable	N	Mean	Ratio	Sd	Min	P50	Max
(1)	WeChat_All	36	103.3	100.00%	97.87	2	72.5	470
(2)	ep_w	36	8.14	7.88%	12.14	0	2	50
(3)	cu_w	36	27.06	26.20%	39.61	0	8.5	171
(4)	em_w	36	24.42	23.64%	40.2	0	5	218
(5)	su_w	36	0.64	0.62%	1.33	0	0	6
(6)	co_w	36	5.17	5.00%	6.35	0	3	27
(7)	go_w	36	8.78	8.50%	14.17	0	3	67
(8)	sh_w	36	29.14	28.21%	36.42	0	18.5	141

4.1.2. Analysis of the Level-2 Environment Subdimension Analysis on Social Media

Table 3 reports the results of the environment level-2 subdimension analysis on social media. Row (1) of Table 3 shows the statistical results of environment level-1 variable ep_w, which are also reported in Row (2) of Table 2. Rows (2) to (13) show the statistics for environment level-2 subdimension variables, including ep1_w (see Table 1 for details). Among them, the top four level-2 subdimensions with the highest proportions in social media environment disclosure are the "Research, development, application, and sale of the environment production and devices" variable ep8_w, the "Environmental protection investment" variable ep4_w, the "Other environment-related" variable ep0_w, and the "Reduce pollution and decrease drain" variable ep10_w, with mean values of 2.19, 1.25, 1.14, and 0.94, respectively, and the last four are the "Forest certification" variable ep6_w, the "Sustainable forest management" variable ep5_w, the "Quantity, kind, and risk to human and environment of toxic or exhaust emission" variable ep7_w, and the "Environmental impact assessment of new investment projects" variable ep2_w. Column (4) shows the proportion of level-2 subdimensions to the total disclosure of environment dimensions, e.g., the social media CSR disclosure of the "Environmental management system (EP1)" dimension accounts for 8.23% of the environmental dimension disclosure. Among the twelve subdimensions, the top three ones with the highest percentages are the "Research, development, application, and sale of the environment production and devices dimension (EP8)", the "Environmental protection investment dimension (EP4)", and the "Other environment-related dimension (EP0)", with 26.90%, 15.36%, and 14.00%, respectively. These three types of level-2 subdimensions account for approximately 60% of the total information in the social media environment disclosure.

Table 3. Results of environment level-2 subdimension analysis on social media.

	Variable	N	Mean	Ratio	Sd	Min	P50	Max
(1)	ep_w	36	8.14	100.00%	12.14	0	2	50
(2)	ep1_w	36	0.67	8.23%	1.2	0	0	5
(3)	ep2_w	36	0.08	0.98%	0.5	0	0	3
(4)	ep4_w	36	1.25	15.36%	3.15	0	0	12
(5)	ep5_w	36	0.14	1.72%	0.59	0	0	3
(6)	ep6_w	36	0.19	2.33%	0.52	0	0	2
(7)	ep7_w	36	0.11	1.35%	0.4	0	0	2
(8)	ep8_w	36	2.19	26.90%	3.4	0	1	12
(9)	ep9_w	36	0.72	8.85%	2.01	0	0	11
(10)	ep10_w	36	0.94	11.55%	2.29	0	0	10
(11)	ep11_w	36	0.28	3.44%	0.74	0	0	3
(12)	ep12_w	36	0.42	5.16%	1.13	0	0	6
(13)	ep0_w	36	1.14	14.00%	2.18	0	0	10

4.1.3. Analysis of the Level-2 Customer Subdimension Analysis on Social Media

Table 4 reports the results of the level-2 customer subdimension analysis on social media. Similarly, we report the level-1 customer variable cu_w in Row (1) of Table 4. Rows (2) to (6) of Table 4 show the statistics of level-2 customer variables, including cu1_w. The highest is the "Information provision of the product and services" variable cu4_w, with a mean value of 23.22. Column (4) shows the proportion of level-2 subdimensions to the total disclosure of customer dimensions, e.g., the "Product quality management system (CU1)" subdimension accounts for 2.99% of the whole customer dimension. Among all level-2 customer subdimensions, the highest percentage belongs to the "Information provision of the product and services subdimension (CU4)", whose percentage is 85.81%, accounting for most of the customer dimension disclosures. The above results show that when disclosing customer dimension information on social media, forestry firms prefer to disclose information under the subdimension "Information provision of the product and services (CU4)".

Table 4. Results of the level-2 customer subdimension analysis on social media.

	Variable	N	Mean	Ratio	Sd	Min	P50	Max
(1)	cu_w	36	27.06	100.00%	39.61	0	8.5	171
(2)	cu1_w	36	0.81	2.99%	1.69	0	0	8
(3)	cu2_w	36	0.39	1.44%	1.4	0	0	8
(4)	cu4_w	36	23.22	85.81%	37.3	0	5	169
(5)	cu6_w	36	0.33	1.22%	1.2	0	0	6
(6)	cu0_w	36	2.31	8.54%	3.54	0	0.5	13

4.1.4. Analysis of the Level-2 Employee Subdimension on Social Media

Table 5 reports the results of the analysis of the level-2 employee subdimension on social media. Row (1) of Table 5 shows the statistics of the level-1 variable em_w. Rows (2) to (8) of Table 5 show seven employee level-2 variables. Among them, the top three variables with the highest scores are "Staff relation management" variable em7_w, "Occupational health and safe producing" variable em6_w, and "Staff development training" variable em5_w, with mean values of 18.25, 3.19, and 2.58, respectively. Column (4) shows the proportion of subdimensions of level-2 disclosure to the total employee dimension disclosure. Among the seven employee subdimensions, the highest disclosure subdimension is the "Staff relation management dimension (EM7)", whose disclosure ratios are 74.73%, accounting for approximately three quarters of the employee dimension disclosure. The above results show that when disclosing employee dimension information on social media, forestry firms prefer to disclose information on the "Staff relation management dimension (EM7)".

Table 5. Results of the analysis of the level-2 employee subdimension on social media.

	Variable	N	Mean	Ratio	Sd	Min	P50	Max
(1)	em_w	36	24.42	100.00%	40.2	0	5	218
(2)	em1_w	36	0.08	0.33%	0.37	0	0	2
(3)	em3_w	36	0.17	0.70%	0.74	0	0	4
(4)	em4_w	36	0.06	0.25%	0.33	0	0	2
(5)	em5_w	36	2.58	10.57%	6.81	0	0	39
(6)	em6_w	36	3.19	13.06%	6.06	0	0	31
(7)	em7_w	36	18.25	74.73%	28.33	0	4.5	148
(8)	em0_w	36	0.08	0.33%	0.37	0	0	2

4.1.5. Analysis of the Level-2 Supplier Subdimension on Social Media

Table 6 reports the results of the analysis of the level-2 supplier subdimension on social media. Row 1 shows the statistics of the level-1 variable su_w. Rows (2) to (6) show the results of level-2 supplier variables, in which the top three highest disclosure positions are the "Other supplier-related" variable su0_w, the "Openness of procurement policy" variable su4_w, and the "Legality of forest product procurement" variable su5_w, with mean (median) values of 0.33 (0), 0.14 (0), and 0.11 (0), respectively. Column (4) shows the percentages of the level-2 supplier dimension. The three highest supplier subdimensions are the "Other supplier-related dimension (SU0)", the "Openness of procurement policy dimension (SU4)", and the "Legality of forest product procurement dimension (SU5)", whose disclosure proportions were 51.56%, 21.88%, and 17.19%, respectively, accounting for approximately 90% of the supplier dimension disclosure. The above results indicate that on social media, forestry firms prefer to disclose information on the "Other supplier-related dimension (SU0)", the "Openness of procurement policy dimension (SU4)", and the "Legality of forest product procurement dimension (SU5)", and that these three subdimensions account for approximately 90% of the supplier dimension cumulatively.

Table 6. Results of the analysis of the level-2 supplier subdimension on social media.

	Variable	N	Mean	Ratio	Sd	Min	P50	Max
(1)	su_w	36	0.64	100.00%	1.33	0	0	6
(2)	su1_w	36	0.03	4.69%	0.17	0	0	1
(3)	su4_w	36	0.14	21.88%	0.42	0	0	2
(4)	su5_w	36	0.11	17.19%	0.4	0	0	2
(5)	su6_w	36	0.03	4.69%	0.17	0	0	1
(6)	su0_w	36	0.33	51.56%	1.01	0	0	5

4.1.6. Analysis of the Level-2 Community Subdimension on Social Media

Table 7 reports the results of the analysis of the level-2 community subdimension on social media. Row (1) of Table 7 shows the level-1 community dimension variable co_w. Rows (2) to (5) of Table 7 show the statistics for level-2 community variables. Among them, the highest score is the "Effect of enterprise operation on community" variable co1_w, whose mean value is 3.31. Column (4) shows the proportion of level-2 subdimensions to the level-1 community dimension. The "Effect of enterprise operation on community dimension (CO1)" accounts for 64.02% of the disclosure of community dimensions. The above results indicate that forestry firms are more likely to disclose information on the subdimension "Effect of enterprise operation on community (CO1)" on social media.

Table 7. Results the analysis of the level-2 community subdimension on social media.

	Variable	N	Mean	Ratio	Sd	Min	P50	Max
(1)	co_w	36	5.17	100.00%	6.35	0	3	27
(2)	co1_w	36	3.31	64.02%	4.08	0	2	14
(3)	co4_w	36	0.56	10.83%	1.83	0	0	10
(4)	co6_w	36	0.67	12.96%	1.41	0	0	6
(5)	co0_w	36	0.64	12.38%	1.64	0	0	9

4.1.7. Analysis of the Level-2 Government Subdimension on Social Media

Table 8 reports the results of the analysis of the level-2 government subdimension. Row (1) of Table 8 shows the statistics of the total government dimension disclosure variable go_w. Rows (2) to (5) of Table 8 show all level-2 subdimensions of government, where the top two highest disclosures are the "Other government-related" variable go0_w and the "Enterprise management abided by rule" variable go1_w, and their mean values are 6.47 and 1.86, respectively. Column (4) shows the proportion of the level-2 subdimensions of the government dimensions. The top two highest disclosure percentages are the "Other government-related dimension (GO0)" and the "Enterprise management abided by rule dimension (GO1)", whose disclosure percentages are 73.69% and 21.18%, respectively, accounting for approximately 95% of the government dimension disclosures.

Table 8. Results of the analysis of the level-2 government subdimension on social media.

	Variable	N	Mean	Ratio	Sd	Min	P50	Max
(1)	go_w	36	8.78	100.00%	14.17	0	3	67
(2)	go1_w	36	1.86	21.18%	3.21	0	0	13
(3)	go2_w	36	0.33	3.76%	0.99	0	0	5
(4)	go4_w	36	0.11	1.25%	0.46	0	0	2
(5)	go0_w	36	6.47	73.69%	11.22	0	1.5	52

4.1.8. Analysis of the Level-2 Shareholder Subdimension on Social Media

Table 9 reports the results for the shareholder dimension. Row (1) of Table 9 shows the statistical results of the total shareholder dimension variable sh_w. Rows (2) to (6) of Table 9 show the level-2 shareholder variables. Among them, the top three highest disclosures are the "Growth potential" variable sh2_w, the "Other shareholder-related" variable sh0_w, and the "Investor relation management" variable sh1_w, with mean values of 12.89, 7.28, and 4.56, respectively. Column (4) shows the proportion of the level-2 subdimensions. Among the five shareholder subdimensions, the top three highest percentages are the "Growth potential dimension" (SH2), "Other shareholder-related dimension" (SH0), and "Investor relation management dimension" (SH1), whose disclosure percentages are 44.23%, 24.98%, and 15.65%, respectively, accounting for approximately 85% of the shareholder dimension disclosure.

Table 9. Results of shareholder level-2 subdimension analysis on social media.

	Variable	N	Mean	Ratio	Sd	Min	P50	Max
(1)	sh_w	36	29.14	100.00%	36.42	0	18.5	141
(2)	sh1_w	36	4.56	15.65%	7.58	0	2.5	42
(3)	sh2_w	36	12.89	44.23%	16.59	0	7.5	68
(4)	sh3_w	36	2.86	9.81%	4.52	0	1	20
(5)	sh4_w	36	1.56	5.35%	3.36	0	0	13
(6)	sh0_w	36	7.28	24.98%	10.06	0	4.5	41

4.1.9. Summary of the Analysis of CSR Disclosures on Social Media

First, forestry firms pay significant attention to the interests of different stakeholders on social media. The three most concerned stakeholders on social media CSR disclosure by forestry firms are shareholders (28.21%), customers (26.20%), and employees (23.64%), and these three categories account for approximately 80% of all disclosures.

Moreover, under the level-1 dimension of each stakeholder, forestry firms also show serious disclosure imbalance for different level-2 subdivisions. Specifically, the three level-2 subdimensions with the highest levels of disclosure of the environmental dimension are the "Research, development, application, and sale of the environment production and devices (EP8)" (26.90%), the "Environmental protection investment dimension (EP4)" (15.36%), and the "Other environment-related dimension (EP0)" (14.00%), which account for approximately 60% of the environmental dimension disclosures. The highest reporting subdimensions of the customer and employee dimensions are the "Information provision of the product and services dimension (CU4) (85.81%) and the "Staff relation management" dimension (EM7) (74.73%). The three highest subdimensions in the supplier dimension are the "Other supplier-related dimension (SU0)" (51.56%), the "Openness of procurement policy dimension (SU4)" (21.88%), and the "Legality of forest product procurement dimension (SU5)" (17.19%), accounting for approximately 90% of the supplier dimension disclosures. The highest community subdimension is the "Effect of enterprise operation on community dimension (CO1)" (64.02%). The highest governmental subdimensions are the "Other government-related dimension (GO0)" (73.69%) and the "Enterprise management abided by rule dimension (GO1)" (21.18%), accounting for approximately 95% of the government dimension. The three highest disclosed subdimensions of the shareholder dimension are the "Growth potential dimension (SH2)" (44.23%), the "Other shareholder-related dimension (SH0)" (24.98%), and the "Investor relation management dimension (SH1)" (15.65%), accounting for approximately 85% of the shareholder dimension.

4.2. Comparative Analysis of CSR Disclosure between Social Media and CSR Reports

4.2.1. Total and Level-1 Dimension Analysis on CSR Reports

Table 10 reports the descriptive statistics of the total and level-1 dimension variables of the CSR report disclosures for the forestry firms using the WeChat official account (11 of 36 listed forestry firms that use WeChat official account issued the CSR reports in 2018 year). Row (1) of Table 10 shows the statistics of the total disclosure variable Report_All with mean and median values of 26.09 and 23, respectively. Rows (2) to (8) of Table 10 show the statistics of the environment dimension variable ep_r, customer dimension variable cu_r, employee dimension variable em_r, supplier dimension variable su_r, community dimension variable co_r, government dimension variable go_r, and shareholder dimension variable sh_r, where the top three level-1 dimension variables are the environmental variable ep_r, the employee variable em_r, and the supplier variable su_r, with mean (median) values of 6.18 (5), 5.46 (5), and 5.27 (6), respectively. Moreover, their percentages of disclosure are 23.69%, 20.91%, and 20.21%, which in total account for approximately 65% of all disclosures of CSR reports.

Table 10. Results of total and level-1 dimension analysis on CSR reports.

	Variable	N	Mean	Ratio	Sd	Min	P50	Max
(1)	Report_All	11	26.09	100%	7.739	20	23	46
(2)	ep_r	11	6.18	23.69%	3.156	2	5	12
(3)	cu_r	11	2.27	8.71%	1.737	0	2	6
(4)	em_r	11	5.46	20.91%	2.339	3	5	10
(5)	su_r	11	1.27	4.88%	1.104	0	1	3
(6)	co_r	11	2.91	11.15%	1.3	1	4	4
(7)	go_r	11	2.73	10.45%	2.37	0	2	6
(8)	sh_r	11	5.27	20.21%	2.453	2	6	8

4.2.2. Comparison between CSR Disclosure under Social Media and CSR Reports

In Figure 1, we compare forestry firms' disclosures on social media and CSR reports by stakeholder dimensions.

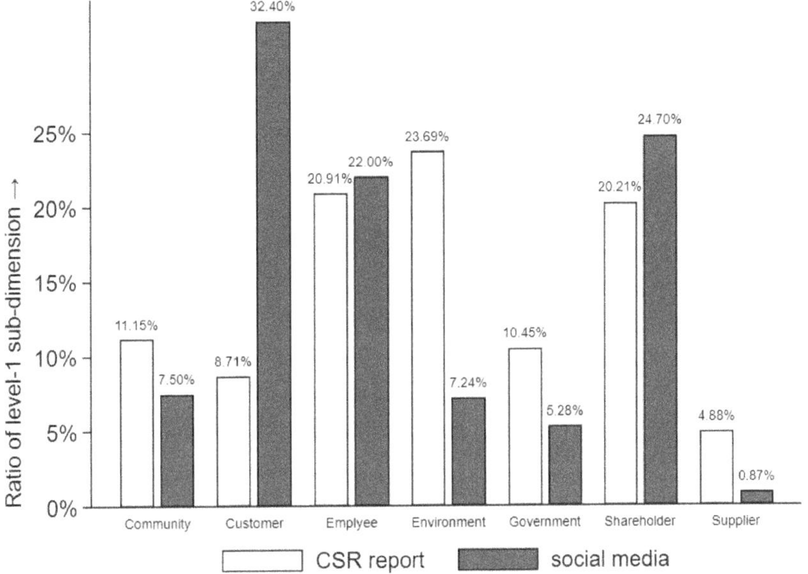

Figure 1. Comparison of CSR disclosure on social media and CSR reports.

First, there are differences in the stakeholder dimensions that forestry firms focus on in different channels. The top three dimensions in CSR report channels are the environment (23.69%), employee (20.91%), and shareholder (20.21%) dimensions, while the top three dimensions in social media channels are the customer (32.40%), shareholder (24.70%), and employee (22.00%) dimensions. The dimension with the highest proportion disclosed in the report is the environment dimension (23.69%), but this dimension only ranks 5th on social media (7.24%), with a difference of more than three times, while the customer dimension, which ranks 6th in the CSR report in terms of proportion (8.71%) and 1st on social media (32.40%), has the same difference of more than three times. We suggest that this difference may be due to the characteristics of the information users focused on by firms in different channels: forestry firms are mostly in heavily polluting and environmentally sensitive industries, so in formal reports (CSR reports), they take more into account the needs of professional information users, such as providing more detailed and rich environmental information to analysts, socially responsible investors (SRIs), fund managers, etc. However, in social media, forestry firms are more likely to have to communicate with non-specialist users, e.g., local customers.

Second, there are also commonalities in the stakeholder dimensions that forestry firms focus on under different channels. For example, in both reporting channels, the lowest proportion of disclosure is the supplier dimension (with CSR reports at 4.88% and social media at 0.87%). This is because forestry firms mostly belong to the upstream of the industry chain or self-supply their raw materials, such as paper, wood supply, furniture manufacturing, and other niche industry firms, and suppliers have little influence on them. In addition, the proportion of community dimension and government dimension is also small, and the proportion in the CSR reports and social media is mostly around 10% or lower. This indicates that for Chinese forestry firms, these two types of stakeholders are not the focus of corporate concern.

5. Discussion and Conclusions

This paper studies the CSR disclosures made by Chinese forestry firms on social media. Based on the framework of Lu, Kozak, Toppinen, D Amato, and Wen [41], we used content analysis to analyze the social media posted by Chinese-listed forestry firms on WeChat in 2018 and found that (1) there are differences for Chinese forestry firms that care about stakeholders in social media and traditional CSR report channels; for example, the top three dimensions present in the social media channel are shareholder, customer, and employee dimensions, while the top three dimensions present in the traditional CSR report channel are environment, employee, and shareholder dimensions; (2) the proportion of disclosed subcontent for each stakeholder on social media also shows a great imbalance; for example, the most concerning subdimension in the disclosure for the shareholder dimension is the "Growth potential" subdimension, which accounted for 44.23% of the shareholder's information. The results of our analysis suggest that the characteristics of social media users lead to distinctive features of CSR disclosures made by forestry firms on social media, with significant differences in the main points of attention from disclosures made through traditional channels (CSR reports). We argue that the main reasons for the differences in the focus of information disclosure between forestry firms in social media and traditional CSR reports are as follows: the target information users of CSR reports are professional users, e.g., analysts, institutional investors, and financial media; however, the general public (e.g., individual consumers, grassroots employees, and community residents) constitute the largest proportion of stakeholders but are less capable of using information, specifically, they lack intuitive judgment of the financial indicators [43] and industrial technical indicators recorded in traditional reports, and are more concerned with intuitive information such as protection of consumer rights and demonstration of corporate strengths, so forestry firms are more interested in displaying such intuitive information in social media than in CSR reports.

The following questions of this paper need future research: first, the WeChat platform cannot fully display all the comments because of its program design (According to the rules of the WeChat platform, users' comments on WOA need to be reviewed by the account owner before they are made public. As a result, negative comments about the firm are filtered out), which leads us to be unable to analyze the interactive communication between firms and information users; second, information activities conducted by firms on the WeChat platform use a large number of pictures and videos, and Kassinis and Panayiotou [44] found that visualization tools play an important role in CSR impression management by firms, which is limited by technical means, making it difficult for us to include these images and videos in our analysis; third, some non-listed forestry firms also use social media to communicate CSR messages, but we are unable to include them in our analysis due to the difficulty of obtaining their firm-level data. We look forward to obtaining additional information through questionnaire analyses, field research, and case studies in the future to provide further answers to the above questions. Furthermore, we will investigate the following in future studies: (1) the motivation and the factors affected on the forestry firms' disclosure differences between social media and CSR report; (2) Social media CSR disclosure of other industries related to carbon emission reduction (e.g., energy industry [45]).

Author Contributions: M.Z.: conceptualization, design, manuscript preparation. F.L.: conceptualization, design, manuscript preparation and supervised this project. Y.Z. and J.C.: data preparation and design, formal analysis, original draft preparation and manuscript preparation sections. Each author contributed to the conceptualization and writing of this paper. All authors have read and agreed to the published version of the manuscript.

Funding: This research was funded by the National Natural Science Foundation of China (Grant number 71902090), Forestry Science and Technology Development Project of China (Grant number KJZXRZ202013), the University Innovation and Technology Project of Shandong Province in China (Grant number 2020RWG004), the High-level Research Foundation of QAU (Grant number 1119710), and the Project Supported by Enterprises and Institutions (Grant number 6602422730).

Institutional Review Board Statement: This study does not involve any ethical issues.

Data Availability Statement: The data of this research are publicly available.

Conflicts of Interest: The authors declare no conflict of interest.

Appendix A

Table A1. CSR content analysis system and examples of social media from WeChat Official Account.

Level-1 Indicators	Definitions	Level-2 Indicators	Definitions	Keywords for Identification	Examples of Scoring
		EP1	Environment management system	Environmental management objectives, environment protection certification for firms, low-carbon green certification and awards, building eco-industrial chains, et al.	Chenming Paper (000488) reported: "Chenming Group was the first in its industry to pass the ISO14001 environmental management system certification in China". We marked 1 point for the EP1.
		EP2	Environmental impact assessment of new investment project	New construction projects are expected to reduce pollution emissions and are expected to improve the environment of a region, et al.	Yuntou Ecology (002200) reported: "After the project is completed, it can effectively collect and treat domestic sewage from the new urban area of Tonghai County, Xiushan Street, Jiulong Street, Sijie Town, and other areas, reducing the pollution caused by domestic sewage discharged directly into Qilu Lake". We marked 1 point for the EP2.
		EP3	Forest biodiversity conservation	n. a.	n. a.
		EP4	Environment protection investment	Amount of investment in environmental protection projects, amount of investment in environmental protection projects, et al.	Chenming Paper (000488) reported: "We have invested more than 8 billion RMB in environmental protection projects, taking the green low-carbon cycle development path". We marked 2 points for the EP4.
		EP5	Sustainable forest management	Conservation of forest resources, et al.	Yueyang Forest and Paper (600963) reported: "We placed the 'first workshop' in the forested hills to feed the forest with paper and promote paper with the forest to achieve the effect of flourishing forest and paper". We marked 1 point for the EP5.
		EP6	Forest certification	Forest stewardship council, et al.	Sun Paper (002078) reported: "FSC-certified 'Happy Sunshine' household paper received positive feedback from the public". We marked 1 point for the EP6.

Table A1. *Cont.*

Level-1 Indicators.	Definitions	Level-2 Indicators	Definitions	Keywords for Identification	Examples of Scoring
EP	Environment Protection	EP7	The quantity, kind, and risk to human and environment of toxic or exhaust emission	Type or number of harmful substances emitted, et al.	Fenglin Group (601996) reported: "Exhaust gas and wastewater emissions are better than national standards, especially the particulate matter emission concentration is about 3.5mg/mL". We marked 2 points for the EP7.
		EP8	Research, development, application, and sale of the environment production and devices	Opening and production of environmentally friendly products, application of environmental technology, renewal of environmental protection equipment, environmental certification of products, production of environmental protection, recycling of old furniture and other forest products, et al.	Chenming Paper (000488) reported: "The company is actively promoting the application of zero-water discharge technology and has visited several countries to learn about wastewater treatment and recycling technology". We marked 1 point for the EP8.
		EP9	Energy resources conservation	Save heavy oil, save water, save electricity, save paper resources, et al.	Chenming Paper (000488) reported: "We are the first one in the industry to put into operation a water reuse project, with a reuse rate of over 40% and water consumption per ton of paper reduced to less than half of the international standard level". We marked 2 points for the EP9.
		EP10	Reduce pollution and decrease drain	Reduction of carbon emissions, reduction of emissions of other gases and substances, reuse of waste, et al.	Yueyang Forest and Paper (600963) reported: "We made construction of deep sewage treatment upgrade project and fluidized bed boiler ultra-low emission project, with a significant reduction in sewage COD and nitrogen oxide emission concentration". We marked 1 point for the EP10.
		EP11	Ecology restoration	Forest restoration, land loss reduction, ecological protection, et al.	Chenming Paper (000488) reported: "Retirement of the forest will have created 3.25 million mu of completed forest land all returned to Huanggang, not cutting a tree in the old areas, production of raw materials is imported wood chips". We marked 2 points for the EP11.
		EP12	Volunteer working for environmental protection	External sanitation and cleaning work, tree planting activities, et al.	Fujian Jinsen (002679) reported: "When the 33rd International Volunteer Day, the Party Committee of Jinsen Group organized volunteers to carry out "clean up the home" environmental protection volunteer work in the co-construction community—Dongmen Community ". We marked 1 point for the EP12.
		EP0	Other environment-related	Green and sustainable development strategies, promotion of green and environmental protection concepts, environmental education, elimination of backward production capacity, et al.	Chenming Paper (000488) reported: "Chenming Group has eliminated 2.72 million tons of backward production capacity, with an elimination rate of 27%". 2 points for the EP0.

Table A1. *Cont.*

Level-1 Indicators.	Definitions	Level-2 Indicators	Definitions	Keywords for Identification	Examples of Scoring
CU	Customers	CU1	Product quality management system	Quality supervision, measurement management system in production department, standardized operation, quality management system certification, FSC-COC chain-of-custody system certification, et al.	Sophia (002572) reported: "Sophia has got ISO 9001 quality management system certification multiple times". We marked 1 point for the CU1.
		CU2	After-sale service system	Content of after-sales service, customer satisfaction, et al.	Sophia (002572) reported: "From the preparation of materials, the plate has been attached to the exclusive QR code, and this QR code will be the identity card of this piece of plate, and after that, sealing, punching, packaging, and other links, until the terminal sales, all the information of the original plate can be tracked at any time. This is more conducive to ensuring stable quality and good after-sales service". We marked 1 point for the CU2.
		CU3	Dispute settlement mechanism	n. a.	n. a.
		CU4	Information provision of the product and services	Product exhibitions, marketing information, display and promotion of products and services and design concepts, product advertising information, et al.	Qumei Home Furnishings (603818) reported: "During the 3rd season of the national sofas sales promotion, Qumei's new fashion series of sofa start from as low as 3999 RMB, with a variety of colors and moods for customers". We marked 2 points for the CU4.
		CU5	Privacy protection of customer	n. a.	n. a.
		CU6	Forest products and other green marketing	Green communication, green marketing activities for forest products, et al.	Qumei Home Furnishings (603818) reported: "The national launch ceremony of Qumei Home furnishings' 'Trade-in' sixth season large green series activities were successfully held at the '2018 Beijing International Home Furnishings Exhibition and Intelligent Life Festival'. Immediately afterward, the green life advocated by 'Trade-in' swept the country with a prairie momentum, and dealers in over 300 cities responded positively". We marked 2 points for the CU6.
		CU0	Other customer-related	Customer relationship management activities, dealer conferences, customer visits and communications, customer evaluations of products or services, and other information on customer-related activities, et al.	Fujian Jinsen (002679) reported: "On the morning of August 3, in the 9th-floor conference room of Hua Hong Technology, the company's fourth logging area timber production and sales tender will be successfully concluded. All the publicized bids were invited, with an area of 1522 m^2 and a timber volume of 12,103 m^3, and the bid amounted to 9.5 million RMB, an increase of 21.3% over the previous year". We marked 2 points for the CU0.

Table A1. *Cont.*

Level-1 Indicators.	Definitions	Level-2 Indicators	Definitions	Keywords for Identification	Examples of Scoring
EM	Employees	EM1	Abidance by rule and laws	Enterprise compliance with labor laws, et al.	Minfeng Special Paper (600235) reported: "The third staff congress of the company was held in the East Conference Room of the Administration Building, and more than 110 staff representatives from all units and departments of the company attended the meeting. The General Assembly considered a new round of Collective Contract. Since the last round of Collective Contract was signed, the Company has been able to strictly implement the content of the terms and conditions of the contract, and no violation of the contract has occurred in the past three years". We marked 2 points for the EM1.
		EM2	Percent of contract signing	n. a.	n. a.
		EM3	Coverage of social insurance	Employee social insurance, et al.	Yutong Technology (002831) reported: "Does the Yuxinyutong Personnel Workers Commercial Insurance take effect immediately after taking out the policy? Sickness death and total disability and critical illness are subject to a 30-day waiting period, and cases occurring within 30 days of enrollment will not be paid. Accidental death and disablement are effective immediately". We marked 1 point for the EM3.
		EM4	Equal employment institution	Male and female employee ratio, et al.	Chenming Paper (000488) reported, "Chenming has more than 3160 female employees, accounting for 31% of the total staff". We marked 2 points for the EM4.
		EM5	Staff development training	Staff training (skill-based), staff training seminars and events, et al.	Yutong Technology (002831) reported: "The Group's Human Resources Management Center brought the course to Vietnam on August 24–August 25. Thirty-six grassroots management cadres (managers, section chiefs and reserve section chiefs) from Yutong Vietnam and Yuzhan Vietnam attended the training, with an attendance rate of 90%". We marked 2 points for the EM5.
		EM6	Occupational health and safe producing	Improvement of work and production environment, employee medical check-ups, safety education and drills, safety hazard inspections, operational and production safety regulations, et al.	Mcc Meili Paper (000815) reported: "From April 2-4, Meili Cloud organized relevant leaders and safety managers, totaling 70 people, to conduct safety qualification certificate review and certification training at the Zhongwei Safety Production Training Center". We marked 2 points for the EM6.
		EM7	Staff relation management	Commendation of employees, recruitment, senior management retirement, employee cultural and entertainment activities, employee care, condolence to employees, corporate culture construction, employee stock ownership, firm internal journals, et al.	Yuntou Ecology (002200) reported: "This staff sports game had 31 participating delegations, more than 860 athletes and 60 referees, as well as more than 100 staff performance teams and 60 volunteer teams, which was a high participation, high quality, and high-level sports event". We marked 2 points for the EM7.

Table A1. *Cont.*

Level-1 Indicators.	Definitions	Level-2 Indicators	Definitions	Keywords for Identification	Examples of Scoring
		EM0	Other employee-related	Promotion of academic qualifications, et al.	Fenglin Group (002078) reported: "Eleven employees of Fenglin factory were happy to get their Adult Education undergraduate certificates, and everyone happily took a group photo in front of the company". We marked 2 points for the EM0.
SU	Suppliers	SU1	Responsibility purchasing system	Responsible purchasing, et al.	Minfeng Special Paper (600235) reported: "The company sent an auditor to conduct a comprehensive audit of the purchasing, production, inventory and sales practices of FSC products produced by Minfeng Special Paper in the past year". We marked 1 point for the SU1.
		SU2	Credit rating	n. a.	n. a.
		SU3	Contradict performance rate	n. a.	n. a.
		SU4	The openness of procurement policy	Number and amount of material procurement, procurement project signing ceremony, et al.	Chenming Paper (000488) reported: "The raw materials for the project are purchased and self-made cellulose-dissolving wood pulp. The purchased portion is procured uniformly in the market through the procurement channel of the group company; the self-made pulp is provided by the existing pulp-making project under construction, which has stable production and can ensure a reliable supply of fiber raw materials for the project". We marked 1 point for the SU4.
		SU5	The legality of forest product procurement	Wood procurement information, paper procurement information, et al.	Mcc Meili Paper (000815) reported, "We expect to import about 15.5 million tons of waste paper for the year, down about 10 million tons from last year". We marked 2 points for the SU5.
		SU6	Supplier qualification evaluation	Selection of high-quality suppliers, et al.	Sophia (002572) reported: "Mr. Jiang Ganjun, Chairman of the Board of Directors, Mr. Ke Jiansheng, President of the Company, and other senior executives presented awards to outstanding suppliers on behalf of the Company". We marked 1 point for the SU6.
		SU0	Other supplier-related	Bidding information, supplier conferences, supplier relationship management, et al.	Sophia (002572) reported: "In 2018 Annual Sophia Home Supplier Conference, nearly 600 supplier representatives attended the conference to discuss how to cooperate closely with the upstream and downstream of the home furnishing industry in the context of the new era for win-win development". We marked 2 points for the SU0.
		CO1	The effect of enterprise operation on community	Promotion of employment, poverty alleviation, disaster relief, product support for local, social public welfare activities, et al.	Chenming Paper (000488) reported, "Zhanjiang Chenming built the largest plant of new wall material in Zhanjiang City, all of which digests solid waste such as ash slag generated from power plants and uses it as raw material to produce lightweight bricks for sale to urban areas, supporting Zhanjiang's urban construction". We marked 1 point for the CO1.

Table A1. *Cont.*

Level-1 Indicators.	Definitions	Level-2 Indicators	Definitions	Keywords for Identification	Examples of Scoring
CO	Community	CO2	Staff localization policy	n. a.	n. a.
		CO3	Localization procurement policy	n. a.	n. a.
		CO4	Donations institution and amount	Donations institution and amount, et al.	Jingxing Paper (002067) reported: "Company Chairman Zhu Zailong donated RMB 600,000 to Zhang Lou Primary School on behalf of the company". We marked 2 points for the CO4.
		CO5	The policy of support for volunteer activity	n. a.	n. a.
		CO6	The data of staff volunteer activity	Employee volunteer activity records, employee donations, and other employee volunteer activities, et al.	Sun Paper (002078) reported: "Chen Wenjun, the company's vice president, took the lead in making donations for the disaster area, and more than 400 Sun people donated in order". We marked 2 points for the CO6.
		CO0	Other community-related	Communication and cooperation with local educational institutions, cooperation with local NGOs, et al.	Sun Paper (002078) reported: "A group of 24 people from the School of Light Industry Science and Engineering of Shaanxi University of Science and Technology came to Shandong Sun Paper Co. for exchange and study". We marked 2 points for the CO0.
GO	Government	GO1	Enterprise management abided by rule	Policy support, response to policy calls, clean and anti-corruption, protection of intellectual property rights, et al.	Yueyang Forest and Paper (600963) reported, "Discipline Inspection Committee of Yueyang Forest and Pape issued the Notice on the Implementation Plan of Yueyang Forest and Paper's 2018 'Anti-Corruption and Integrity Propaganda and Education Month' Activities ". We marked 1 point for the GO1.
		GO2	Tax payment	Tax recognition, tax payment amount, et al.	Yinhua Lifestyle Technology (600978) reported: "Taxes paid are 2.1 billion RMB". We marked 2 points for the GO2.
		GO3	Employment security policy	n. a.	n. a.
		GO4	Employment amount over the report periods	Number of employees in the enterprise, et al.	UE Furniture (603600) reports, "The company employs nearly 4500 people". We marked 2 points for the GO4.
		GO0	Other government-related	Participation in local or national People's Congress, communication and cooperation with the government, governmental officer visits, party organization building, party member activities, et al.	Yuntou Ecology (002200) reported, "On June 30, the Party Committee of Yuntou Ecology organized party education for all party members and cadres of the company to further strengthen their ideal beliefs and awareness of purpose and to remember the responsibilities and obligations of party members by watching the red movie 'Unforgettable Years' and revisiting the historical stories of party building during the Yan'an period". We marked 1 point for the GO0.

Table A1. *Cont.*

Level-1 Indicators.	Definitions	Level-2 Indicators	Definitions	Keywords for Identification	Examples of Scoring
SH	Shareholders	SH1	Investor relation management	Holding of senior management meetings, executive appointments, general meeting of stockholders, corporate investment activities, mergers and acquisitions or creation of subsidiaries, formulation of corporate strategies, strategic cooperation between other enterprises, changes in corporate shares, changes on shareholder holdings, annual planning, financing activities such as issuance of bond, the shareholding of distributors, dividends, et al.	Chenming Paper (000488) reported: "Chenming will acquire 1.369 billion shares of Nanyue Bank with 2.546 billion RMB, accounting for 14.55% of the total share capital of Nanyue Bank, through a series of combined operations such as 'Subscription to privately issued shares + Public purchase' by its subsidiary". We marked 2 points for the SH1.
		SH2	Growth potential	Project information, production capacity, sales growth, new production lines, technological breakthroughs, description of company status, company development history, talent reserves, et al.	Chenming Paper (000488) reported: "Zhanjiang Chenming invested a total of 26.5 billion yuan to build these four integrated pulp and paper production lines, of which the first phase production line and the fourth phase production line are the cultural paper production lines and white cardboard production lines with the widest paper width, the fastest speed, and the highest single-machine capacity in the world". We marked 2 points for the SH2.
		SH3	Profitability	Financial indicators such as profit and operating income, brand value, et al.	Chenming Paper (000488) reported: "The project will achieve annual sales revenue of approximately RMB 9904 million and net profit of approximately RMB 1016 million upon completion". We marked 2 points for the SH3.
		SH4	Safety	Safety of financing, safety of operation, Safety of production materials such as inventory, et al.	Guangdong Ganhua (000576) reported: "The diversified business model would also strengthen the company's financial soundness, enhance the company's anti-risk capability, and help protect the interests of the shareholders, especially the small and medium-sized shareholders". We marked 1 point for the SH4.
		SH0	Other shareholder-related	Hold or participate in academic activities, external publicity of corporate culture, social interviews of executives, et al.	Yueyang Forest and Paper (600963) reported: "Yueyang Forest and Paper, upholding the essence of history and culture and shouldering the burden of sustainable development, gathered with nearly 400 partners to talk about the future". We marked 2 points for the SH0.

Appendix B

Table A2. Firms sampled in this study.

ID	Firms	WeChat Name	Websites of the CSR Report
000488	Chenming Paper Group Co., Ltd.	晨鸣集团	
000576	Guangdong Ganhua Co., Ltd.	广东甘化000576	
000815	Mcc Meili Cloud Computing Industry Investment Co., Ltd.	中冶美利云产业投资股份有限公司	http://www.cninfo.com.cn/new/disclosure/detail?stockCode=000815&announcementId=1206091798&orgId=gssz0000815&announcementTime, accessed on 25 April 2019.
000833	Guangxi Guitang (Group) Co., Ltd.	粤桂股份	
002067	Zhejiang Jingxing Paper Co., Ltd.	景兴纸业002067	http://www.cninfo.com.cn/new/disclosure/detail?stockCode=002067&announcementId=1206113242&orgId=9900000601&announcementTime, accessed on 27 April 2019.
002078	Shandong Sun Paper Co., Ltd.	山东太阳纸业股份有限公司	http://www.cninfo.com.cn/new/disclosure/detail?stockCode=002078&announcementId=1206020135&orgId=9900001223&announcementTime, accessed on 16 April 2019.
002200	Yuntou Ecology Co., Ltd.	云投生态	
002228	Yueyang Forest and Paper Co., Ltd.	合兴包装ips	
002235	Anne Co., Ltd.	安妮股份	
002303	Meiyingsen Group Co., Ltd.	美盈森集团	
002521	Qifeng New Material Co., Ltd.	齐峰新材	
002572	Sophia Household Co., Ltd.	索菲亚家居	http://www.cninfo.com.cn/new/disclosure/detail?stockCode=002572&announcementId=1205874312&orgId=9900019037&announcementTime, accessed on 5 March 2019.
002679	Fujian Jinsen Forestry Co., Ltd.	福建金森集团	
002798	D&O Home Collection Group Co., Ltd.	帝王洁具monarch	
002831	Shenzhen YUTO Packaging Technology Co., Ltd.	裕同科技	
002853	Guangdong Piano Home Co., Ltd.	皮阿诺家居	
600076	Kangxin New Materials Co. Ltd.	康欣新材料股份有限公司	
600235	Minfeng Special Paper Co., Ltd.	民丰特纸	
600321	Rightway Holding Co., Ltd.	正源股份	
600356	Mudanjiang Hengfeng Paper Co., Ltd.	牡丹江恒丰纸业股份有限公司	http://www.cninfo.com.cn/new/disclosure/detail?stockCode=600356&announcementId=1206117239&orgId=gssh0600356&announcementTime, accessed on 27 April 2019.
600433	Guangdong Guanhao High-Tech Co., Ltd.	广东冠豪高新技术股份有限公司	http://www.cninfo.com.cn/new/disclosure/detail?stockCode=600433&announcementId=1205903842&orgId=gssh0600433&announcementTime, accessed on 18 March 2019.
600963	Yueyang Forest and Paper Co., Ltd.	百年岳纸千载文章	
600978	Yinhua Lifestyle Technology Co., Ltd.	宜华生活创享优悦	
601996	Guangxi Fenglin Wood Industry Co., Ltd.	丰林集团	http://www.fenglingroup.com/shzrbg/info_31.aspx?itemid=3137, accessed on 18 March 2019.
603022	Shanghai XTL Packaging Co., Ltd.	新通联xtl	
603165	Rongsheng Environmental Protection Technology Co., Ltd.	荣晟环保	http://www.cninfo.com.cn/new/disclosure/detail?stockCode=603165&announcementId=1205931681&orgId=9900030004&announcementTime, accessed on 23 March 2019.

Table A2. *Cont.*

ID	Firms	WeChat Name	Websites of the CSR Report
603180	Gold kitchen cabinet Home Technology Co., Ltd.	金牌厨柜官方号	
603208	Oupai Group Co., Ltd.	欧派	
603226	Vohringer Home Technology Co., Ltd.	菲林格尔vohringer	CSR report provided by China Forest Products Industry Association
603313	Mlily Furniture Co., Ltd.	mlily 梦百合	
603326	OLO Furniture Co., Ltd.	olo 我乐家居	
603389	A-Zenith Furniture Co., Ltd.	a-zenith 亚振	
603600	UE Furniture Co., Ltd.	永艺股份	
603801	ZBOM Furniture Co., Ltd.	志邦家居	
603816	KUKA Home Co., Ltd.	顾家家居	http://www.cninfo.com.cn/new/disclosure/detail?stockCode=603816&announcementId=1206054348&orgId=9900027317&announcementTime, accessed on 19 April 2019.
603818	Qumei Home Furnishings Group Co., Ltd.	曲美家居	http://www.qumei.com/upload/files/2020/3/b259ecb270526e84.pdf, accessed on 19 April 2019.

References

1. UN. Available online: https://www.un.org/zh/desa/un-adopts-new-global-goals (accessed on 1 June 2022).
2. Reinhard, A. *Innovation Management and Corporate Social Responsibility*; Springer Nature: Cham, Switzerland, 2019.
3. Ghardallou, W.; Alessa, N. Corporate Social Responsibility and Firm Performance in GCC Countries: A Panel Smooth Transition Regression Model. *Sustainability* **2022**, *14*, 7908. [CrossRef]
4. Ghardallou, W. Corporate Sustainability and Firm Performance: The Moderating Role of CEO Education and Tenure. *Sustainability* **2022**, *14*, 3513. [CrossRef]
5. Campbell, J.L. Why would corporations behave in socially responsible ways? an institutional theory of corporate social responsibility. *Acad. Manag. Rev.* **2007**, *32*, 946–967. [CrossRef]
6. Schröter, D.; Cramer, W.; Leemans, R.; Prentice, I.C.; Araújo, M.B.; Arnell, N.W.; Bondeau, A.; Bugmann, H.; Carter, T.R.; Gracia, C.A.; et al. Ecosystem Service Supply and Vulnerability to Global Change in Europe. *Science* **2005**, *310*, 1333–1337. [CrossRef] [PubMed]
7. Davies, A.L. Long-term approaches to native woodland restoration: Palaeoecological and stakeholder perspectives on Atlantic forests of Northern Europe. *Forest Ecol. Manag.* **2011**, *261*, 751–763. [CrossRef]
8. D Amato, D.; Li, N.; Rekola, M.; Toppinen, A.; Lu, F. Linking forest ecosystem services to corporate sustainability disclosure: A conceptual analysis. *Ecosyst. Serv.* **2015**, *14*, 170–178. [CrossRef]
9. Korhonen, J.; Zhang, Y.; Toppinen, A. Examining timberland ownership and control strategies in the global forest sector. *Forest Policy Econ.* **2016**, *70*, 39–46. [CrossRef]
10. Lu, F.; Wen, Z.; D'Amato, D.; Toppinen, A. Corporate Social Responsibility in the Wood-Based Panel Industry: Main Strategies from Four Enterprises in China. *Forest Prod. J.* **2018**, *68*, 163–171. [CrossRef]
11. Amin, M.H.; Mohamed, E.K.A.; Elragal, A. CSR disclosure on Twitter: Evidence from the UK. *Int. J. Account. Inf. Syst.* **2021**, *40*, 100500. [CrossRef]
12. She, C.; Michelon, G. Managing stakeholder perceptions: Organized hypocrisy in CSR disclosures on Facebook. *Crit. Perspect. Account.* **2019**, *61*, 54–76. [CrossRef]
13. Elliott, W.B.; Grant, S.M.; Hodge, F.D. Negative News and Investor Trust: The Role of $Firm and #CEO Twitter Use. *J. Account. Res.* **2018**, *56*, 1483–1519. [CrossRef]
14. Prokofieva, M. Twitter-Based Dissemination of Corporate Disclosure and the Intervening Effects of Firms' Visibility: Evidence from Australian-Listed Companies. *J. Inform. Syst.* **2014**, *29*, 107–136. [CrossRef]
15. Ettredge, M.; Richardson, V.J.; Scholz, S. Dissemination of information for investors at corporate Web sites. *J. Account. Public Pol.* **2002**, *21*, 357–369. [CrossRef]
16. Rahman, A.R.; Debreceny, R.S. Institutionalized Online Access to Corporate Information and Cost of Equity Capital: A Cross-Country Analysis. *J. Inform. Syst.* **2013**, *28*, 43–74. [CrossRef]
17. Xiao, J.Z.; Yang, H.; Chow, C.W. The determinants and characteristics of voluntary Internet-based disclosures by listed Chinese companies. *J. Account. Public Pol.* **2004**, *23*, 191–225. [CrossRef]
18. Lyon, T.P.; Montgomery, A.W. Tweetjacked: The Impact of Social Media on Corporate Greenwash. *J. Bus. Ethics* **2013**, *118*, 747–757. [CrossRef]

19. Zheng, B.; Liu, H.; Davison, R.M. Exploring the relationship between corporate reputation and the public's crisis communication on social media. *Public Relat. Rev.* **2018**, *44*, 56–64. [CrossRef]

20. Blankespoor, E.; Miller, G.S.; White, H.D. The Role of Dissemination in Market Liquidity: Evidence from Firms' Use of Twitter™. *Account. Rev.* **2013**, *89*, 79–112. [CrossRef]

21. Yang, J.H.; Liu, S. Accounting narratives and impression management on social media. *Account. Bus. Res.* **2017**, *47*, 673–694. [CrossRef]

22. Ang, J.S.; Hsu, C.; Tang, D.; Wu, C. The Role of Social Media in Corporate Governance. *Account. Rev.* **2020**, *96*, 1–32. [CrossRef]

23. Kesavan, R.; Bernacchi, M.D.; Mascarenhas, O.A. Word of mouse: CSR communication and the social media. *Int. Manag. Rev.* **2013**, *9*, 58–66.

24. Manetti, G.; Bellucci, M. The use of social media for engaging stakeholders in sustainability reporting. *Account. Audit. Account. J.* **2016**, *29*, 985–1011. [CrossRef]

25. Khanal, A.; Akhtaruzzaman, M.; Kularatne, I. The influence of social media on stakeholder engagement and the corporate social responsibility of small businesses. *Corp. Soc. Responsib. Environ. Manag.* **2021**, *28*, 1921–1929. [CrossRef]

26. Lee, K.; Oh, W.; Kim, N. Social Media for Socially Responsible Firms: Analysis of Fortune 500's Twitter Profiles and their CSR/CSIR Ratings. *J. Bus. Ethics* **2013**, *118*, 791–806. [CrossRef]

27. Ahmad, N.; Naveed, R.T.; Scholz, M.; Irfan, M.; Usman, M.; Ahmad, I. CSR Communication through Social Media: A Litmus Test for Banking Consumers' Loyalty. *Sustainability* **2021**, *13*, 2319. [CrossRef]

28. Ali, I.; Jiménez-Zarco, A.I.; Bicho, M. *Using Social Media for CSR Communication and Engaging Stakeholders*; Emerald Group Publishing Limited: Bingley, UK, 2015; Volume 7, pp. 165–185. [CrossRef]

29. Araujo, T.; Kollat, J. Communicating effectively about CSR on Twitter. *Internet Res.* **2018**, *28*, 419–431. [CrossRef]

30. Fatma, M.; Ruiz, A.P.; Khan, I.; Rahman, Z. The effect of CSR engagement on eWOM on social media. *Int. J. Organ. Anal.* **2020**, *28*, 941–956. [CrossRef]

31. Dutot, V.; Lacalle Galvez, E.; Versailles, D.W. CSR communications strategies through social media and influence on e-reputation. *Manag. Dec.* **2016**, *54*, 363–389. [CrossRef]

32. Pizzi, S.; Moggi, S.; Caputo, F.; Rosato, P. Social media as stakeholder engagement tool: CSR communication failure in the oil and gas sector. *Corp. Soc. Responsib. Environ. Manag.* **2021**, *28*, 849–859. [CrossRef]

33. Russo, S.; Schimperna, F.; Lombardi, R.; Ruggiero, P. Sustainability performance and social media: An explorative analysis. *Meditari Account. Res.* **2022**, *30*, 1118–1140. [CrossRef]

34. Vidal, N.G.; Kozak, R.A. The recent evolution of corporate responsibility practices in the forestry sector. *Int. For. Rev.* **2008**, *10*, 1–13. [CrossRef]

35. Colaço, R.; Simão, J. Disclosure of corporate social responsibility in the forestry sector of the Congo Basin. *Forest Policy Econ.* **2018**, *92*, 136–147. [CrossRef]

36. D'Amato, D.; Korhonen, J.; Toppinen, A. Circular, Green, and Bio Economy: How Do Companies in Land-Use Intensive Sectors Align with Sustainability Concepts? *Ecol. Econ.* **2019**, *158*, 116–133. [CrossRef]

37. Wang, L.; Juslin, H. Corporate Social Responsibility in the Chinese Forest Industry: Understanding Multiple Stakeholder Perceptions. *Corp. Soc. Responsib. Environ. Manag.* **2013**, *20*, 129–145. [CrossRef]

38. Hansen, E.; Panwar, R.; Vlosky, R. The Global Forest Sector: Changes, Practices, and Prospects. *Anesthesiology* **2013**, *21*, 322.

39. Panwar, R.; Hansen, E.; Kozak, R. Evaluating Social and Environmental Issues by Integrating the Legitimacy Gap With Expectational Gaps: An Empirical Assessment of the Forest Industry. *Bus. Soc.* **2012**, *53*, 853–875. [CrossRef]

40. Li, N.; Toppinen, A.; Tuppura, A.; Puumalainen, K.; Hujala, M. Determinants of sustainability disclosure in the global forest industry. *Electron. J. Bus. Ethics Organ. Stud.* **2011**, *16*, 33–40.

41. Lu, F.; Kozak, R.; Toppinen, A.; D Amato, D.; Wen, Z. Factors Influencing Levels of CSR Disclosure by Forestry Companies in China. *Sustainability* **2017**, *9*, 1800. [CrossRef]

42. Lu, F.; Wang, Z.; Toppinen, A.; D Amato, D.; Wen, Z. Managerial Risk Perceptions of Corporate Social Responsibility Disclosure: Evidence from the Forestry Sector in China. *Sustainability* **2021**, *13*, 6811. [CrossRef]

43. Deng, D.; Liu, L.Y.J.; Wen, S. The Impact of Accountability-Oriented Control Aspects of Variance Investigation on Budgetary Slack and Moderating Effect of Moral Development. *Front. Psychol.* **2020**, *11*, 583643. [CrossRef]

44. Kassinis, G.; Panayiotou, A. Visuality as Greenwashing: The Case of BP and Deepwater Horizon. *Organ. Environ.* **2017**, *31*, 25–47. [CrossRef]

45. Deng, D.; Li, C.; Zu, Y.; Liu, L.Y.J.; Zhang, J.; Wen, S. A Systematic Literature Review on Performance Evaluation of Power System From the Perspective of Sustainability. *Front. Environ. Sci.* **2022**, *10*, 914. [CrossRef]

 *forests* 

Article

Estimation of Extreme Daily Rainfall Probabilities: A Case Study in Kyushu Region, Japan

Tadamichi Sato [1,*] and Yasuhiro Shuin [2]

[1] Graduate School of Bioresource and Bioenvironmental Sciences, Kyushu University, Fukuoka 819-0382, Japan
[2] Faculty of Agriculture, Kyushu University, Fukuoka 819-0382, Japan
* Correspondence: sato.tadamichi.343@s.kyushu-u.ac.jp

Abstract: Extreme rainfall causes floods and landslides, and so damages humans and socioeconomics; for instance, floods and landslides have been triggered by repeated torrential precipitation and have caused severe damage in the Kyushu region, Japan. Therefore, evaluating extreme rainfall in Kyushu is necessary to provide basic information for measures of rainfall-induced disasters. In this study, we estimated the probability of daily rainfall in Kyushu. The annual maximum values for daily rainfall at 23 long-record stations were normalized using return values at each station, corresponding to 2 and 10 years, and were combined by the station-year method. Additionally, the return period (RP) was calculated by fitting them to the generalized extreme value distribution. Based on the relationship between the normalized values of annual maximum daily rainfall and the RP, we obtained a regression equation to accurately estimate the RP up to 300 years by using data at given stations, considering outliers. In addition, we verified this equation using data from short-record stations where extreme rainfall events triggering floods and landslides were observed, and thereby elucidated that our method was consistent with previous techniques. Thus, this study develops strategies of measures for floods and landslides.

Keywords: extreme value analysis; daily rainfall; floods; rainfall-induced landslides; regional frequency analysis; station-year method; Kyushu region

Citation: Sato, T.; Shuin, Y. Estimation of Extreme Daily Rainfall Probabilities: A Case Study in Kyushu Region, Japan. *Forests* **2023**, *14*, 147. https://doi.org/10.3390/f14010147

Academic Editor: Filippo Giadrossich

Received: 9 December 2022
Revised: 10 January 2023
Accepted: 11 January 2023
Published: 12 January 2023

1. Introduction

Extreme rainfall is an inducing factor for floods and landslides, which cause severe damage to humans and socioeconomics [1–3]. Hagon et al. [1] investigated six types of climate- and weather-related disasters (flood, storm, hydrometeorological landslide, wildfire, extreme temperature, and drought) worldwide between 1960 and 2020 and showed that floods affect more people globally than any other disaster. Ushiyama and Yokomaku [3] evaluated six types of disasters induced by heavy rainfall (storm surge, strong wind, flood, landslide, water accident in rivers excluding flood, and others) in Japan between 2004 and 2011. They showed that the number of deaths due to landslides was the highest and due to floods was the second highest [3]. In addition, increases in precipitation influenced by climate change are likely to increase floods and landslides [4–6].

Japan is within the East Asian monsoon region where torrential rainfall is frequent during the summer monsoon [7], and so floods and rainfall-induced landslides occur commonly and often cause damage [8]. In the Kyushu region, floods and landslides have occurred repeatedly [9–12] and may have been affected by climate change in recent years [13,14]; for example, an extreme rainfall event in the northern part of Kyushu in July 2017 induced severe damage from landslides, driftwood, and floods [12]. Thus, to mitigate the damage from floods and landslides, it is necessary to evaluate rainfall characteristics that may cause these disasters.

Rainfall frequency analysis statistically evaluates the magnitude of rainfall characteristics and provides basic information for the planning, design, and management of hydraulic structures (e.g., check dam and culvert) [15–18]. Similarly, rainfall frequency analysis is required in case studies of catastrophic disasters [15,19]. Hence, many researchers proposed theories for evaluating extreme rainfall (e.g., [20–23]). However, the rainfall record in the individual site is generally less than the return period (RP) required to design hydraulic structures [19]. Moreover, the record at one station may not include extreme rainfall because extreme rainfall is low frequency and has spatial heterogeneity [24]. Consequently, the results of rainfall frequency analysis may include uncertainty.

Regional frequency analysis (e.g., [25–28]) addresses these problems by using rainfall data observed at all stations in the region. The station-year method [15,16] combines the rescaled data from all stations into a single sample and fits a distribution by treating the combined sample as a single random sample [27]. Hosking and Wallis [27] mentioned that this method was rarely used because it is not appropriate to treat the rescaled data as a single random sample in many cases. Nevertheless, the station-year method has been used even in recent years (e.g., [29–31]) due to its simplicity.

In Japan, Suzuki and Kikuchihara [24] applied the station-year method to the annual maximum daily rainfall data from 137 stations and estimated the probability of extreme daily rainfall. They showed that daily rainfall for RP up to 1000 years was estimated successfully in the region using the return values of 2 and 10 years estimated by the plotting position formula (Hazen's formula) at a given station [24]. However, there was no comparison between the RPs estimated by their technique and by other processes at individual stations where extreme rainfall events occurred, that is, the usefulness and validation of their method were not sufficiently evaluated. In addition, rainfall data were not examined to the same standard because the periods for calculating the return values were different for each station.

The purpose of this study is to estimate regional daily rainfall in Kyushu by improving the previous method [24]. Additionally, we also examine the usefulness and limitations of our method in terms of development measures against floods and landslides. In this paper, we first describe the procedures for normalizing and combining daily rainfall at 23 stations in Kyushu, respectively (Section 2). Next, we examine the relationship between the RP and the normalized daily rainfall and propose an empirical dependence for estimating daily rainfall (Section 3.4). Then, we validate our method using daily rainfall data from short-record stations, including extreme events (Section 3.5). Lastly, we discuss the usefulness and limitations of our method (Section 3.6).

2. Materials and Methods

2.1. Analysis Procedure

The method of this study is divided into four steps. First, the data for the annual maximum daily rainfall in the Kyushu region are collected. Next, the spatial correlation of rainfall data between the two stations is investigated. Following this, the annual maximum daily rainfall is normalized using quantiles, corresponding to the non-exceedance probability of 50% and 90% of the generalized extreme value (GEV) distribution. The second and third steps are performed to apply the station-year method. Then, the normalized rainfall data are combined and converted to the RP.

2.2. Data Collection

The annual maximum value for daily rainfall at 23 meteorological observatories operated by the Japan Meteorological Agency (JMA) in Kyushu until 2020 was used to estimate the RP of daily rainfall because long-term records are available (Figure 1 and Table 1). The longest record is 143 years at Nagasaki, the shortest is 59 years at Fukue, and the average for the 23 sites is 98 years. The highest elevation at the rainfall station is 677.5 m at Unzen, the lowest is 2.5 m at Fukuoka, and the average is 20.6 m. The distance between the stations is 25.0 km at the shortest and 441.8 km at the longest.

In addition, the annual maximum value for daily rainfall until 2020 from the Automated Meteorological Data Acquisition System (AMeDAS) Izumi, Morotsuka, Aso-

Otohime, and Asakura operated by the JMA was used to validate our method (Figure 1). These stations were selected because they have short-period records (approximately 40 years) and observed extreme rainfall events that triggered floods and landslides (Table 2) [12,32–34]. Specifically, a rainfall event that triggered a deep-seated landslide and debris flow was observed at AMeDAS Izumi in July 1997. At AMeDAS Morotsuka, a rainfall event that triggered deep-seated landslides was observed in September 2005. At AMeDAS Aso-Otohime, rainfall events that triggered floods and debris flow were observed in July 1990 and July 2012, respectively. At AMeDAS Asakura, an extreme rainfall event that triggered shallow landslides and debris flow including driftwood was observed in July 2017.

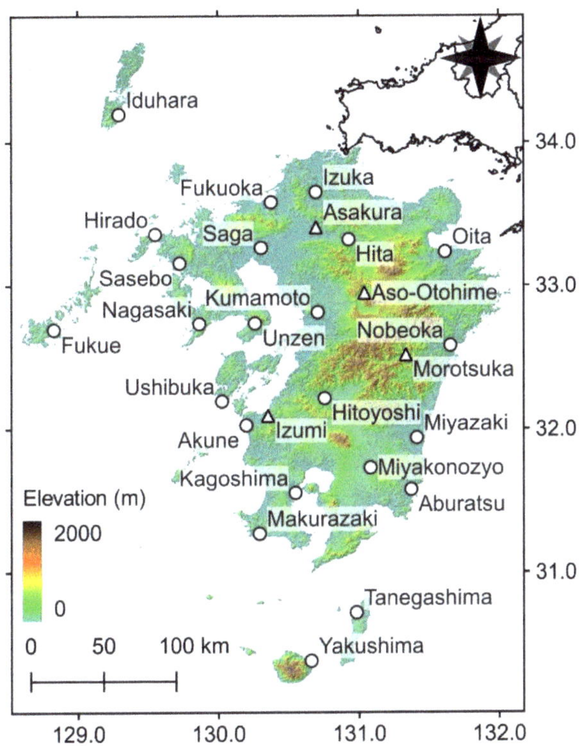

Figure 1. Location of rainfall stations. White circles and triangles indicate meteorological observatories and AMeDAS, respectively.

Table 1. Specifications of meteorological observatories in Kyushu.

Name	Location			Observation Period (Start Year)
	X	Y	Z (m)	
Fukuoka	130°22.5′ E	33°34.9′ N	2.5	131 (1890~)
Izuka	130°41.6′ E	33°39.1′ N	37.5	86 (1935~)
Oita	131°37.1′ E	33°19.3′ N	4.6	134 (1887~)
Hita	130°55.7′ E	33°19.3′ N	82.9	79 (1942~)
Saga	130°18.3′ E	33°15.9′ N	5.5	131 (1890~)
Hirado	129°33.0′ E	33°09.5′ N	57.8	81 (1940~)
Sasebo	129°43.6′ E	33°09.5′ N	3.9	75 (1946~)
Nagasaki	129°52.0′ E	32°44.0′ N	26.9	143 (1978~)

Table 1. *Cont.*

Name	Location			Observation Period (Start Year)
	X	**Y**	**Z (m)**	
Unzen	130°15.7′ E	32°44.2′ N	677.5	97 (1924~)
Fukue	128°49.6′ E	32°41.6′ N	25.1	59 (1962~)
Induhara	129°17.5′ E	34°11.8′ N	3.7	135 (1886~)
Kamamato	130°42.4′ E	32°11.8′ N	37.7	131 (1890~)
Ushibuka	130°01.6′ E	32°11.8′ N	3.0	72 (1949~)
Hitoyoshi	130°45.3′ E	32°13.0′ N	145.8	78 (1943~)
Nobeoka	131°39.4′ E	32°34.9′ N	19.2	60 (1961~)
Aburatsu	131°22.4′ E	31°34.7′ N	2.9	72 (1949~)
Miyakonojo	131°04.9′ E	31°43.8′ N	153.8	79 (1942~)
Miyazaki	131°24.8′ E	31°56.3′ N	9.2	135 (1886~)
Akune	130°12.0′ E	32°01.6′ N	40.1	82 (1939~)
Kagoshima	130°32.8′ E	31°33.3′ N	3.9	138 (1883~)
Makurazaki	130°17.5′ E	31°16.3′ N	29.5	98 (1923~)
Yakushima	130°39.5′ E	30°23.1′ N	37.3	84 (1937~)
Tanegashima	130°58.9′ E	30°43.2′ N	24.9	73 (1948~)

Table 2. Specifications of AMeDAS.

AMeDAS	Location			Occurrence of Disasters (Year)	References
	X	**Y**	**Z (m)**		
Asakura	130°41.7′ E	33°24.4′ N	38.0	2017	Takahashi et al. (2021) [12]
Aso-Otohime	131°02.4′ E	32°56.8′ N	487.0	1990 2012	Ishikawa and Shida (1990) [32] Yang et al. (2015) [34]
Izumi	130°21.1′ E	32°05.6′ N	11.0	1997	Sassa (1998) [33]
Morotsuka	131°20.1′ E	32°31.0′ N	150.0	2005	Chigira (2005) [9]

2.3. Investigation of Spatial Correlation of Annual Maximum Value for Daily Rainfall

The Kendall rank correlation coefficient (Kendall's τ) [35] of the annual maximum daily rainfall between two stations was investigated because the station-year method assumes the spatial independence of stations [17]. In previous studies [17,28,30,36], spatial independence was investigated using Pearson's correlation coefficient, but Kendall's τ was used in this study since the annual maximum daily rainfall between two stations was not assumed to be Gaussian distribution.

Kendall's τ is a non-parametric method for testing the dependence between two variables based on an ordinal association between two measured quantities. Kendall's τ is given as:

$$\tau = \frac{\sum_{i<j} sign(x_i - x_j) sign(y_i - y_j)}{n(n-1)/2} \tag{1}$$

where $sign\,(\cdot)$ is the sign function. To conduct this investigation, common period data from all stations between 1981 and 2010 were used.

2.4. Normalization of Daily Rainfall Data

Quantiles corresponding to the non-exceedance probability of 50% and 90% were used to normalize the daily rainfall data; in other words, the values corresponding to the RP of 2 and 10 years (2- and 10-year values). These indices were less variable since they were calculated by interpolation at each station and were considered suitable for normalization. Suzuki and Kikuchihara [24] also used 2- and 10-year values to normalize the daily rainfall data. The data for the annual maximum daily rainfall were normalized following Suzuki and Kikuchihara [24]:

$$y_T = \frac{x_j - x_2}{x_{10} - x_2} \tag{2}$$

where y_T is normalized daily rainfall and x_j is the annual maximum daily rainfall. x_2 and x_{10} are the 2- and 10-year values, respectively.

2.5. Extreme Value Analysis

The 2- and 10-year values were calculated using parameters in the GEV distribution [22], estimated by the L-moment method [21]. The data for the annual maximum daily rainfall between 1981 and 2010 were used due to the need to unify the period for calculating the GEV parameters. The GEV cumulative distribution function and the quantile of the GEV corresponding to the non-exceedance probability are given as, respectively:

$$F(x) = exp\left\{-\left(1 - k\frac{x - c}{a}\right)^{\frac{1}{k}}\right\} \text{ for } k \neq 0 \tag{3}$$

$$x_p = c + \frac{a}{k}[1 - \{-\ln(p)\}^k] \tag{4}$$

where k is the shape parameter, c is the scale parameter, a is the location parameter, and p is the non-exceedance probability. The parameters of the GEV are given as:

$$\begin{cases} k = 7.8590d + 2.9554d^2 \\ a = \frac{k\lambda_2}{(1 - 2^{-k})\Gamma(1 + k)} \\ c = \lambda_1 - \frac{a}{k}\{1 - \Gamma(1 + k)\} \end{cases} \tag{5}$$

$$d = \frac{2\lambda_2}{\lambda_3 + 3\lambda_2} - \frac{\ln(2)}{\ln(3)} \tag{6}$$

where λ_{1-3} are sample L-moments and Γ the gamma function. λ_{1-3} are given as:

$$\begin{cases} \lambda_1 = \beta_0 = \frac{1}{N}\sum_{j=1}^{N} x_{(j)} \\ \lambda_2 = \beta_1 = \frac{1}{N(N-1)}\sum_{j=1}^{N}(j - 1)x_{(j)} \\ \lambda_3 = \beta_2 = \frac{1}{N(N-1)(N-2)}\sum_{j=1}^{N}(j - 1)(j - 2)x_{(j)} \end{cases} \tag{7}$$

where $x_{(j)}$ is the j-th value from the smallest when the sample is arranged in increasing order.

The standard least-squares criterion (SLSC) [37,38] was used to evaluate the goodness of fit between the observed rainfall and the probability distribution. The SLSC compares the goodness of fit across distributions, and their smaller values imply better fits [37]. In this study, we judged a good fit when the SLSC value was below 0.04 in accordance with the JMA [39]. The SLSC value is given as:

$$SLSC = \frac{\sqrt{\frac{1}{N}\sum_{j=1}^{N}\left\{s\left(x_{(j)}\right) - s^*\left(p_{(j)}\right)\right\}^2}}{|s_{0.99} - s_{0.01}|} \tag{8}$$

$$s(x_{(j)}) = -\ln\left[\left\{\left(1 - k\frac{x_{(j)} - c}{a}\right)^{\frac{1}{k}}\right\}\right] \tag{9}$$

$$s^*\left(p_{(j)}\right) = -\ln\left[-\ln\left\{p_{(j)}\right\}\right] \tag{10}$$

where $s(x_{(j)})$ is the standardized variate by GEV parameters and $s^*(p_{(j)})$ is the standardized variate corresponding to the non-exceedance probability calculated by the plotting position formula [40]. $s_{0.99}$ and $s_{0.01}$ are the standardized variates corresponding to the non-exceedance probability of 1% and 99%, respectively. The plotting position formula is given as:

$$p_{(j)} = F\left(x_{(j)}\right) = \frac{j - \alpha}{N + 1 - 2\alpha} \tag{11}$$

where α is a constant; we used Cunnane's formula ($\alpha = 0.4$) [41] to give p_i following the JMA [39].

3. Results and Discussion

3.1. Spatial Correlation of Annual Maximum Value Daily Rainfall between 1981 and 2010

Figure 2 shows Kendall's τ of the annual maximum daily rainfall between two stations plotted against distance. As shown in Figure 2, Kendall's τ of the annual maximum daily rainfall between two stations tended to decrease as the distance between the two stations increased. In general, the correlation of rainfall decreases [17,28,30,36]; for example, Kuzuha et al. [34] examined the spatial correlation structure of precipitation using rainfall data at AMeDAS in Japan and showed that Pearson's correlation coefficient of daily rainfall decreased exponentially as the distance between the two stations increased. Hence, our result agreed with previous studies.

Furthermore, Kendall's τ of the annual maximum daily rainfall between two stations was 0.57 at its maximum and was generally small (Figure 2). Overeem et al. [17] examined the spatial dependence to apply the station-year method by using Pearson's correlation coefficient of daily rainfall between the two stations. They assumed the spatial dependence even if the Pearson's correlation coefficient was around 0.60 [17]. Although the Kendall's τ and Pearson's correlation coefficient were not simply comparable, we believed that the data for the annual maximum daily rainfall at 23 stations were spatially independent considering a previous study [17].

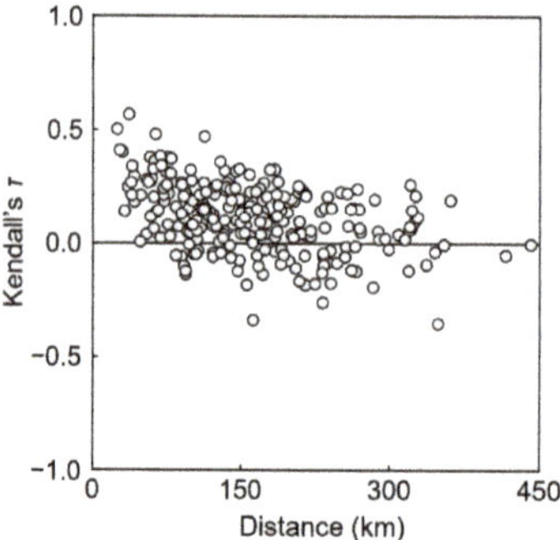

Figure 2. Kendall's τ of annual maximum daily rainfall between stations plotted against distance.

3.2. Spatial Distribution of 2- and 10-Year Values

Table 3 shows the 2- and 10-year values and the SLSC for each station. As shown in Table 3, the SLSC values at all stations were below the JMA criterion (0.04) [39]; therefore, the observed rainfall and GEV distribution were a good fit.

Figures 3 and 4 show the spatial distribution of the 2- and 10-year values, respectively. As shown in Figures 3 and 4, the 2- and 10-year values tended to be lower in the northern plains and were larger in the south, especially in the east of the Kyushu Mountains. At the Fukuoka Regional Headquarters, the JMA [42] investigated the relationship between the spatial distribution of annual precipitation and the topographical conditions in northern Kyushu Island. They showed that the annual precipitation on the plain in the north was smaller than in the mountainous areas, the eastern hillsides near the Kyushu Mountains had more precipitation due to the influence of typhoons, and the annual precipitation at 32–33° N was higher than that of around 34° N because this area was affected by typhoons and Baiu precipitation [42]. Moreover, the JMA [39] examined the spatial distribution of probable rainfall for 30, 50, 100, and 200 years using the annual maximum daily rainfall between 1901 and 2006 at 51 sites nationwide and showed that the probable rainfall in the Kyushu region (Fukuoka, Oita, Nagasaki, Kumamoto, Miyazaki, and Kagoshima) tended to be smaller in the north, but larger in the south. Thus, the 2- and 10-year values represented the regional rainfall characteristics and were good indexes for normalization.

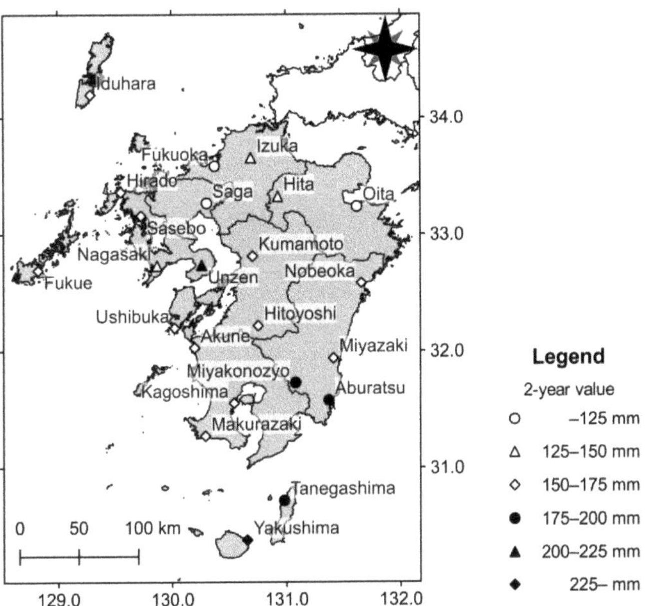

Figure 3. Spatial distribution of the two-year value.

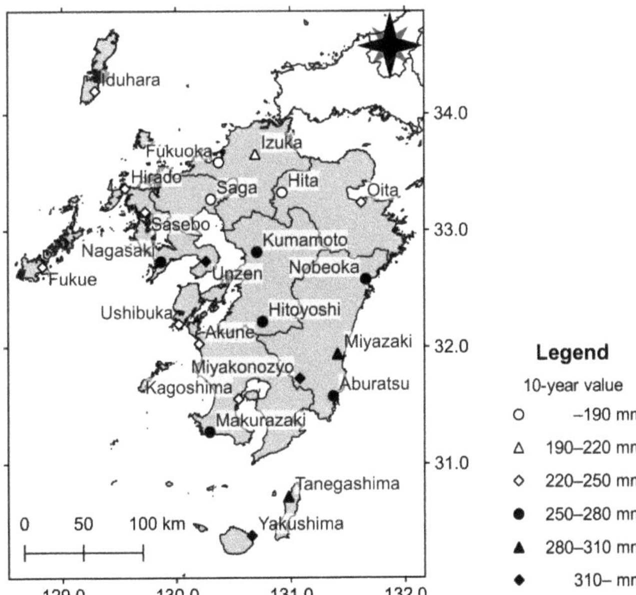

Figure 4. Spatial distribution of the 10-year value.

Table 3. The 2- and 10-year values at long-term record stations.

Name	2-Year Value	10-Year Value	SLSC
Fukuoka	121.6	182.7	0.019
Izuka	131.1	213.6	0.026
Oita	124.4	240.2	0.031
Hita	126.1	178.6	0.018
Saga	124.4	186.8	0.022
Hirado	159.4	239.0	0.006
Sasebo	150.2	242.7	0.024
Nagasaki	130.4	250.6	0.033
Unzen	207.3	330.8	0.021
Fukue	152.7	273.7	0.028
Induhara	160.2	243.9	0.024
Kamamato	166.2	276.6	0.023
Ushibuka	146.2	246.1	0.024
Hitoyoshi	161.2	254.0	0.027
Nobeoka	172.2	265.3	0.021
Aburatsu	176.1	271.6	0.022
Miyakonojo	180.7	310.4	0.028
Miyazaki	166.8	293.1	0.028
Akune	144.2	256.1	0.021
Kagoshima	158.8	234.4	0.022
Makurazaki	155.9	253.6	0.030
Yakushima	241.0	341.8	0.022
Tanegashima	176.5	284.5	0.019

3.3. Normalized Values of Annual Maximum Daily Rainfall at Each Station

Figure 5 shows a boxplot for the normalized values of the annual maximum daily rainfall. As shown in Figure 5, there are wide variations in the data from all stations because the annual maximum value fluctuates by year. On the other hand, there are few differences in the maximum, minimum, interquartile range, and median between stations (Figure 5). As a result, the data exhibited the same probability distributions at all stations and were combined using the station-year method.

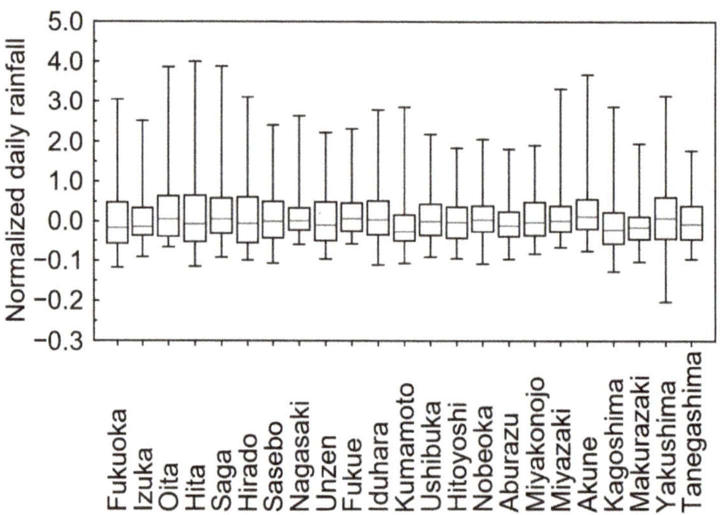

Figure 5. Boxplots for normalized values of annual maximum daily rainfall at 23 stations. Whiskers of the box show 25th (lower) and 75th (upper) percentile values. The gray line is the median value.

3.4. Relationship between Normalized Daily Rainfall and the RP

Figure 6 shows the quantile–quantile plot (Q-Q plot). As shown in Figure 6, the SLSC value was below the JMA criterion (0.04) [39]. Due to this, the standardized variate and the GEV distribution were a good fit.

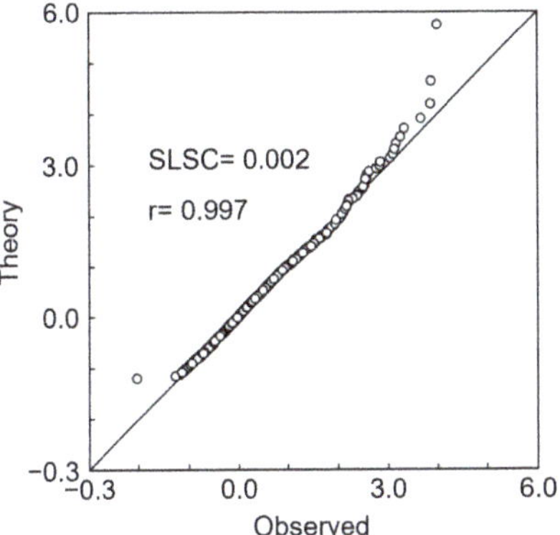

Figure 6. Quantile–quantile plot.

Figure 7 shows the relationship between the RP and the normalized values of the annual maximum daily rainfall. As shown in Figure 7, the RP increased exponentially with the increase in the normalized values of the annual maximum daily rainfall when the RP was greater than 1.5 years. Furthermore, this relation indicates that an exponential approximation was obtained by the least-squares method.

$$RP = 2.156e^{1.525y_T} \tag{12}$$

This equation accurately estimates the RP of daily rainfall up to 300 years (Figure 7) by using the 2- and 10-year values at the given stations in Kyushu.

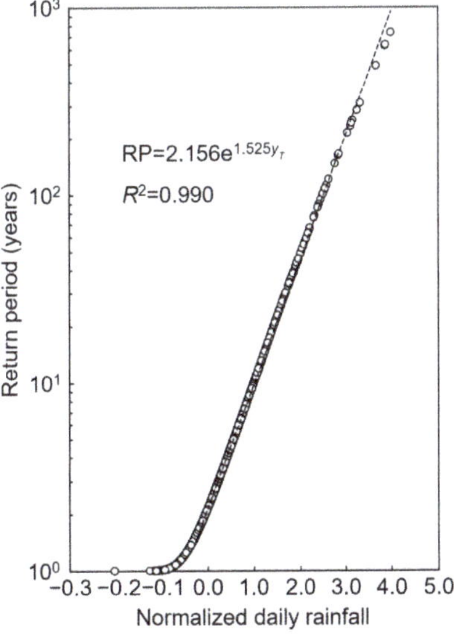

Figure 7. Relationship between normalized values of annual maximum daily rainfall and the RP.

3.5. Verification of Our Method at Short-Record Stations

Figure 8 shows the RP of the annual maximum values for daily rainfall estimated by our method and previous methods. The black and white circle indicates the RP estimated by fitting the GEV distribution and by the plotting position formula, using data from each station, respectively. As shown in Figure 8, the RP estimated by our method was consistent with that calculated by other methods, excluding the AMeDAS Asakura. At the AMeDAS Asakura, the RP might be not consistent with the result of the plotting position formula because a record-breaking regional heavy rainfall event occurred in 2017 [14], and large values for daily rainfall occurred consecutively after periods when the GEV parameters were estimated (Figure 9).

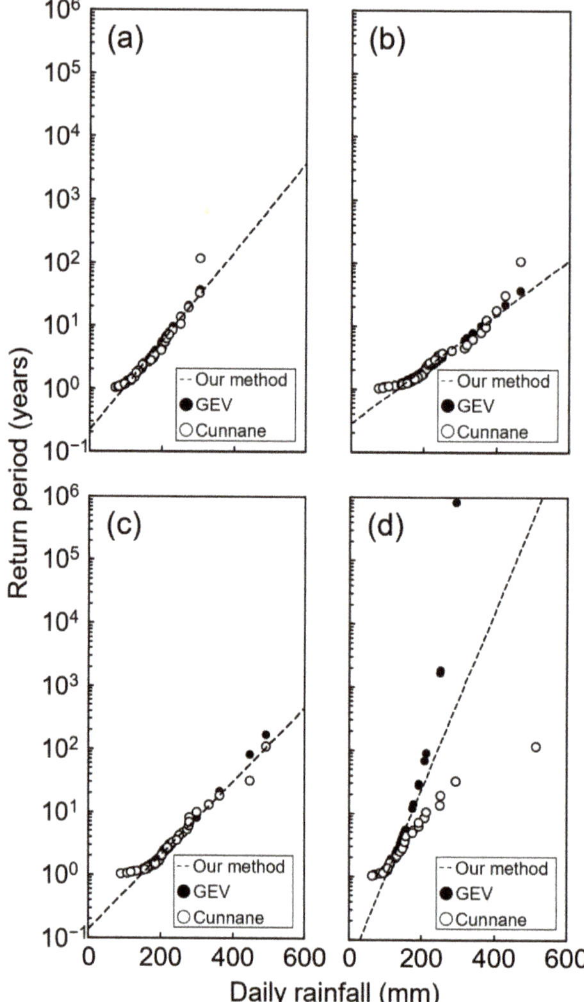

Figure 8. Comparison of RP estimated by the proposed regression equation, RP estimated by the GEV, and RP by the empirical method (Cunnane [39]) at AMeDAS Izumi (**a**), Morotsuka (**b**), Aso-Otohime (**c**), and Asakura (**d**).

3.6. Usefulness and Limitation of Our Method

Our method estimates the RP of daily rainfall up to 300 years in Kyushu using the 2- and 10-year values at a given station (Figure 7). Extreme daily precipitation is a trigger for floods, landslides, and debris flows (e.g., [43,44]), and estimating them is required to plan, design, and manage hydraulic structures against these disasters [15–18,45,46]. For example, the Sabo Planning Division, Sabo Department, NILIM, MLIT [46] summarized the methods of the Sabo master plan for preventing damage triggered by debris flows, including driftwood, in Japan. They decided that daily rainfall, corresponding to an exceedance probability of 100 years, can be used to design the scale of measures against debris flow and driftwood [46]. Hence, our method may be applied to estimate the extreme rainfall for developing measures against floods and landslides.

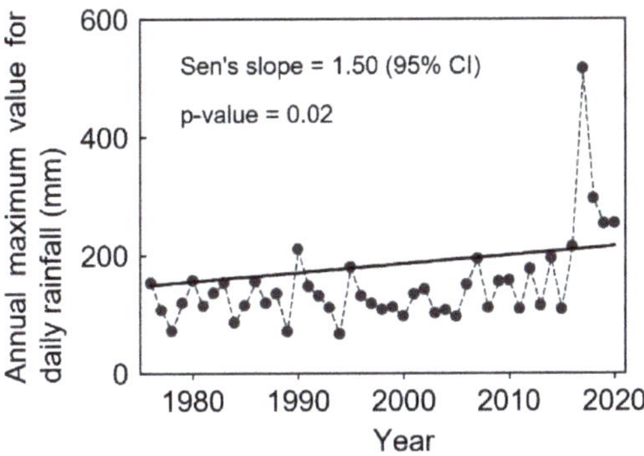

Figure 9. Annual maximum daily rainfall at the AMeDAS Asakura between 1976 and 2020. Gray circles indicate annual maximum value. Bold line indicates Sen's slope (Hipel and McLeod 1996 [47]; Sen 1968 [48]) ($p < 0.05$).

Extreme daily precipitation is generally estimated by extrapolation because rainfall records at individual stations are often short (e.g., [19]). By contrast, our technique only uses the 2- and 10-year values, which can be calculated by interpolation at many points. Kikuchihara and Suzuki [24] estimated the probability of daily precipitation using a similar approach. Meanwhile, they have not verified their applicability and limitations. In the current study, an empirical dependence (Figure 7) was verified using short-record data (approximately 40 years) at four stations where extreme rainfall was observed. As a result, the RP estimated by our method was consistent with that obtained by other means at most sites (Figure 8a–c); however, it was difficult to apply in stations with an increasing trend in daily rainfall (Figures 8d and 9). This problem may be solved to modify periods for calculating 2- and 10-year values, considering rainfall trends at each station. Overall, we concluded that our method estimates the probability of daily rainfall up to 300 years using data at given stations and contributes to developing strategies for measures against floods and landslides.

4. Conclusions

This study estimated the probability of daily rainfall in the Kyushu region, Japan. The data for the annual maximum values of daily rainfall were obtained from 23 long-record stations and were normalized by quantiles, corresponding to the non-exceedance probability of 50% and 90%. The normalized rainfall at all stations was combined by the station-year method, and the RP was calculated using GEV parameters estimated by the L-moment method. Then, a regression equation linking the normalized values of annual maximum daily rainfall and the RP was obtained; this accurately estimates the RP for up to 300 years. In addition, this relation was verified using the data at short-record stations that observed extreme rainfall events triggering floods and landslides. As a result, the probability of daily precipitation estimated by our approach was consistent with the results of previous techniques. In contrast, our method reduced the uncertainty of extrapolation by using parameters estimated by interpolation. Nevertheless, our method was difficult to apply at sites overserving an increasing trend in daily rainfall; therefore, trends in daily rainfall should be examined to use them. In conclusion, our technique estimates the RP up to 300 years for daily precipitation in Kyushu using data from the given stations, considering outliers. Hence, our findings help to develop measures for floods and landslides.

Author Contributions: Conceptualization: T.S. and Y.S.; data curation: T.S.; formal analysis: T.S.; funding acquisition: Y.S.; methodology: T.S.; validation: T.S.; visualization: T.S.; writing—original draft: T.S.; writing—review and editing: T.S. and Y.S. All authors have read and agreed to the published version of the manuscript.

Funding: This work was supported by the Grant-in-Aid for Scientific Research, JSPS (project number 21H01581).

Data Availability Statement: Daily rainfall data in Kyushu are available from the JMA website "https://www.data.jma.go.jp/gmd/risk/obsdl/index.php (accessed on 26 November 2022)".

Conflicts of Interest: The authors declare no conflict of interest.

References

1. Hagon, K.; Turmine, V.; Pizzini, G.; Freebairn, A.; Singh, R. Hazards everywhere-climate and disaster trends and impacts. In *World Disasters Report 2020*; IFRC, Ed.; IFRC: Geneva, Switzerland, 2020; pp. 34–115.
2. Hilker, N.; Badoux, A.; Hegg, C. The Swiss flood and landslide damage database 1972–2007. *Nat. Hazards Earth Syst. Sci.* **2009**, *9*, 913–925. [CrossRef]
3. Ushiyama, M.; Yokomaku, S. Location characteristics of victims caused by recent heavy rainfall disasters. *J. Disaster Inf. Stud.* **2013**, *11*, 81–89, (In Japanese with English Abstract). [CrossRef]
4. Hirabayashi, Y.; Mahendran, R.; Koirala, S.; Konoshima, L.; Yamazaki, D.; Watanabe, S.; Kim, H.; Kanae, S. Global flood risk under climate change. *Nat. Clim. Chang.* **2013**, *3*, 816–821. [CrossRef]
5. McBride, C.M.; Kruger, A.C.; Dyson, L. Changes in extreme daily rainfall characteristics in South Africa: 1921–2020. *Weather. Clim. Extrem.* **2022**, *38*, 100517. [CrossRef]
6. Shinohara, Y.; Kume, T. Changes in the factors contributing to the reduction of landslide fatalities between 1945 and 2019 in Japan. *Sci. Total Environ.* **2022**, *827*, 154392. [CrossRef]
7. Matsumoto, J. Heavy rainfalls over east Asia. *Int. J. Climatol.* **1989**, *9*, 407–423. [CrossRef]
8. Oguchi, T.; Saito, K.; Kadomura, H.; Grossman, M. Fluvial geomorphology and paleohydrology in Japan. *Geomorphology* **2001**, *39*, 3–19. [CrossRef]
9. Chigira, M. September 2005 rain-induced catastrophic rockslides on slopes affected by deep-seated gravitational deformations, Kyushu, southern Japan. *Eng. Geol.* **2009**, *108*, 1–15. [CrossRef]
10. Jitousono, T.; Igura, M.; Ue, H.; Ohishi, H.; Kakimoto, T.; Kitou, K.; Koga, S.; Sakai, Y.; Sakasnhima, T.; Shinohara, Y.; et al. The july 2020 rainfall-induced sediment disasters in Kumamoto prefecture, Japan. *Int. J. Eros. Control Eng.* **2021**, *13*, 93–100. [CrossRef]
11. Sidle, R.C.; Chigira, M. Landslides and debris flows strike Kyushu, Japan. *EOS Trans. Am. Geophys. Union* **2004**, *85*, 145–151. [CrossRef]
12. Takahashi, T.; de Jong, W.; Kakizawa, H.; Kawase, M.; Matsushita, K.; Sato, N.; Takayanagi, A. New frontiers in Japanese Forest Policy: Addressing ecosystem disservices in the 21st century. *Ambio* **2021**, *50*, 2272–2285. [CrossRef] [PubMed]
13. Fujiwara, K.; Kawamura, R. Appearance of a Quasi-Quadrennial Variation in Baiu Precipitation in Southern Kyushu, Japan, after the Beginning of This Century. *SOLA* **2022**, *18*, 181–186. [CrossRef]
14. Imada, Y.; Kawase, H.; Watanabe, M.; Arai, M.; Shiogama, H.; Takayabu, I. Advanced risk-based event attribution for heavy regional rainfall events. *NPJ Clim. Atmos. Sci.* **2020**, *3*, 37. [CrossRef]
15. Buishand, T.A. Extreme rainfall estimation by combining data from several sites. *Hydrol. Sci. J.* **1991**, *36*, 345–365. [CrossRef]
16. Koutsoyiannis, D.; Baloutsos, G. Analysis of a Long Record of Annual Maximum Rainfall in Athens, Greece, and Design Rainfall Inferences. *Nat. Hazards* **2000**, *29*, 29–48. [CrossRef]
17. Overeem, A.; Buishand, A.; Holleman, I. Rainfall depth-duration-frequency curves and their uncertainties. *J. Hydrol.* **2007**, *348*, 124–134. [CrossRef]
18. Stewart, E.J.; Reed, D.W.; Faulkner, D.S.; Reynard, N.S. The FORGEX method of rainfall growth estimation I: Review of requirement. *Hydrol. Earth Syst. Sci.* **1999**, *3*, 187–195. [CrossRef]
19. Svensson, C.; Jones, D.A. Review of rainfall frequency estimation methods. *J. Flood Risk Manag.* **2010**, *3*, 296–313. [CrossRef]
20. Gumbel, E.G. *Statistics of Extremes*; Columbia University Press: New York, NY, USA, 1958; pp. 1–375.

21. Hosking, J.R.M. L-moments: Analysis and estimation of distributions using linear combinations of order statistics. *J. Roy. Stat. Soc. Ser. B (Methodol.)* **1990**, *52*, 105–124. [CrossRef]

22. Jenkinson, A.F. The frequency distribution of the annual maximum (or minimum) values of meteorological elements. *Q. J. R. Meteorol. Soc.* **1955**, *81*, 158–171. [CrossRef]

23. Jou, P.H.; Mirhashemi, S.H. Frequency analysis of extreme daily rainfall over an arid zone of Iran using Fourier series method. *Appl. Water Sci.* **2023**, *13*, 16. [CrossRef]

24. Suzuki, A.; Kikuchihara, H. Method for estimating the return period of annual maximum daily precipitation considering about extreme values. *Tenki* **1984**, *31*, 179–189. (In Japanese)

25. Clarke-Halstad, K. A statistical method for estimating the reliability of a station-year rainfall-record. *EOS Trans. Am. Geophys. Union* **1938**, *19*, 526–529. [CrossRef]

26. Clarke-Hafstad, K. Reliability of Station-Year Rainfall-Frequency Determinations. *Trans. Am. Soc. Civ. Eng.* **1942**, *107*, 633–652. [CrossRef]

27. Hosking, J.R.M.; Wallis, J.R. *Regional Frequency Analysis: An Approach Based on L-Moments*; Cambridge University Press: Cambridge, UK, 1997; 224p.

28. Buishand, T.A. Bivariate extreme-value data and the station-year method. *J. Hydrol.* **1984**, *69*, 77–95. [CrossRef]

29. Daksiya, V.; Mandapaka, P.V.; Lo, E.Y.M. Effect of climate change and urbanisation on flood protection decisionmaking. *J. Flood Risk Manag.* **2021**, *14*, e12681. [CrossRef]

30. Olsson, J.; Södling, J.; Berg, P.; Wern, L.; Eronn, A. Short-duration rainfall extremes in Sweden: A regional analysis. *Hydrol. Res.* **2019**, *50*, 945–960. [CrossRef]

31. Mohammadi, M.; Finnan, J.; Baker, C.; Sterling, M. The potential impact of climate change on oat lodging in the UK and Republic of Ireland. *Adv. Meteorol.* **2020**, *2020*, 4138469. [CrossRef]

32. Ishikawa, Y.; Shida, T. Mud flow-floating log Disasters in Ichinomia Town, Kumamoto Pref. on July 2nd, 1990. *Jpn. Soc. Eros. Control* **1990**, *43*, 63–66. [CrossRef]

33. Sassa, K. Mechanisms of Landslide Triggered Debris Flows. In *Environmental Forest Science*; Sassa, K., Ed.; Springer: Dordrecht, The Netherlands, 1998; pp. 499–518. [CrossRef]

34. Yang, H.; Wang, F.; Vilímek, V.; Araiba, K.; Asano, S. Investigation of rainfall-induced shallow landslides on the northeastern rim of Aso caldera, Japan, in July 2012. *Geoenviron. Disasters* **2015**, *2*, 20. [CrossRef]

35. Kendall, M.G. A new measure of rank correlation. *Biometrika* **1938**, *30*, 81–93. [CrossRef]

36. Kuzuha, Y.; Tomosugi, K.; Kishii, T. Spatial correlation structure of precipitation. *Proc. Hydraul. Eng.* **2002**, *46*, 127–132. (In Japanese with English Abstract). [CrossRef]

37. Takara, K. History and Perspectives of Hydrologic Frequency Analysis in Japan. In *Pioneering Works on Extreme Value Theory*; Hoshino, N., Mano, S., Shimura, T., Eds.; Springer: Singapore, 2021; pp. 113–134. [CrossRef]

38. Takasao, T.; Takara, K.; Shimizu, A. A basic study on frequency analysis of hydrologic data in the lake biwa basin. *Annu. Disaster Prev. Res. Inst. Kyoto Univ.* **1986**, *29*, 57–171. (In Japanese with English Abstract). Available online: http://hdl.handle.net/2433/71967 (accessed on 22 August 2022).

39. Japan Meteorological Agency. Extreme Weather Risk Map 2006. Available online: https://www.jma.go.jp/jma/press/0703/28b/riskmap18.pdf (accessed on 8 November 2022). (In Japanese).

40. Hazen, A. Storage to be provided in impounding municipal water supply. *Trans. Am. Soc. Civ. Eng.* **1914**, *77*, 1539–1640. [CrossRef]

41. Cunnane, C. Unbiased plotting positions—A review. *J. Hydrol.* **1979**, *37*, 205–222. [CrossRef]

42. Fukuoka Regional Headquarters, JMA. Information on Meteorology, Earthquakes, Volcanoes, and Oceans—About Climate. Available online: https://www.jma-net.go.jp/fukuoka/kaiyo/tenkou_main.html (accessed on 8 November 2022). (In Japanese).

43. Chaithong, T. Influence of changes in extreme daily rainfall distribution on the stability of residual soil slopes. *Big Earth Data* **2022**, 1–25. [CrossRef]

44. Gaume, E.; Livet, M.; Desbordes, M.; Villeneuve, J.P. Hydrological analysis of the river Aude, France, flash flood on 12 and 13 November 1999. *J. Hydrol.* **2004**, *286*, 135–154. [CrossRef]

45. Sabo Planning Division Sabo Department, NILIM, MLIT. Manual of Technical Standard for establishing Sabo master plan for debris flow and driftwood. *Tech. Note Natl. Inst. Land Infrastruct. Manag.* **2016**, *904*, 1–77. (In Japanese)

46. Sabo Planning Division Sabo Department, NILIM, MLIT. Manual of Technical Standard for designing Sabo facilities against debris flow and driftwood. *Tech. Note Natl. Inst. Land Infrastruct. Manag.* **2016**, *905*, 1–78. (In Japanese)

47. Hipel, K.W.; McLeod, A.I. *Time Series Modelling of Water Resources and Environmental Systems*; Elsevier: Amsterdam, The Netherlands, 1994; 1012p.

48. Sen, P.K. Estimates of the regression coefficient based on Kendall's tau. *J. Am. Stat. Assoc.* **1968**, *63*, 1379–1389. [CrossRef]

Article

Rain-Driven Failure Risk on Forest Roads around Catchment Landforms in Mountainous Areas of Japan

Masaru Watanabe [1], Masashi Saito [2,*], Kenichiro Toda [3] and Hiroaki Shirasawa [4]

[1] United Graduate School of Agricultural Sciences, Iwate University, Morioka 020-8550, Japan
[2] Agri-Innovation Center, Faculty of Agriculture, Iwate University, Morioka 020-8550, Japan
[3] Geo Forest Co., Ltd., Minamiminowa 399-4511, Japan
[4] Forestry and Forest Products Research Institute, Tsukuba 305-8687, Japan
* Correspondence: msaito@iwate-u.ac.jp

Abstract: Although the causes of and impacts against forest road failure differ according to the type of damage that occurs, the statistical understanding of the trends in the type of failure is insufficient. In this study, we collected data on 526 forest road failures due to heavy rainfall during 2006–2010 in the mountainous regions of Japan and statistically analyzed the characteristics. The forest roads covered in this study include those used primarily for timber extraction as well as those used for public purposes. Forest road segments were classified into four categories: streamside, stream crossings, zero-order basin, and others, and comparisons were made regarding the length of damage, the relative probability of occurrence, repair costs, and induced rainfall intensity in each category. Streamside segments accounted for only 15% of the total length of routes analyzed but 42% of all damaged segments. Furthermore, the relative risk of the streamside segments was about 6.0 times higher than that of the other categories of segments, indicating that they were the most likely to be damaged in this analysis. It is clear that the most important issue in the target area is to prevent damage to streamside segments.

Keywords: forest road failure; streamside; stream crossing; zero-order basin; relative risk

Citation: Watanabe, M.; Saito, M.; Toda, K.; Shirasawa, H. Rain-Driven Failure Risk on Forest Roads around Catchment Landforms in Mountainous Areas of Japan. *Forests* **2023**, *14*, 537. https://doi.org/10.3390/f14030537

Academic Editor: Chong Xu

Received: 15 February 2023
Revised: 7 March 2023
Accepted: 8 March 2023
Published: 9 March 2023

1. Introduction

A properly maintained network of forest roads provides the forest accessibility necessary for the sustainable use of forest resources, including forest maintenance, timber harvesting, hunting management, and recreational activities [1]. On the other hand, forest roads gradually lose their functionality due to erosion caused by use and exposure to rainfall after establishment, and if the degree of erosion becomes severe enough, they are damaged and become dysfunctional. The forms and factors of erosion and damage vary. Accumulation of sediment and organic matter in culverts crossing forest roads reduces the drainage capacity of culverts [2], causing gully of the road surface due to overflow of stream water [3], culvert failure, and road body failure at stream crossings. Overflow of road surface runoff water causes road body failure by significantly changing the shape of the rutted area itself. Sediment accumulation on the road surface due to soil avalanches from the cut slope or its upper slope makes the road impassable for vehicular traffic.

Soil erosion, in which forest roads are directly or indirectly involved, is an important source of sediment to streams in forested watersheds, affecting the hydrologic system of the forested watershed [4]. Sediment inputs to forested stream systems can have adverse effects on water quality, such as turbidity, increased nutrient concentrations, and reduced water clarity. Therefore, studies have examined the effects of rainfall, vegetation, and season on sediment runoff generated by the existing road network [5,6], as well as studies on the interaction between the road network and the river network [7].

Facilities for proper drainage are of paramount importance in road design [8]. Therefore, studies have been conducted on the layout and dimensioning of appropriate drainage facilities, taking into account the topography and hydrological factors of the hillsides and the road structure [2,9]. From the viewpoint of minimizing the amount of road surface erosion, guidelines have been given for appropriate cross-sectional trench spacing according to the longitudinal gradient [10]. In recent years, forest road design and maintenance methods have been proposed that take into account the impact on the forest environment as well as economic efficiency; Aruga et al. [11] attempted to reduce sediment deposition to rivers and the total cost of forest road alignment planning; Dodson et al. [12] studied the environmental impact on rivers, the cost both environmental and economic to the river, the occurrence of road failures that may lead to both environmental and economic costs, and the total cost of owning a forest road, they examined a method for scheduling maintenance and improvement activities on forest road segments to minimize these three costs. Saito et al. [13] studied an automatic forest road design model that considers shallow landslides using LiDAR data. Pellegrini and Grigolato [14] examined how the integrated use of GIS and Analytical Hierarchy Process (AHP) analysis can be used to determine priorities in road network maintenance work to minimize sediment generation from road surfaces and maximize the social value of roads.

In Japan, studies have analyzed the risk of forest road failure from the perspective of topography and forest road structure. Located on the eastern edge of monsoon Asia, one of the world's rainiest regions, Japan has an average annual precipitation of 1718 mm, which is about twice the world average (880 mm) [15]. The rainy season in June and typhoons in autumn bring torrential rains. Geologically, the terrain is complex and steep due to its location in an orogenic belt where the Pacific Plate is subducting toward the continental plate, and there are many mountainous areas with active erosion. Therefore, the placement of forest roads on unstable slopes may cause slope failure. Kondo and Kamiya [16] surveyed 100 forest road failures that occurred on three forest road routes over a 13-year period from 1977 to 1989 and analyzed the characteristics of damage-prone stream crossings using Quantification II. Surface failures on hillsides occur most frequently in concave landforms (zero-order basin) where seepage water and weathered sediments are concentrated and have not yet developed in the primary valley [17]. Yoshimura and Kanzaki [18] analyzed slope failure factors by analysis of variance using only topographic factors legible from topographic maps. In the analysis, inclination, cross-section slope, turning point inclination, and catchment area were used as the factors estimating the risk of slope failure. They reported that the risk of failure is particularly high on concave slopes. Yoshimura et al. [19] also examined forest road alignment geometry in slope failure areas and reported that the risk of failure was significantly higher on inner curves where water had been concentrated, followed by a higher risk of failure at outer curve inflection points (the starting and ending points of a curve on a curve located on a ridge). Kondo and Kamiya [16] and Yoshimura et al. [19] suggest that the topography and structural configuration of forest road sections have some influence on the risk of forest road failure in Japan, although they do not consider the layout of drainage facilities. In Japanese forest road engineering textbooks, the rising water level of the stream parallel to the forest road has also been cited as a cause of damage on the valley side of forest roads [20].

Forest road failures in Japan are more likely to occur in forest road segments with topography, such as stream crossings, stream sides, and zero-order basins [16–20]. However, all of the existing studies in Japan are only the results of a few lines of investigation in some areas. Thus, for example, Kondo and Kamiya [16] surveyed 100 forest road failures that occurred on three forest road routes over a 13-year period from 1977 to 1989. They reported that failures at stream crossings accounted for 72% of the total in the research site and reported a characteristic form of damage, but it is not known to what extent this trend applies in Japan. Wemple et al. [21] reported that of 103 sediment transport events that occurred near forest roads in the western Cascade Range region of Oregon due to heavy rainfall in February 1996, three-quarters were mass movements, debris flows, cut slope slides, fill slope slides, and Phillips [22] investigated 116 landslides that occurred during two heavy rainfall events on the east coast of the North Island and found 17 landslides associated with forest roads. Sidle et al. [23] studied mass wasting along a road in Yunnan, China, to determine the extent to which landslide erosion was caused by cut-and-fill slope failure or dry ravel. The results show that the landslide erosion was caused by a cut slope, embankment slope failure, and dry ravel. However, only one or two heavy rainfall events were reported in each case.

Factors and countermeasures for forest road failure differ depending on the type of failure that occurs, but the statistical understanding of the type of damage events is not sufficient because of the limited study area in the existing studies. Since we do not know the macroscopic occurrence trends of failures, it is also impossible to determine whether there are sufficient guidelines and research on the important countermeasures for failures that occur in the target area. In addition to stream crossings, other failure hazards have been identified in concave terrain, but it is not known to what extent the susceptibility to failure differs. If there is a difference in the likelihood of failure even in the same category of damage-prone areas, the weight should be changed when selecting routes and scheduling maintenance and management. If there is a pattern of failure that tends to require high repair costs, then the most important issue from the viewpoint of reducing repair costs is to prevent the occurrence of such a pattern of failure. Since the sensitivity to rainfall intensity may differ depending on the damage pattern, it is important to understand the relationship between the failure pattern and rainfall intensity in order to consider the maintenance of forest roads under climate change. By understanding the overall trend of forest road failure, more strategic maintenance may be possible. In Japan, forest road failures occur frequently due to heavy rains caused by typhoons and rainy season fronts, which predominantly occur from July to October. The total number of damaged segments of forest roads and the total amount spent on repair in Japan in the last 3 years has been 8181 (19.3 billion yen) in the 2017 fiscal year, 13,241 (39.8 billion yen) in the 2018 fiscal year, and 12,448 (34.1 billion yen) in the 2019 fiscal year [24]. However, the latest Basic Plan for Forestry and Forest Products states that the desired length for forest roads is 250,000 km, compared to the current 190,000 km [25]. If the frequency of heavy rainfall and the length of forest roads increase in the future, the number of failures caused by heavy rainfall and the cost of repair is very likely to increase [26]. In recent years, there has been a shortage of forest engineers who can formulate repair plans [24]. If the number of failures is not reduced, it may not be possible to provide the labor required for failure repair in the future. If the number of failures increases, the time required for failure recovery may become longer, with forest roads remaining impassable for a longer period and forestry activities potentially being severely hampered. As the intensity and frequency of rainfall events increase, it is important to avoid planning routes in unstable terrain and to take appropriate precautionary measures in hazardous areas on existing routes. This is particularly important to promote appropriate forest road maintenance in the context of limited human and economic resources. To achieve this, it is necessary to identify what types of forest road failure are important to address.

Therefore, in order to provide statistical information on forest road failures, this study collected data on a large number of forest road failures that occurred in mountainous areas in Japan and statistically analyzed the characteristics of each topographic type at the points where failures occurred. Specifically, forest road segments were classified into four categories: stream crossings, streamside, zero-order basin, and others, and comparisons were made regarding the length of forest road failure, the relative probability of occurrence, repair costs, and induced rainfall intensity in each category.

The terms "erosion" and "failure" to forest roads are not strictly defined. In this study, the difference between the two terms is the degree of erosion, and the term "failure" is used when a forest road is so eroded that it cannot fulfill its design purpose and becomes dysfunctional (Figure 1). In this study, "failure" includes shallow landslides, slope failures, road surface erosion, and road body failure.

Figure 1. Image of forest road failure. A culvert at a stream crossing was destroyed by heavy rain, and the road body was eroded by stream water, making it inaccessible to vehicles.

2. Materials and Methods

2.1. Dataset

The difficulty in collecting forest road failure data over a wide area is that the larger the field survey area, the more labor is required. Therefore, it was difficult to obtain a forest road failure data set with a sample size sufficient for statistical analysis. On the other hand, in Japan, for the purpose of applying for government subsidies, the location and details of repair work are recorded in an administrative document ("Forest Road Facility Failure Assessment Document"). The failure is subject to certain conditions, such as there being a repair cost of more than 400,000 yen/point (As of 2006–2010, USD 2942 as of 7 March 2023) and the failure being caused by a rainfall event with 24 h of rainfall of 80 mm or more. The data in this document include the date of occurrence of the failure and its triggers, the distance from the start of the forest road to the damaged place, the repair cost, and the length of the segment of the road where the repair was required (Table 1).

Table 1. Items listed in the Forest Road Facility Failure Assessment Document and examples of how to complete the document.

Item	Example of Description
Identification number of the failures	No.1
Causes of failure	Typhoon No. 4 on 14–15 July 2007
Maximum hourly rainfall	11.0 mm
Maximum 24 h rainfall	112.5 mm
Failure extension	32 m
Repair cost	13,369,000 yen (USD 98,328 as of 7 March 2023)
Distance from beginning point	4500 m

However, since this document was created for administrative purposes only, it was supposed to be discarded after a five-year retention period and has not been used for research on forest road failures. In this study, we were able to obtain the "Forest Road Facility Failure Assessment Document" prepared in Nagano Prefecture between 2006 and 2010. Based on this document, an inventory of forest road failures was compiled. Among the forest road failures in the document, 10 failures caused by factors other than heavy rainfall (snowmelt) were excluded from the data. The forest roads covered in this study include those used primarily for timber extraction as well as those used for public purposes.

Nagano Prefecture is in the central region of Japan's Honshu Island and has mountainous characteristics, with forests covering 78% of its total area (approximately 1.06 million ha) [27] (Figure 2). In the Nagano Prefecture, 1962, forest roads totaling approximately 4904 km in length have been established as of the 2019 fiscal year, and the average forest road density is 7.1 m/ha [27]. Forest road counting is based on road identification codes. At least 207 roads, comprising approximately 10.7% of all roads, have experienced failures due to rainfall events during 2006–2010 (Figure 3a). The total number of damaged forest roads due to rainfall events during 2006–2010 was 701. Of these, 526 were identified as failure locations according to the forest road vector data (Figure 3b). The analysis in this study covers the 526 failures that occurred along these 207 routes. The median and mean number of the damaged locations (locations/line) for each forest road were 1 and 3, respectively. In 79% of the lines, the number of damaged points per line was less than 3 (Figure 4).

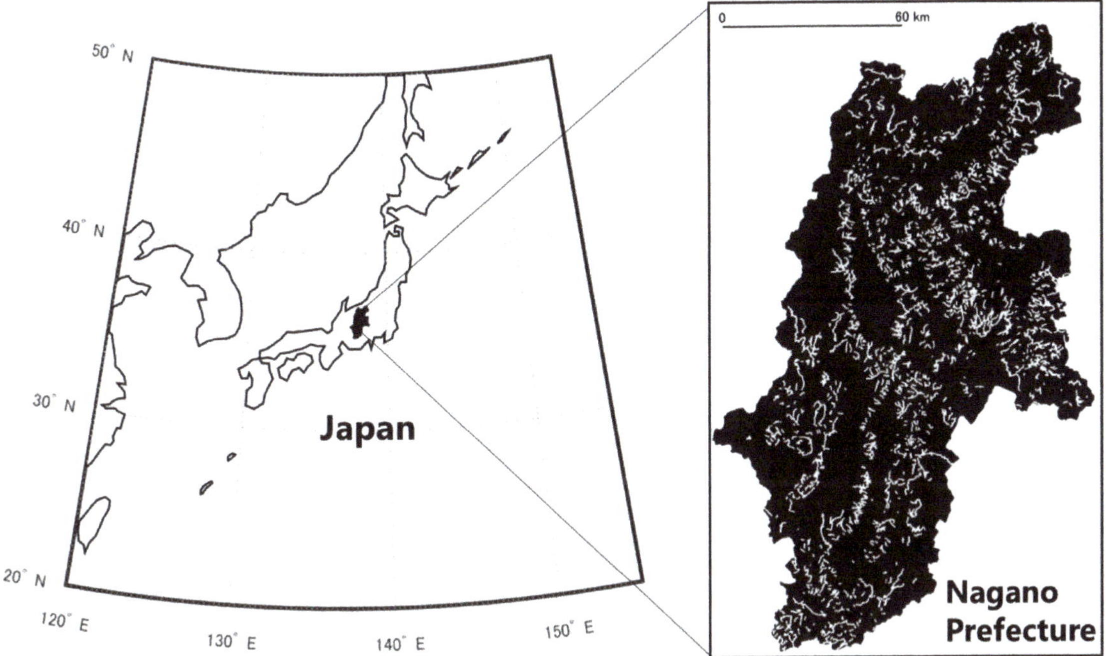

Figure 2. Map showing the research site. White lines indicate all forest roads in the Nagano Prefecture. In the figure on the right, no public roads other than forest roads are shown, so the forest roads are shown as if they exist independently. The forest road vector data shown in the figure was created by Nagano Prefecture.

2.2. Classification of Forest Road Segments

The Forest Road Facility Failure Assessment Document obtained for this study does not record the type of failure. In other words, it is not possible to determine whether the failure is the road surface, cut slope, fill, or road body. However, since the location data of the failures are recorded, the topographical features of failures can be clarified by analysis. Since it is possible to infer to some extent the causes of road failure from the topographical features of the failures, we decided to classify failures from the viewpoint of topographical features.

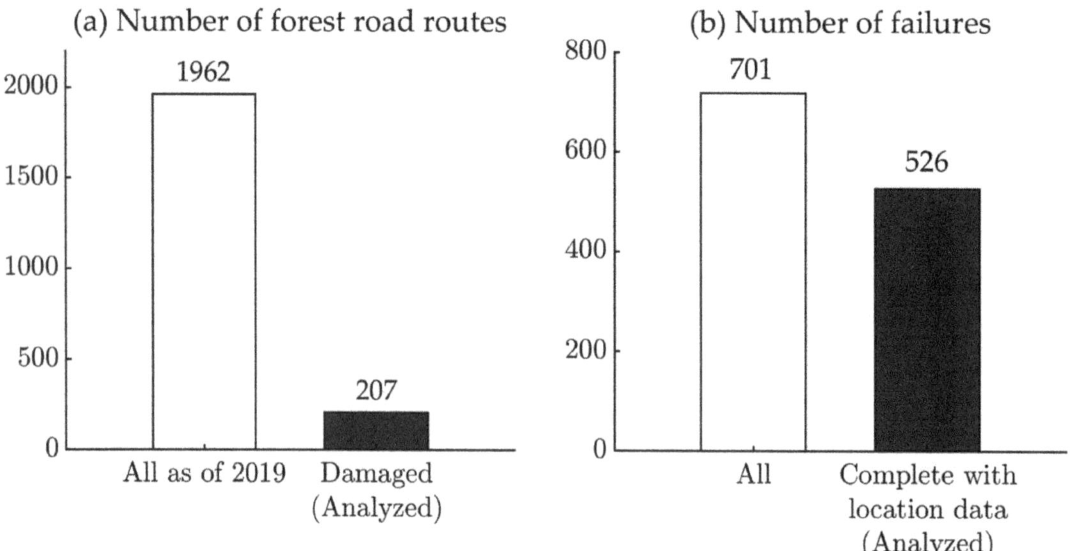

Figure 3. Number of forest road routes and failures analyzed. (**a**) Two hundred-seven routes where the failure occurred were included in the analysis. (**b**) Failures for which location information on both the failure and the forest road route has been developed are included in the analysis.

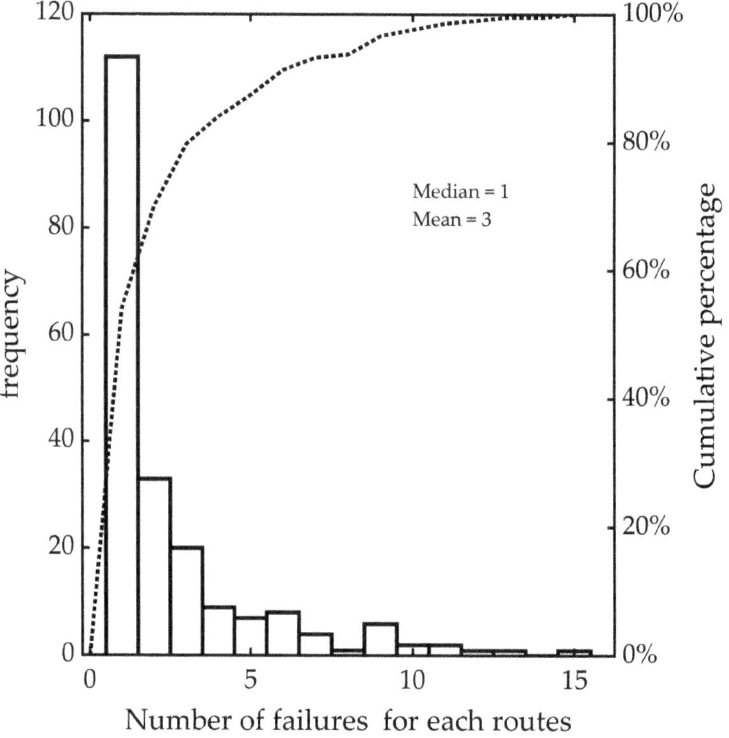

Figure 4. Distribution of the number of failures according to route.

To accurately assess the risk of failure, it is desirable to consider the presence or absence of structures, such as drainage facilities, as well as their types and functional status. However, in order to conduct an analysis that takes structures into account, it is necessary to record the presence and absence of structures, their types, and their functional status prior to the occurrence of a failure. In this study, due to the availability of data, it is not possible to obtain information on structures prior to the occurrence of the failure since the study deals with forest road failure that occurred from 2006 to 2010. Therefore, the structural information was not considered in the analysis of the failure in this study. Even if the above structural information could be obtained and failure classification was performed considering both structures and topography, it is expected that the results would be complicated because a single failure belongs to multiple categories. Since the purpose of this study is to statistically understand the morphological characteristics of the elucidated failure events rather than to strictly describe the risk of forest road failure, we believe that a concise approach to classifying failure based on topographical factors alone is effective. Even without considering structural information, the identification of topographic features that are prone to forest road failure can, paradoxically, identify the characteristics of areas where structures are not functioning properly.

Referring to previous studies [16–20], forest road segments were classified into three categories, namely streamsides, including stream crossings (Figure 5), around concave landforms such as zero-order basins (Figure 6), and "others" that do not fall into either of these categories. A category for the slope was also considered. However, steep slopes were not included in the categories because they are a more universal landform type and could be mixed with other landform types, complicating categorization. Lithology and geology were not included for the same reason. In Japan, the Geospatial Information Authority of Japan provides spatial information on landslide landforms, geology, and faults. However, not all landslide landforms and faults have been databased, and it is possible that some small-scale landslides, especially in short sections of forest roads, have not been extracted, so they were not used in the evaluation categories. However, since a high-resolution DEM was used in this study, it is likely that the characteristics of erosional landforms, such as fault zones and landslide landforms, were also taken into account as erosion heights, which will be discussed later.

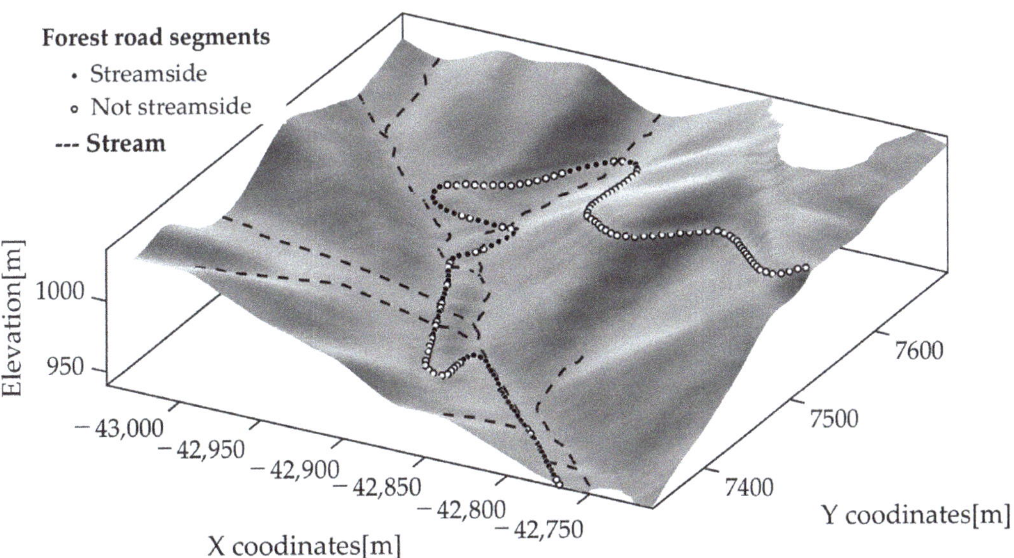

Figure 5. Image of a forest road segment that falls into the streamside category. Note: The points in the figure represent the center coordinates of the segment.

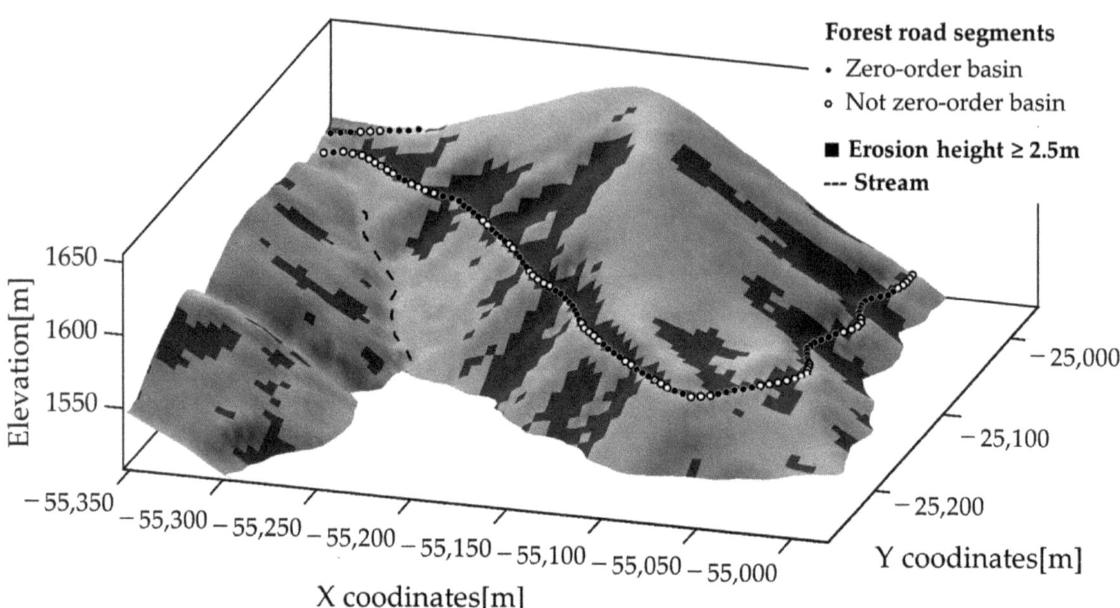

Figure 6. Image of a forest road segment that falls into the zero-order basin category. Note: The points in the figure represent the center coordinates of the segment.

First, each forest road was equally divided into 5 m intervals. The division process is performed by creating a new apex every 5 m from the beginning point of the forest road. The apex of the division' refers to each vertex created by this process. In this study, the apex of the division represented the center of the forest road segment. The presence or absence of forest road segment failure was determined based on the positional relationship between the forest road segment feature and the failure feature. First, line features (vertex interval: 5 m) were created from The Forest Road Facility Failure Assessment Document to indicate the forest road segments where forest road failure occurred. Next, a nearest neighbor search was conducted from each vertex of the failure line feature to the forest road segment center point set. The forest road segment that corresponded to the nearest neighbor was defined as the damaged segment.

Next, the forest road segments were classified into three categories using a Digital Elevation Model (DEM). The DEM used was a 1 m resolution DEM created from an aerial laser survey conducted by Nagano Prefecture in 2013, which was resampled to a 5 m resolution.

2.2.1. Streamside

The raster data representing the water system lines were converted into vector data, which were equally divided into 5 m intervals, as was the case with forest roads. The nearest waterline vertex was searched from each forest road vertex, and its distance was obtained. The forest road vertices within 15 m of the waterline were defined as forest road segments along the stream. In this study, stream features were created from a DEM with a 5-m cell size. Therefore, we used 15 m, the distance of 3 cells in the DEM, as a guide to represent the adjacent range of the stream.

2.2.2. Stream Crossing

Stream line features and forest road line features were determined to intersect, and forest road segments that intersected were defined as stream crossings.

2.2.3. Zero-Order Basin

As a method for estimating zero-order basins, Shirasawa et al. [28] developed and validated a method using three landform quantities; erosion height, uneroded height, and erosion rate were considered to represent the degree of erosion. The erosion height is calculated as the difference between the summit level map and the base level map, which contains the highest and lowest elevation points in an area [29]. The amount of topography represented by the erosion height depends on the sampling grid size of the DEM. Kühni and Pfiffner [29] analyzed the morphology of valleys incised by major rivers in the Swiss Alps by using a 10 km square smoothing filter. In this study, erosion heights were determined using a smoothing filter of 30 m square to evaluate erosion due to surface failure, following the method of Shirasawa et al. [28]. Shirasawa et al. [28] determined the threshold of erosion heights representing zero-order basins by examining erosion heights on shallow landslides. The erosion height was used to simplify the method of estimating the zero-order basins. Some studies have used curvature to assess the risk of surface failure, but the assessment model uses multiple factors other than curvature [30]. Since it has been suggested that the erosion height calculated from the tangent peak surface and the tangent valley surface could be used to estimate the zero-order basin with a single indicator, we used the erosion height instead of the curvature in this study.

However, in this study, it was not possible to collect a sufficient sample of damaged slope areas. Therefore, cells with erosion heights of 2.5 m or more were treated as zero-order basins with reference to the relative risk values. The erosion height tends to be high around the waterline, and the area around the waterline is also defined as a concave landform according to the defining parameters used in this study. In this study, we wanted to make a clear distinction between streamside and zero-order basins in terms of topographic scale. Therefore, we masked the erosion heights of cells located 15 m around the waterline to avoid confusion between the two landforms. Zero-order basins defined in this study represent unchanneled landforms with a catchment area of 1 ha or less. Even under these conditions, forest road segments located along a stream with a zero-order basin on the backslope fall into both the along-stream and around-concave landform categories.

At each forest road apex, a concave cell was examined for the presence of a concave cell at 20 m in the transverse direction. The 20 m cell was examined to avoid evaluating the forest road slope as a concave cell. The vertices where concave cells were present were defined as the forest road segment around the concave cell. Although a simplified method was used in this study, there is room for further study of the evaluation method for the area zero-order basin.

2.3. Extension, Relative Risk, Repair Cost, and Rainfall at Failure by Terrain Type

First, by totaling the extensions of forest road segments by topographic type, we obtained the percentage of topographic types on an extension basis. The percentage of the landforms that were damaged was also calculated by adding up the extensions of the damaged segments by landform type. By comparing these data, the characteristics of failure in each landform type were examined in terms of the extension of failure.

Next, we calculated the relative risk of each landform type with respect to the other categories in order to examine the extent to which the susceptibility to failure differs by landform type. For example, the relative risk of the streamside category is calculated as the ratio of the probability of failure in the streamside category to the probability of failure in the other categories. Since the relative risk of each landform category represents the likelihood of failure relative to the other categories, it is also possible to compare the likelihood of failure among landforms. In order to clarify the relationship between topographic form categories and failure rates, a test of independence was conducted using the χ-square distribution. The null hypothesis of the χ-squared test is that the two variables, i.e., a given landform category and loss rate, are independent. If the null hypothesis is rejected, it suggests that there is some relationship between the two variables, i.e., the category may have an effect on the susceptibility to damage. Finally, we examined whether the cost of repair and rainfall at the time of failure differed by terrain type. While repair costs and rainfall were recorded on a per-loss basis from the Forest Road Facility Failure Assessment Document, the topographic configuration differed on a per forest road segment basis. Therefore, depending on the length of the failure, a single failure may consist of segments of multiple terrain categories. In this case, the category with the highest number of segments was treated as the topographic category of the failure in question. The Kruskal–Wallis test was performed to determine whether the samples were drawn from the same population (or different populations with the same distribution) by comparing the median repair cost and rainfall at the time of failure for each topographic category group. If the null hypothesis is rejected, it suggests that not all samples may have been obtained from the same continuously distributed population. In this case, a further multiple comparison test is performed to determine whether there is a significant difference between any of the groups.

The rainfall at the time of failure recorded in the Forest Road Facility Failure Assessment Document was obtained from the nearest meteorological station (Automated Meteorological Data Acquisition System). These meteorological stations are located at intervals of about 17 km, and 45 of them are located in Nagano Prefecture (Figure 7) [31].

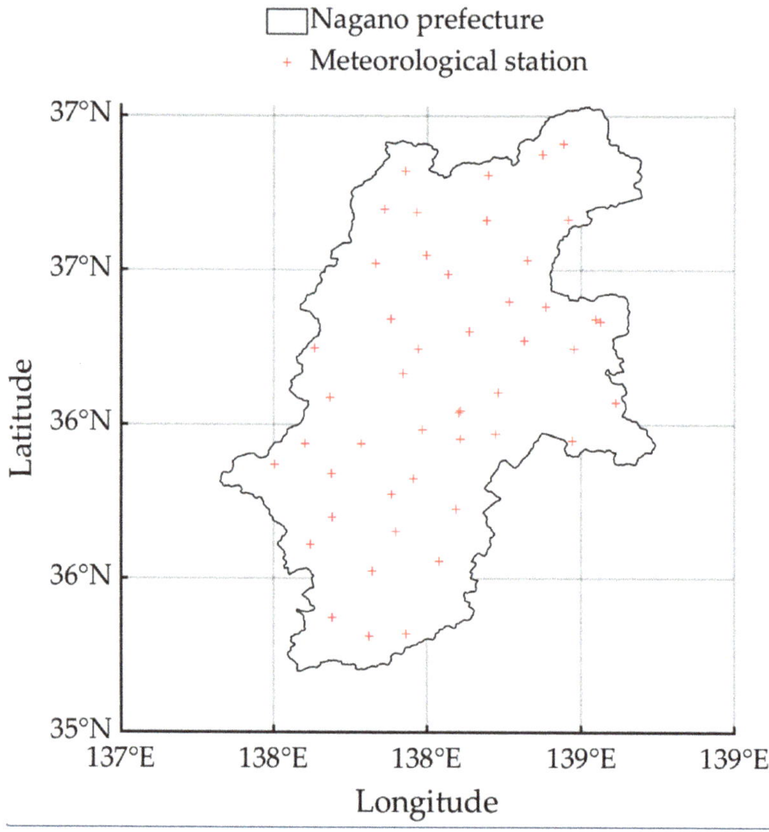

Figure 7. Distribution of meteorological stations established in Nagano Prefecture.

3. Results

3.1. Percentage of Forest Roads Damaged by Segment Categories

Figure 8 shows the percentage of all forest road segments and damaged forest road segments by terrain type category.

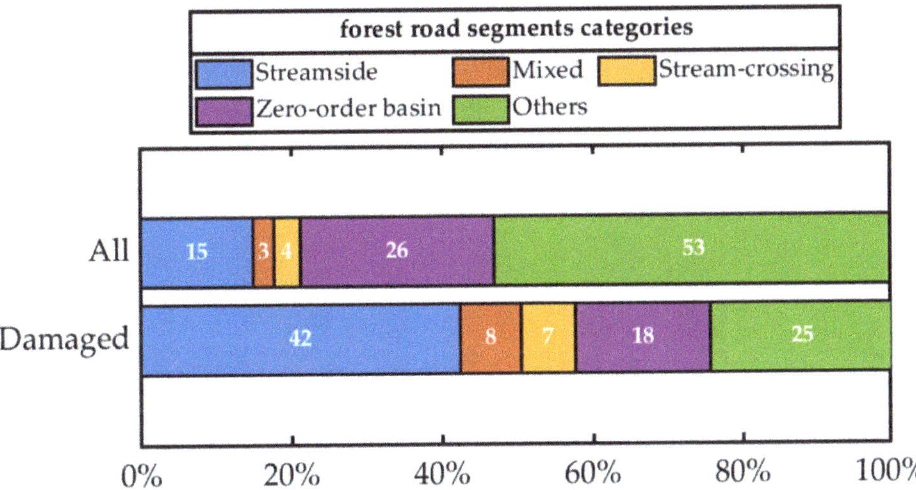

Figure 8. Percentage of all forest road segments and damaged forest road segments by terrain type category." Mixed" indicates segments that fall into both streamside and zero-order basin landforms.

The bar graph in the upper row of Figure 8 shows the percentage of forest road segment categories calculated for the 207 routes; 15% was streamside, 26% zero-order basin, 3% in both categories ("mixed"), 4% was stream crossing, and 53% was of other types.

The bar graph in the lower row of Figure 8 shows the percentage of forest road segment categories calculated for the failure segments; 42% was streamside, 18% zero-order basin, 8% in both categories, 7% was stream crossing, and 25% was of other types. By extension, 75% of the forest road failures that occurred in the target area occurred along streams or stream crossing or zero-order basins, indicating that forest road failures along streams are most common.

3.2. Relative Risk of Forest Road Segment Categories

The number of damaged and undamaged segments was recorded for each forest road segment category, and the relative risk to forest road segments in the "other" category was calculated (Figure 9). The χ^2 test results showed that the ratio of the number of damaged to undamaged forest road segments in the "other" category was significantly higher than that in the "other" category for all categories. The relative risk for forest road segments in the "other" category was higher along streams and zero-order basins, in that order. The streamside was about 6.0 times more likely to be damaged, stream crossings were about 4.6 times more likely to be damaged, and zero-order basins were about 1.9 times more likely to be damaged than the other categories.

3.3. Failure Characteristics by Forest Road Segment Category
3.3.1. Failure Repair Costs

Figure 10 shows the distribution of repair costs for the 287 damaged areas where repair costs were available (USD 1 = JPY 135.96 as of 7 March 2023). Only a small percentage of the failures were applicable to stream crossings (failures in which stream crossing segments were in the majority), so they were excluded from the analysis. The median and mean repair cost for all the damaged sites was approximately 2.33 million yen and 2.94 million yen, respectively (USD 16,946 and 21,584, respectively, as of 7 March 2023). The Kruskal–Wallis test was conducted for the three forest road segment categories, and the null hypothesis that all samples were derived from the same distribution with a 5% risk rate was not rejected. It is considered necessary to consider factors other than topographic morphology in analyzing the causes of damage that lead to high repair costs.

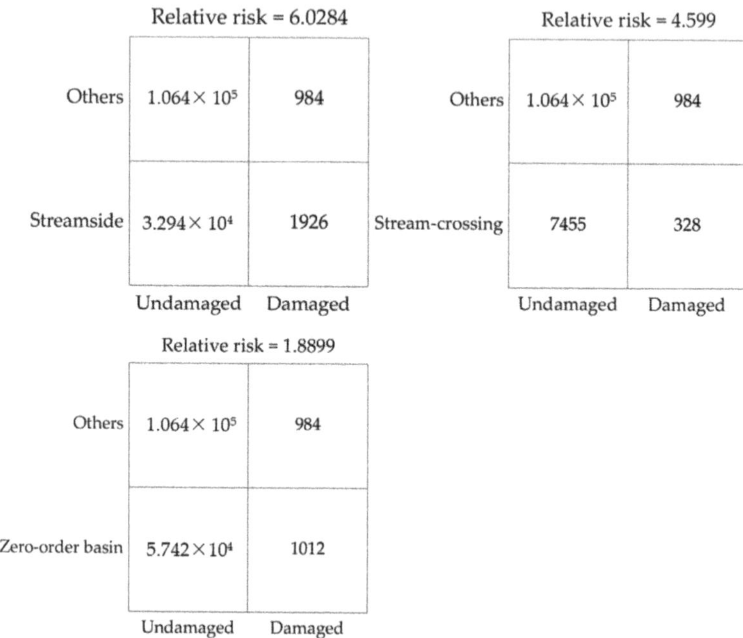

Figure 9. Cross tabulation table of forest road segment topography form and damage status. Rows correspond to the topographic form of the forest road segment, and columns correspond to the damage status.

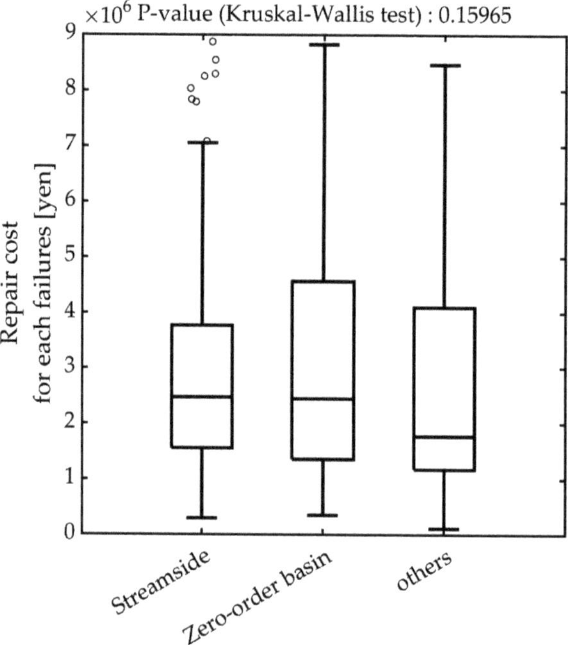

Figure 10. Repair cost by risk category for each area damaged by failure. The line in each box represents the median of the sample; the upper and lower ends of each box represent the upper and lower quartiles, respectively.

3.3.2. Amount of Rainfall

The distribution of the maximum 1 h rainfall and maximum 24 h rainfall for the rainfall events that triggered the forest road failures are shown in Figures 11 and 12, respectively. The maximum and minimum values for 1 h potential rainfall [32] and 24 h potential rainfall [33] at the target sites are plotted in the figures as reference values. The median maximum 1 h rainfall for all the damaged locations was 15.5 mm/h. The Kruskal–Wallis test was performed for the three forest road segment categories. However, the null hypothesis that all samples were derived from the same distribution with a 5% risk rate was not rejected. The quartile range of maximum 1 h rainfall at the damaged sites being located between the maximum and minimum values of 1-year potential rainfall suggests that forest road failures can occur even when the intensity of the maximum 1 h rainfall is not extreme. The median maximum 24 h rainfall intensity for all damaged sites was 175 mm/24 h. The Kruskal–Wallis test was conducted for the three forest road segment categories. The null hypothesis that all samples were derived from the same distribution with a 5% risk rate was rejected. Multiple comparisons showed that the mean maximum 24-h rainfall at the time of damage for the streamside category was significantly higher than the mean for the other categories at a 5% risk rate. The first quartile of the streamside category was higher than the other categories and exceeded the 30-year potential rainfall in some areas.

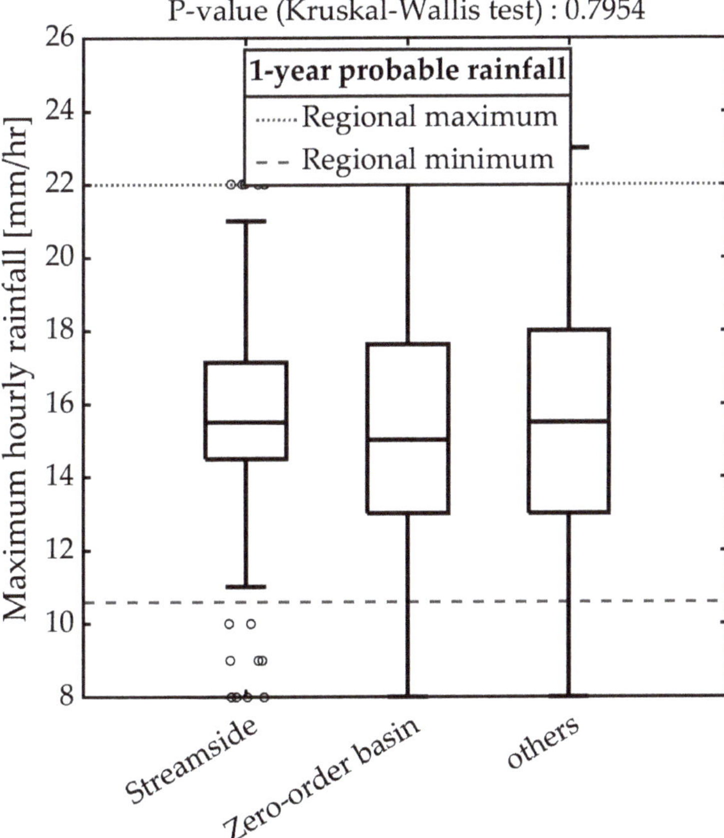

Figure 11. Distribution of maximum hourly rainfall in the damaged area. The line in each box represents the median of the sample; the upper and lower ends of each box represent the upper and lower quartiles, respectively.

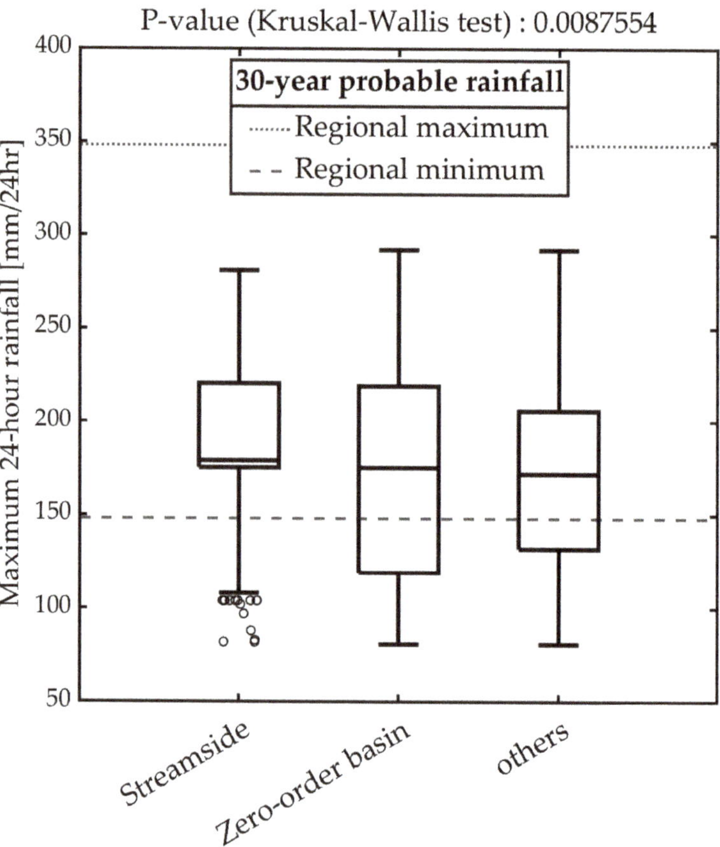

Figure 12. Distribution of maximum 24 h rainfall in the damaged area. The line in each box represents the median of the sample; the upper and lower ends of each box represent the upper and lower quartiles, respectively.

4. Discussion

Streamside forest road segments accounted for only 15% of the total length of forest roads analyzed, but 42% of the road segments that were damaged. Furthermore, the relative risk of the streamside forest road segment was about 6.0 times higher than that of the other categories of forest road segments, indicating that it was the most likely terrain type to be damaged in this analysis. From the perspectives of both the length of failure and the relative risk, it is clear that the most important issue in the target area is the prevention of failure on the streamside forest road segments. The Forestry Agency supports the planning and improvement of trunk forest roads important for failure prevention and conducts projects to promote the strengthening of forest roads [34]. Countermeasures along stream segments will be particularly important for the resilience of Japan's forest roads against failures.

In previous studies evaluating the environmental impact of forest roads, distance to streams [35,36] and avoidance of stream crossings [37] have been used as one evaluation index. This study found that proximity to a stream is also an important evaluation indicator for forest road maintenance in terms of the susceptibility of forest roads to damage. Jing et al. [38] studied the spatial relationship between road and river networks in central China and noted that road and river networks were closely related spatially, with the density of high-standard roads increasing the closer they were to rivers. In Japan, there is a history of forest roads being located along streams because of the preference for timber haul-out routes using progressive gradients before the 1960s, when the driving performance of timber transport vehicles was low [39]. Especially when forest roads located along streams play a key role in the road network, they need to be subject to maintenance and improvement because of the significant disruption that would result from their destruction. The mean maximum 24-h rainfall at the time of the failure of a forest road segment along a stream was significantly higher than the mean for other geomorphic categories at a 5% risk rate. The frequency of extreme precipitation events per degree of warming also increased, and these trends were reported to be greater in Europe and Japan than in the United States and Australia [40]. Future changes in rainfall patterns may further increase the risk of damage, especially along streamside forest roads, and the relationship between failure and rainfall should continue to be investigated.

Even in terrain forms that were considered to be at high risk of failure in previous studies, the ease of failure in terms of relative risk differed by several fold. In studies of forest road evaluation based on multi-criteria evaluation, the AHP method has been used to determine the importance of different criteria [12,35–37]. Through interviews with experts and others, evaluations have been made, for example, that Criterion A is twice as important as Criterion B. However, these evaluations are based on the subjective opinions of experts. With regard to the weight of the criteria for failure, a more objective evaluation may be possible by accumulating statistical knowledge.

The topographic morphology of the forest road segments analyzed in this study could not explain the difference in failure repair costs. According to Watanabe et al. [41], who studied the past repair costs of 1504 forest roads in Nagano Prefecture, Japan, the standard deviation of the repair cost of one forest road failure amounts to about 26.2 million yen. Therefore, it is considered necessary to first investigate the characteristics of the failure in detail, especially for the failure with high repair costs.

In this study, we were able to obtain five years of inventory data on forest road failure by the prefectural government, which allowed us to statistically analyze the characteristics of forest road failure. In Japan, such inventory data are temporarily maintained by administrative departments but are discarded after a certain period of time because of a lack of familiarity with how to utilize them. As revealed in this study, some knowledge can be obtained only by collecting data on forest road failure on a scale of several hundred routes. Therefore, it is important to establish a system to compile a database of forest road failure inventories in the future. In this study, we focused on forest road failure that occurred more than 10 years ago, so we were unable to consider the presence, location, and condition of structures (drainage facilities, retaining walls, road surface pavement) that existed before the damage occurred. In Japan, it is rare to record information on such forest road structures in normal times because the value of their use is not well known. Since these structures are essential for a more accurate assessment of the risk of forest road failure, it is desirable to investigate their location and other information prior to the occurrence of a failure. Recently, studies have evaluated the degree of damage to failure-risk forest road surfaces by remote sensing [42–44]. Although it is labor-intensive to survey all forest roads immediately, it is also important to combine techniques to determine the functional status of forest road structures in a labor-saving manner, focusing on forest roads composed of segments at high risk of failure, as identified in this study and to build an inventory data of the structures.

There are various types of failure to forest roads, and it is necessary to construct an evaluation model of failure risk and prioritize countermeasures according to the cause and risk of failure. For example, in a forest road segment that runs alongside a stream, the nourishing force of the stream is considered to be the dominant failure factor, and reinforcement of the fill slope may be a more important failure risk reduction method than the placement of cross-drainage ditches. In this study, forest road segments were classified in terms of topography, and comparisons were made regarding the length of forest road failure, the relative probability of occurrence, repair cost, and induced rainfall intensity in each category. The results showed the importance of responding to forest road failures along streams, which was revealed by the analysis based on a large number of data for a wide area of the prefecture.

5. Conclusions

In this study, we statistically analyzed the characteristics of each topographic type at the points where failures occurred. Specifically, forest road segments were classified into four categories: stream crossings, streamside, zero-order basin, and others, and comparisons were made regarding the length of forest road failure, the relative probability of occurrence, repair costs, and induced rainfall intensity in each category. We were able to obtain the "Forest Road Facility Failure Assessment Document" prepared in Nagano Prefecture between 2006 and 2010. Based on this document, an inventory of forest road failures was compiled. The analysis in this study covers the 526 failures that occurred along these 207 routes. There are various types of failure to forest roads, and it is necessary to construct an evaluation model of failure risk and prioritize countermeasures according to the cause and risk of failure. Streamside forest road segments accounted for only 15% of the total length of forest roads analyzed, but 42% of the road segments that were damaged. Furthermore, the relative risk of the streamside forest road segment was about 6.0 times higher than that of the other categories of forest road segments, indicating that it was the most likely terrain type to be damaged in this analysis. From the perspectives of both the length of failure and the relative risk, it is clear that the most important issue in the target area is the prevention of failure on the streamside forest road segments. The results showed the importance of responding to forest road failures along streams, which was revealed by the analysis based on a large amount of data for a wide area of the prefecture.

Author Contributions: Conceptualization, M.W. and M.S; formal analysis, M.W.; resources, K.T. and H.S.; data curation, K.T. and H.S.; writing—original draft preparation, M.W.; writing—review and editing, M.W., M.S., K.T. and H.S.; funding acquisition, M.S. All authors have read and agreed to the published version of the manuscript.

Funding: This research was funded by the Japan Society for the Promotion of Science (JSPS) KAKENHI (Grant Number: 22H03800, 21K05665, 21H03672, 19H02991) by M.S.

Institutional Review Board Statement: Not applicable.

Informed Consent Statement: Not applicable.

Data Availability Statement: The data are not publicly available due to confidentiality reasons.

Acknowledgments: We would like to thank Nagano Prefecture and the municipalities in Nagano Prefecture for providing valuable data for this study.

Conflicts of Interest: The authors declare no conflict of interest.

References

1. Demir, M. Impacts, Management and Functional Planning Criterion of Forest Road Network System in Turkey. *Transp. Res. Part Policy Pract.* **2007**, *41*, 56–68. [CrossRef]
2. Minematsu, H.; Akita, O. New Design Criterion for Forest-Road. *J. Jpn. Soc.* **1987**, *69*, 489–491. [CrossRef]
3. Weaver, W.; Weppner, E.; Hagans, D. *Handbook for Forest, Ranch and Rural Roads: A Guide for. Planning, Designing, Constructing, Reconstructing, Upgrading, Maintaining and Closing Wildland Roads*; Mendocino County Resource Conservation District: Ukiah, CA, USA, 2014.
4. Kastridis, A. Impact of Forest Roads on Hydrological Processes. *Forests* **2020**, *11*, 1201. [CrossRef]
5. Akay, A.E.; Erdas, O.; Reis, M.; Yuksel, A. Estimating Sediment Yield from a Forest Road Network by Using a Sediment Prediction Model and GIS Techniques. *Build. Environ.* **2008**, *43*, 687–695. [CrossRef]
6. Martínez-Zavala, L.; Jordán López, A.; Bellinfante, N. Seasonal Variability of Runoff and Soil Loss on Forest Road Backslopes under Simulated Rainfall. *Catena* **2008**, *74*, 73–79. [CrossRef]
7. Jones, J.A.; Swanson, F.J.; Wemple, B.C.; Snyder, K.U. Effects of Roads on Hydrology, Geomorphology, and Disturbance Patches in Stream Networks. *Conserv. Biol.* **2000**, *14*, 76–85. [CrossRef]
8. Schiechti, H.M. FAO–SFM Tool Detail: FAO Watershed Management Field Manual–Vegetative and Soil Treatment Measures Chapter 4. Available online: https://www.fao.org/sustainable-forest-management/toolbox/tools/tool-detail/en/c/263442/ (accessed on 16 January 2023).

9. Zhao, Q.; Wang, A.; Jing, Y.; Zhang, G.; Yu, Z.; Yu, J.; Liu, Y.; Ding, S. Optimizing Management Practices to Reduce Sediment Connectivity between Forest Roads and Streams in a Mountainous Watershed. *Remote. Sens.* **2022**, *14*, 4897. [CrossRef]

10. Kochenderfer, J.N. *Erosion Control on Logging Roads in the Appalachians*; Northeastern Forest Experiment Station, Forest Service, US Department of Agriculture: Upper Darby, PA, USA, 1970; Volume 158. Available online: https://www.fs.usda.gov/research/treesearch/23767 (accessed on 29 January 2023).

11. Aruga, K.; Sessions, J.; Miyata, E.S. Forest Road Design with Soil Sediment Evaluation Using a High-Resolution DEM. *J. For. Res.* **2005**, *10*, 471–479. [CrossRef]

12. Dodson, E.; Sessions, J.; Wing, M. Scheduling Forest Road Maintenance Using the Analytic Hierarchy Process and Heuristics. *Silva Fenn.* **2006**, *40*, 143–160. [CrossRef]

13. Saito, M.; Goshima, M.; Aruga, K.; Matsue, K.; Shuin, Y.; Tasaka, T. Study of Automatic Forest Road Design Model Considering Shallow Landslides with LiDAR Data of Funyu Experimental Forest. *Croat. J. Eng.* **2013**, *34*, 1–15.

14. Pellegrini, M.; Grigolato, S.; Cavalli, R. Spatial Multi-Criteria Decision Process to Define Maintenance Priorities of Forest Road Network: An Application in the Italian Alpine Region. *Croat. J. Eng.* **2013**, *34*, 31–42.

15. Ministry of Land, Infrastructure, Transport and Tourism. Chapter 3: Causes of Flood and Landslide Disasters and Changes in Social Structures. Available online: https://www.mlit.go.jp/river/pamphlet_jirei/bousai/saigai/kiroku/suigai/suigai_3-1-1.html (accessed on 29 January 2023).

16. Kondo, K.; Kamiya, N. The Analysis on the Factors of Forest Road Disasters in the Area of Southern Akaishi Mountains. *J. Jpn. Eng. Soc* **1995**, *10*, 205–212. [CrossRef]

17. Tsukamoto, Y. Study on the Growth of Stream Channel (1) -Relationship between Stream Channel Growth and Landslides Occurring during Heavy Storm. *Shin. Sabo* **1973**, *25*, 4–13. [CrossRef]

18. Yoshimura, T.; Kanzaki, K. Method of Planning a Forest-Road Network Using the Degrees of Slope Failure Potentials (I) Estimation of the Degrees of Slope Failure Potentials with Topographic Maps. *J. Jpn. For. Soc.* **1995**, *77*, 1–8. [CrossRef]

19. Yoshimura, T.; Akahane, H.; Kanzaki, K. The Evaluation of Potential Slope Failure of Forest Roads Using the Fuzzy Theory. *J. Jpn. Eng. Soc.* **1995**, *10*, 195–204. [CrossRef]

20. Kamiizaka, M.; Kobayashi, N. New Forestry Civil Engineering. Available online: https://www.kinokuniya.co.jp/f/dsg-01-9784254440133 (accessed on 29 January 2023).

21. Wemple, B.C.; Swanson, F.J.; Jones, J.A. Forest Roads and Geomorphic Process Interactions, Cascade Range, Oregon. *Earth Surf. Process. Landf.* **2001**, *26*, 191–204. [CrossRef]

22. Phillips, C. Geomorphic Effects of Two Storms on the Upper Waitahaia River Catchment, Raukumara Peninsula, New Zealand. *J. Hydrol.* **1988**, *27*, 99–112.

23. Sidle, R.C.; Furuichi, T.; Kono, Y. Unprecedented Rates of Landslide and Surface Erosion along a Newly Constructed Road in Yunnan, China. *Nat. Hazards* **2011**, *57*, 313–326. [CrossRef]

24. Forestry Agency. Study Group on Future Road Network Development 1st Meeting. Available online: https://www.rinya.maff.go.jp/j/seibi/sagyoudo/attach/pdf/kentokai-20.pdf (accessed on 29 January 2023).

25. Forestry Agency. Basic Plan for Forest and Forestry (Cabinet Decision on 15 June 2021). Available online: https://www.rinya.maff.go.jp/j/kikaku/plan/attach/pdf/index-10.pdf (accessed on 29 January 2023).

26. Muneoka, H.; Shirasawa, H.; Zushi, K.; Suzuki, H. Frequency of Forest Road Collapse with Various Rainfall Intensities. *J. Jpn. Eng. Soc* **2021**, *36*, 43–50. [CrossRef]

27. Nagano Prefecture Forestry Statistics. Available online: https://www.pref.nagano.lg.jp/rinsei/toukei/ringyotoukeisyo.html (accessed on 27 January 2022).

28. Shirasawa, H.; Saito, M.; Toda, K.; Tada, Y.; Daimaru, H. Shallow Landslide Hazard Assessment Using High-Resolution DEM for Application to Road Alignment Design. *J. Jpn. Eng. Soc* **2018**, *33*, 123–131. [CrossRef]

29. Kühni, A.; Pfiffner, O.A. The Relief of the Swiss Alps and Adjacent Areas and Its Relation to Lithology and Structure: Topographic Analysis from a 250-m DEM. *Geomorphology* **2001**, *41*, 285–307. [CrossRef]

30. Shirzadi, A.; Bui, D.T.; Pham, B.T.; Solaimani, K.; Chapi, K.; Kavian, A.; Shahabi, H.; Revhaug, I. Shallow Landslide Susceptibility Assessment Using a Novel Hybrid Intelligence Approach. *Environ. Earth Sci.* **2017**, *76*, 60. [CrossRef]

31. Nagano District Meteorological Observatory | Outline of Weather Observation Services and AMeDAS. Available online: https://www.data.jma.go.jp/nagano/shosai/kansoku/kansoku.html (accessed on 29 January 2023).

32. Probable Precipitation List by Location (AMeDAS) Western Hokuriku. Available online: https://www.pref.nagano.lg.jp/kasen/infra/kasen/keikaku/koukyodo280401.html (accessed on 30 January 2023).

33. Rainfall Intensity Formula in Nagano Prefecture. Available online: https://www.data.jma.go.jp/cpdinfo/riskmap/rtpd/listRTPD_tb_211.html (accessed on 30 January 2023).

34. Forestry Agency Summary of the FY2022 Forestry Agency Budget. Available online: https://www.rinya.maff.go.jp/j/rinsei/yosankesan/pdf/R4_k3.pdf (accessed on 29 January 2023).

35. Mohammadi Samani, K.; Hosseini, S.A.; Lotfian, M.; Najafi, A. Planning Road Network in Mountain Forests Using GIS and Analytic Hierarchical Process (AHP). *Casp. J. Environ. Sci.* **2010**, *8*, 151–162.

36. Hayati, E.; Majnounian, B.; Abdi, E.; Sessions, J.; Makhdoum, M. An Expert-Based Approach to Forest Road Network Planning by Combining Delphi and Spatial Multi-Criteria Evaluation. *Environ. Monit. Assess.* **2013**, *185*, 1767–1776. [CrossRef]

37. Norizah, K.; Mohd Hasmadi, I. Developing Priorities and Ranking for Suitable Forest Road Allocation Using Analytical Hierarchy Process (AHP) in Peninsular Malaysia. *Sains Malaysa* **2012**, *41*, 1177–1185.
38. Jing, Y.; Zhao, Q.; Lu, M.; Wang, A.; Yu, J.; Liu, Y.; Ding, S. Effects of Road and River Networks on Sediment Connectivity in Mountainous Watersheds. *Sci. Total Environ.* **2022**, *826*, 154189. [CrossRef]
39. Forestry Road Division of the Forestry Agency, 20th Anniversary Publication Group ed. *Trajectory and Development of Forest Roads*; Tokyo Antiquarian Booksellers Cooperative Association: Tokyo, Japan, 1972.
40. Myhre, G.; Alterskjær, K.; Stjern, C.W.; Hodnebrog, Ø.; Marelle, L.; Samset, B.H.; Sillmann, J.; Schaller, N.; Fischer, E.; Schulz, M.; et al. Frequency of Extreme Precipitation Increases Extensively with Event Rareness under Global Warming. *Sci. Rep.* **2019**, *9*, 16063. [CrossRef]
41. Watanabe, M.; Saito, M.; Toda, K.; Shirasawa, H.; Ueki, T. The Elucidation of Actual Disaster Restoration Project Costs by Compiling the Forest Road Registers. *J. Jpn. Eng. Soc.* **2022**, *37*, 155–160. [CrossRef]
42. Hrůza, P.; Mikita, T.; Janata, P. Monitoring of Forest Hauling Roads Wearing Course Damage Using Unmanned Aerial Systems. *Acta Univ. Agric. Silvic. Mendel. Brun.* **2016**, *64*, 1537–1546. [CrossRef]
43. Akgul, M.; Yurtseven, H.; Akburak, S.; Demir, M.; Cigizoglu, H.K.; Ozturk, T.; Eksi, M.; Akay, A.O. Short Term Monitoring of Forest Road Pavement Degradation Using Terrestrial Laser Scanning. *Measurement* **2017**, *103*, 283–293. [CrossRef]
44. Hrůza, P.; Mikita, T.; Tyagur, N.; Krejza, Z.; Cibulka, M.; Procházková, A.; Patočka, Z. Detecting Forest Road Wearing Course Damage Using Different Methods of Remote Sensing. *Remote Sens.* **2018**, *10*, 492. [CrossRef]

 *forests*

Article

Verification of Structural Strength of Spur Roads Constructed Using a Locally Developed Method for Mountainous Areas: A Case Study in Kochi University Forest, Japan

Yasushi Suzuki [1,*], Shouma Hashimoto [2], Haruka Aoki [3], Ituski Katayama [1] and Tetsuhiko Yoshimura [4]

[1] Faculty of Agriculture and Marine Science, Kochi University, Nankoku 783-8502, Japan
[2] Tokushima Prefectural Government, Tokushima 770-8570, Japan
[3] Okayama-No-Mori-Seibi-Kousha (Okayama Forest Management Public Corporation), Tsuyama 708-0013, Japan
[4] Faculty of Life and Environmental Sciences, Shimane University, Matsue 690-8504, Japan
*   Correspondence: ysuzuki@kochi-u.ac.jp

Abstract: Owing to steep terrain and complicated geology, constructing spur roads with low cost and sufficient strength is crucial for sustainable forest management in Japan. The Shimanto method was developed for making narrow spur roads robust against collapse around the 2000s in the Shimanto geology belt area, where the strata were slanted because of an accretion wedge. Kochi University Forest adopted this method and constructed some routes of spur roads in the 2010s. In the present study, we assessed the performance of this method in terms of the roadbed strength and bearing capacity. Two routes were selected, namely Sites 1 and 2, constructed in 2013–2016 and 2019–2021, and tested in 2017 and 2021, respectively. The roadbed strength was measured up to a depth of 100 cm using a handy dynamic cone penetrometer with a rammer of 5 kg. The results showed that the roadbed strength of the embankment side was weaker than that of the cut slope side, although the method was supposed to compact the roadbeds equally over the road width. However, most of the roadbeds had sufficient strength; the younger ones tended to have lower strength than the older ones, and the same tendency was observed for the bearing capacity. It was suggested that the soil under the road width should be excavated more widely toward the cut slope side before compaction.

Keywords: bearing capacity; road construction method; roadbed strength; soil; spur road

Citation: Suzuki, Y.; Hashimoto, S.; Aoki, H.; Katayama, I.; Yoshimura, T. Verification of Structural Strength of Spur Roads Constructed Using a Locally Developed Method for Mountainous Areas: A Case Study in Kochi University Forest, Japan. *Forests* **2023**, *14*, 380. https://doi.org/10.3390/f14020380

Academic Editor: Rodolfo Picchio

Received: 19 January 2023
Revised: 9 February 2023
Accepted: 10 February 2023
Published: 13 February 2023

1. Introduction

The forested land in Japan occupies an area of 25 million ha, which is 68% of the land, and most of it is located in mountainous areas: i.e., 24.5% on less than $15°$ of slope, 39.1% on 15–$30°$, 13.5% on 30–$35°$, and 22.8% on more than $35°$ [1]. The geology features of the Japanese islands are complicated and fragile because the islands are on the edge of the Eurasian tectonic plate, under which the Pacific and the Philippines tectonic plates move gradually downward, creating an accretionary wedge [2–4]. The Japan Forestry Agency recommends the following classification for road networks in the forested area. One classification is forest roads with widths of 3.0–5.0 m and, if necessary, retaining structures, such as concrete walls. Standard-sized log trucks (10-t trucks) transport logs on forest roads, which also witness public use. The other classification is spur roads or "forest operation roads" with a road width of around 3.0 m for logging operations; log transportation is achieved by smaller-sized log trucks [5–8]. This is because the low standards of spur roads, such as narrower road width, higher allowable gradient, and smaller curve radius, are suitable for steep and complicated terrain. Most of the forest roads in Japan allow public traffic and, if necessary, are paved while this is not the case in other countries [3,9–12]. The construction of road networks in forested areas in Japan has shifted from forest roads to spur roads. In 2012, the total lengths of forest roads and spur roads were 19×10^4 km and 12×10^4 km, respectively. In 2020, they were 20×10^4 km and 20×10^4 km, respectively, equivalent to 8.0 m/ha and 8.0 m/ha for forested areas in Japan [13,14]. Therefore, constructing spur roads with low cost and sufficient strength is crucial for sustainable forest management in Japan. Different terms are used for such roads in Japan, i.e., spur road, strip road, forest operation road, forestry operation road, operation road, etc. Hereafter, the term spur road is used in this paper.

Among the various forest practitioners who have developed spur road construction methods for their local properties of the terrain, geology, soil, and forest, Mr. Keizabrou Ohashi, or Oohashi, is the most famous practitioner, whose contribution includes his method of spur road construction and forest management [3,4,15–19]. The essence of his method, termed the Ohashi-type spur road construction method, is as follows: a minimum road width of around 2.0 m for small-sized trucks, 3-t to 5-t class excavators, and 2-t trucks; detailed route selection against collapse concerning terrain, geology, soil type, forest type, etc.; and use of small, round wood material as the retaining material for the fill slope toe and road shoulder [3,4,15–19]. He developed the method in the late 20th century, mainly for the Chichbu geology belt areas combined with decomposed granite areas. The Chichibu belt collapsed, and once it was deposited below sea level, the geologic stratum was not formed and the soil is fine-grained [3]. Owing to fine-grained soils, large landslides with long runouts can result from heavy precipitation [3]. Because fine-grained soils limit the forest road cut slope height and the ability to construct fill slopes [3], the Ohashi method often employs retaining materials for fill slopes.

The method witnessed widespread use in Japan around the 1990s and 2000s, with some modifications to fit the geology and soil of other areas [20,21]. One such variant is the Shimanto spur road construction method developed around the 2000s by a municipal officer of the forestry section in order to make narrow spur roads robust against collapse, even in steep terrain. The Shimanto belt is relatively young and exhibits typical features of accretionary wedges composed of sedimentary rocks [3]. The slopes facing the north or the Japan Sea are generally dip slopes, and those facing the south or the Pacific Ocean are opposite slopes [3]. There are numerous circular slips and deep-seated landslides on the north-facing slopes, and it is risky to construct roads on the dip slopes [3]. For the inclined stratum in the Shimanto belt, which is composed of sedimentary rocks, deep excavation has been conducted for preparing the structural foundation of the road [4]. After deep excavation to ensure the depth required for the road width, ~30-cm-thick blocks of compacted subsoil were piled up [4]. This method is suitable for gravelly soil, which is the main soil type in the Shimanto belt, while a small portion of clay serves as the bondage medium [4]. The surface soil of the natural slope was maintained during construction and used for the surface of the filling slope to provide an early green surface because it contained buried seeds [4]. In some cases, stumps are placed at the fill slope toes with thick roots embedded under the embankment [6]. The road width is limited to 2.5 m considering the usage of 3 t–5 t class excavators and mini-forwarders [22]. A mini-forwarder is a popular machine for small-scale forestry in Japan. It is propelled by rubber tracks and equipped with a winch with a typical loading capacity of 500–1000 kg [23].

Including the co-related principals of these methods, a training manual [24] was compiled by forest engineers in 2010 for on-the-job training of spur road construction operators in government-aided training programs. The book introduces various local methods developed to fit the geological and terrain features of the area. Umeda et al. [21] pointed out that the designated strength of the roadbed cannot always be obtained in certain areas if a method developed for another area is adopted unless the operator completely understands the concept of the method.

Regarding the proper compaction of roadbed construction, Sawaguchi et al. [25] found that the maximum compaction strength can be obtained by less than 10 round trips of an excavator when it is used as a compaction device. Although more retaining walls are required for forest roads, which have higher design standards than spur roads, on steep terrain [26], the following problem has recently arisen: many sections of improperly constructed spur roads on steep slopes have collapsed one or two years after construction [27]. This problem should be taken into account in construction supervision. Furthermore, companies involved in such unsustainable forest management should be punished by law. Concerning the age-related change effect of road bed strength and bearing capacity, Umeda et al. [21] reported that in areas with low compaction ability soils, there is a spur road construction method that requires three years for completion. This method seems to effectively utilize the age-related change effect of the road strength.

Kochi University Forest adopted the Shimanto method and constructed some routes of spur roads in the 2010s. The forest is in the same prefecture for which the method was developed but in a different geology belt. The present study assesses the performance of the method in terms of the roadbed strength and bearing capacity. We verify the hypothesis that the spur road construction method was properly used to achieve sufficient mechanical strength.

2. Materials and Methods

2.1. Study Site

2.1.1. Location

The study was carried out at two sites in Kochi University Forest, which is located at the northwest boundary of Kami city, Kochi prefecture, Japan (133°36′32.0″ E, 33°42′12.8″ N). It has an area of 124 ha and is composed of eight compartments, which are divided into two blocks: the west and the east blocks (Figure 1). The altitude ranges from 660–1045 m. Coniferous planted forests and natural broad-leaved forests account for 60% and 40% of the vegetation, respectively, which is dominated by *Cryptomeria japonica* and *Chamaecyparis obtusa*, and *Quercus acuta* and *Q. serrata*, respectively [28]. All the forest roads have a width of 3.0 m, including shoulders, and are unpaved. The spur roads were constructed for forest management and logging operation performed by an 8-t class excavator-based machine (CAT 308D CR) and a 3-t class forwarder (IWAFUJI U3) [29]. The average annual precipitations from 2013 to 2021 at the two nearby meteorological stations, Motoyama town (6100 m to NNW) and Shigeto in Kami city (7900 m to ESE), were 3096 ± STD 637 mm and 3778 ± STD 757 mm, respectively [30].

Two routes were selected for the study: Sites 1 and 2. Site 1 was constructed in FY (fiscal year) 2013–2016 and Site 2 was constructed in FY 2019–2021. Field investigations of Sites 1 and 2 were carried out in 2017 and 2021, respectively. The coordinates of the routes were surveyed using a survey compass with a transit telescope for Site 1 and using a single band GNSS receiver [31] for Site 2.

The entire Kochi University Forest is in the Chichibu geology belt, which consists of an accretionary wedge sedimented from the Carboniferous period to the Jurassic period and is located north, adjacent to the Shimanto geology belt. The detailed geological features are as follows: geological period, Mesozoic–Early Cretaceous–Albian period to Cenozoic–Paleogene–Oligocene–Selandian period; lithology and muddy millstone or metamorphic chert; metamorphism, high P/T wide-area metamorphic rocks, and green mudstone belt [32]. Although many areas are prone to landslides in the Sanbagawa geology belt in Kochi prefecture, which is north adjacent to the Chichibu belt, there are few such areas in both Chichibu and Shimanto geology belts in Kochi prefecture [33]. From the terrain analysis of Yamasaki et al. [34], the average slope gradient is 25–30°, the average height difference within a 500-m circle is 300–400 m, and the contour-round-about factor is moderate for the Kochi University Forest area, resulting in moderate to highly difficult terrain within the Kochi prefecture [34].

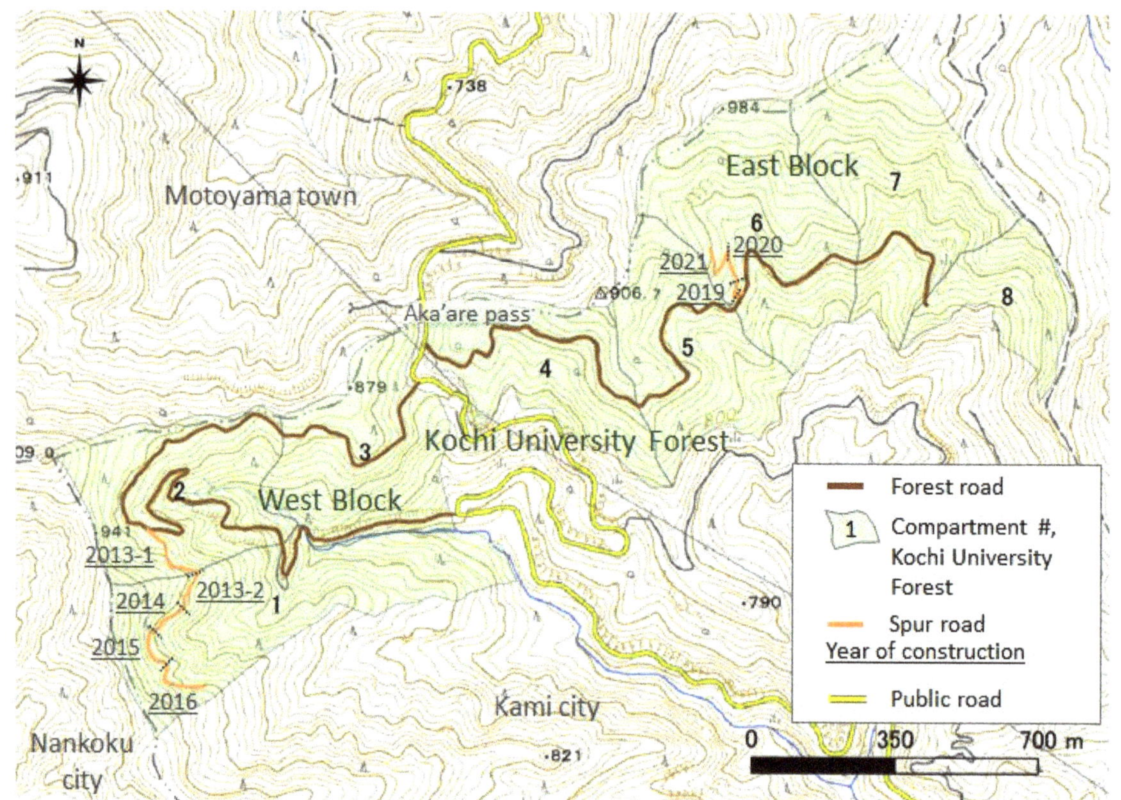

Figure 1. Study site. Site 1: A spur road constructed in FY 2013–2016 in the west block, Site 2: A spur road constructed in FY 2019–2021 in the east block of Kochi University Forest.

2.1.2. Spur Road Construction Method

The spur roads were constructed by three university forest officers using an 8-t class excavator (CAT 308D CR; bucket capacity, 0.28 m^3). They had 24, 8, and 6 years of experience at the time of Site 1 spur road construction in 2013. They learned the Shimanto method in some training courses held outside the university and then commenced spur road construction within the university forest on their own around the 2010s.

Figure 2 shows the spur road construction in Kochi University Forest using the Shimanto method. The soil is dug over once with additional depth, then compacted with 20–30-cm-thick layers. An example of the cross-section of the middle-grade slope in Figure 2a is 20° of natural ground. On a steeper natural ground slope (Figure 2b, e.g., 30°), the embankment volume and fill slope length tend to be larger. In such cases, the center line of the spur road is shifted to the uphill side in order to decrease the embankment volume (Figure 2c). The cut slope can be constructed as vertical if its maximum height is less than the limit height of 1.4 m.

Figure 3 shows a practical example of spur road construction in Kochi University Forest during a training course held for university forest officers of other universities in 2012. The excavator dug over the soil for extra depth (Figure 3a). Then, the loose soil was mixed using the bucket, spread over the designated road width (Figure 3b), and compacted by the crawler (Figure 3c). The compaction was carried out for a few layers with a thickness of 20–30 cm each up to the level of the existing road height (Figure 3d).

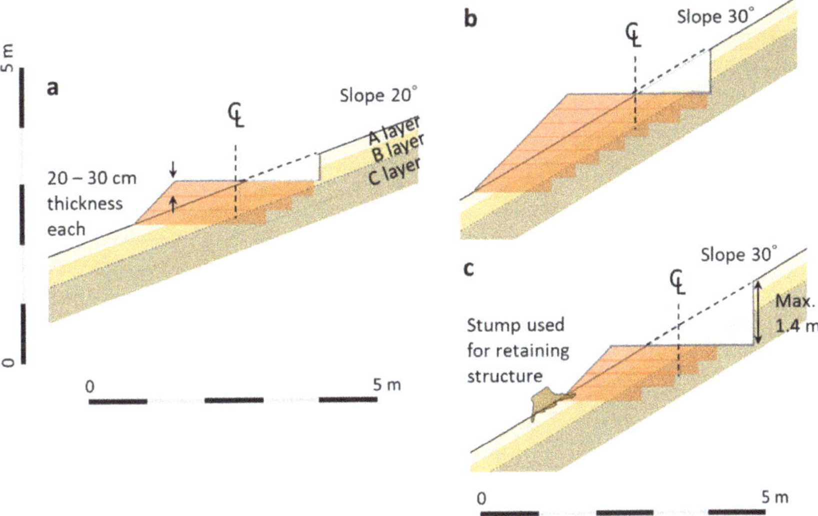

Figure 2. Spur road construction in Kochi University Forest using the Shimanto method. The soil was dug over once with additional depth, and then compacted with layers of 20–30 cm thickness. (**a**) Example of the cross-section of the middle-grade slope (e.g., 20°) natural ground. (**b**) On a steeper natural ground slope (e.g., 30°), the embankment volume and fill slope length tend to be larger. (**c**) In such cases, the center line of the spur road is shifted to the uphill side in order to decrease the embankment volume. The cut slope can be constructed as vertical if its maximum height is less than the limit height of 1.4 m.

Figure 3. Example of spur road construction in Kochi University Forest during a training course. The excavator dug over the soil for extra depth (**a**). Then, the loose soil was mixed using the bucket, spread over the designated road width (**b**), and compacted by the crawler (**c**). The compaction was carried out for a few layers with a thickness of 20–30 cm each up to the level of the existing road height (**d**).

2.2. Field Survey

2.2.1. Measurement of Roadbed Strength

The roadbed soil strength was tested using a simple dynamic cone penetrometer. There are some types of dynamic cone penetrometers for in situ soil strength measurement in Japan [35], varying in terms of the cone size, hammer mass, and height of a single strike with a hammer. Using a dynamic cone penetrometer, the strength of the soil is expressed as the number of hammer strikes required to penetrate a certain depth of the soil. Umeda et al. [36] standardized the strength values of three penetrometers. Yoshinaga et al. [37] concluded that the in situ density of underground soil can be estimated with the measured strength of a simple dynamic cone penetrometer called the Doken-type penetrometer. Gotou et al. [38] verified the roadbed strength of spur roads and forest roads using a Doken-type simple dynamic cone penetrometer. In the present study, a Doken-type simple dynamic cone penetrometer was used for the measurement of the roadbed strength. The Doken-type penetrometer was developed for in situ measurement on steep slopes with a lighter hammer (5 kg) for convenience rather than the other types [39]. Its measuring procedure has been standardized by the Japan Geotechnical Society, Committee for Standard of Geotechnical Survey, (Japanese Industrial Standards)-1433-2012 [39].

Figure 4 shows the standardized design of the Doken-type dynamic penetrometer (Figure 4a) [39]. The soil strength is measured by striking the hammer at a height of 500 mm (Figure 4b,c) with the road kept vertical. By measuring the penetration depth of each strike or a few strikes, the penetrated depth is converted into an N_d (or N_c) value, which represents the number of strikes required for a penetration depth of 10 cm.

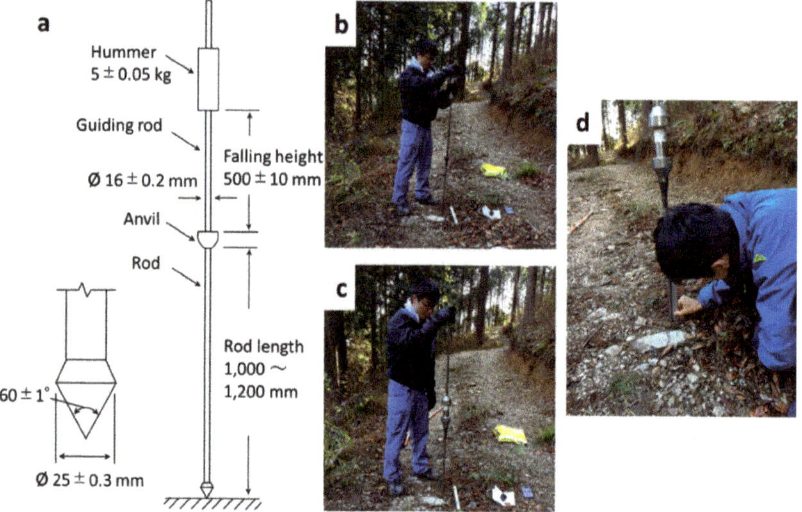

Figure 4. Doken-type dynamic penetrometer. Standardized design of the Doken-type dynamic penetrometer (**a**) (the Japan Geotechnical Society, Committee for Standard of Geotechnical Survey 2014 [36]). The soil strength is measured by striking the hammer at a height of 500 mm (**b**–**d**) with the road kept vertical. By measuring the penetration depth by each strike or a few strikes, the penetrated depth is converted into an N_d value, which represents the number of strikes required for a depth of penetration of 10 cm.

The N_d value is defined as follows:

$$N_d = 10\,(N/\Delta h),\tag{1}$$

where Δh is the penetrating depth of N strikes. In general, Δh is 10 cm; however, for softer soil, Δh was set to one strike. and for harder soil, Δh was set to less than 10 cm. The maximum penetrating depth was set to 100 cm because some research reports have pointed out that the friction of the rod cannot be ignored when the penetrating depth exceeds 100 cm [38–40]. In some cases, the N_d value was too large, implying that there was a large rock just under the penetrating point. In such cases, we moved the penetrating point 20–0 cm apart from the original point.

Thresholds of the N_d values were determined for judging the strength requirements as follows. Okimura and Tanaka [41] predicted the potential sliding layer depth in forested areas of sandy soil caused by heavy rain as $N_{10} < 12$. The N_{10} value is measured using a handy dynamic cone penetrometer termed the Kobe University type, which provides nearly the same value as the Doken-type penetrometer. Sugiyama et al. [42] reported that the average N_d value of a collapsed railway embankment was 5.4. Ogawa [43] studied the relationship between slope surface failure and water content variation on the forested slope and found that most of the sliding depth was 0.5–1.5 m, of which the N_d values ranged from 5–10. Hiramatsu and Bitoh [44] analyzed a landslide area on the Chichibu geology belt in Yusuhara town, Kochi prefecture, and proposed a threshold N_d value of 9 to estimate the slippage depth of the soil layer. Koyama [45] conducted a series of field investigations on spur roads using a simpler dynamic penetrometer called the Tottori FK-type penetrometer and found that cracks appeared at the road shoulders if the N_d equivalent values were less than 1.4 with a penetrating depth of up to 25 cm. For these preceding achievements, we set the following three thresholds: $N_d = 1.4$ as "very weak" for clacks on shoulders, $N_d = 5.4$ as "weak" for the collapse of the embankment, and $N_d = 10.0$ as "moderate" for natural slope failure.

2.2.2. Measurement of Bearing Capacity

The California bearing ratio (CBR) is a widely used parameter for assessing the bearing capacity of a road subsurface. While proper CBR values are obtained by testing standardized sample soils using appropriate instruments in a laboratory, in situ CBR values are obtained using a purpose-built instrument. Therefore, although the CBR value was originally for the construction of public roads, it is often employed for assessing the bearing capacity of unpaved forest roads and spur roads using in situ CBR testers [46–49].

Kobayashi and Fukuda [46] measured in situ CBR values of spur roads using a sphere-shaped rammer type in situ CBR tester and obtained the average value of 18.9%. Kobayashi et al. [47] investigated two spur road routes constructed using the Ohashi method with an in situ CBR tester called CASPFOL, and obtained average values of 14.7% and 23.4%. Suzuki et al. [48] evaluated the total strength of a spur road constructed using the Shimanto method in a volcanic ash soil area. For the bearing ratio, they used a CASPFOL tester and obtained in situ CBR values of 2.5–9.8% for ruts on the cut slope sides and 8.0–12.5% for ruts on the fill slope sides. Sawaguchi et al. [49] investigated the age change of the bearing capacity for spur roads in Neogene period gravelly soils for the Tohoku area, northern Japan, using a CASPFOL tester. They found that the CBR values increased from 1.4% to 4.0% at the shoulder, from 1.7 to 5.6% at the center, and from 2.7 to 7.8% at the ruts over 10 years, even if there was nearly no traffic during the gap years [49].

In the present study, we used an in situ CBR tester called CASPFOL (Figure 5a) [50,51]. The soil strength is measured by striking the rummer at a height of 500 mm with the road kept vertical through the built-in acceleration meter (Figure 5b,c). The acceleration of the rummer is measured with the instrument when the rummer falls and strikes the soil. The standard measuring method requires five measurements around a designated point with a radius of 20 cm (Figure 5d).

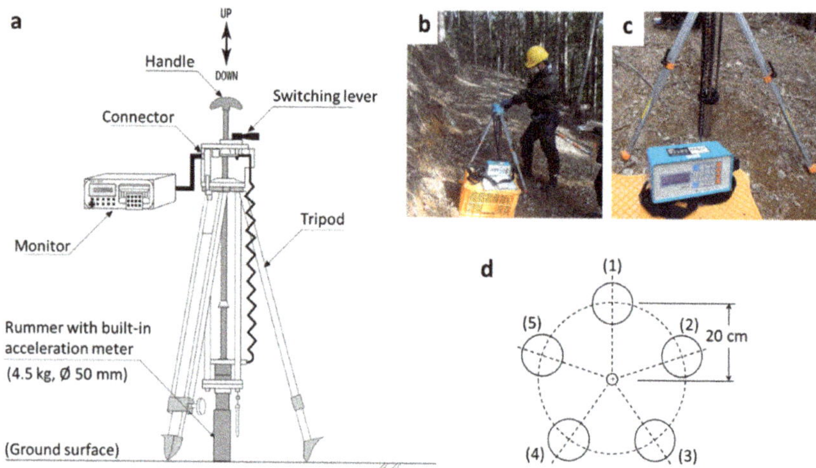

Figure 5. In-situ CBR measuring instrument (CASPFOL). Schematic diagram of CASPFOL (**a**) [51]. The acceleration of the rummer is measured by the instrument when the rummer falls and strikes the soil (**b**,**c**). The soil strength is measured by striking the rummer at a height of 500 mm with the road kept vertical through the built-in acceleration meter. The standard measuring method requires five measurements around a designated point with a 20 cm radius (**d**).

2.2.3. Soil Property

The grain size distribution has significant effects on soil properties, especially mechanical features, such as compatibility [21]. Soil samples were obtained from a cut slope at two points each for the two sites. The samples were tested for grain size analysis by following the procedure of JIS-A-1202 [52]. Soil property as engineering materials is classified as one of 24 engineering soil types through a standard classification system of the Japanese Geotechnical Society using the parameters derived by the grain size distribution and accompanying tests for the grain size analysis procedures. An additional property, i.e., wideness of size distribution, is assessed using two parameters, the uniformity coefficient U_c and coefficient of curvature U_c', which are defined as follows:

$$U_c = D_{60}/D_{10}, \text{ and} \tag{2}$$

$$U_c' = (D_{30})^2/(D_{10} \times D_{60}), \tag{3}$$

where D_{10}, D_{30}, and D_{60} are the grain sizes of the accumulated grain mass percentages of 10%, 30%, and 60%, respectively [52]. If $U_c \geq 10$ and $1 < U_c' < 3$, then the soil is supposed to have wide grain size distribution, which means that it is suitable for compaction [52].

2.2.4. Experimental Design and Analysis

The first series of the investigation was conducted for Site 1 for the measurement of the roadbed strength. To check the equality of the compacted roadbed soil strength around the road width, Factor A was set as a relative position within the road width (Figure 6) with five levels: A1, downhill side apart from the shoulder at least 30 cm; A2, rut of downhill side; A3, center; A4, rut of uphill side; and A5, uphill side apart from the toe of the cut slope at least 30 cm. Factor B was set to the fiscal year of construction in order to check the age-related change in the strength. The spur road of Site 2 was constructed from 2013 to 2016. Because the constructed length of FY 2013 was nearly twice that of the other years, the FY 2013 constructed section was separated at its mid-point into two sections. Therefore, the levels of Factor B are B1, FY2013-1; B2, FY2013-2; B3, FY2014; B4, FY2015; and B5, FY2016. The spur roads were constructed from fall to winter, and Site 1 was investigated in the winter of FY 2017. Hence, the number of years elapsed were 4.2, 4, 3, 2, and 1 for B1, B2, B3, B4, and B5, respectively. It should be noted that the number of years elapsed could not be treated as the effect of time on each road segment in a precise sense. However, in the analysis, it was assumed that the number of years elapsed had a meaning of elapsed time by assuming that each road segment has a homogeneous property. Nevertheless, the effect of the number of years elapsed should be carefully evaluated in this regard.

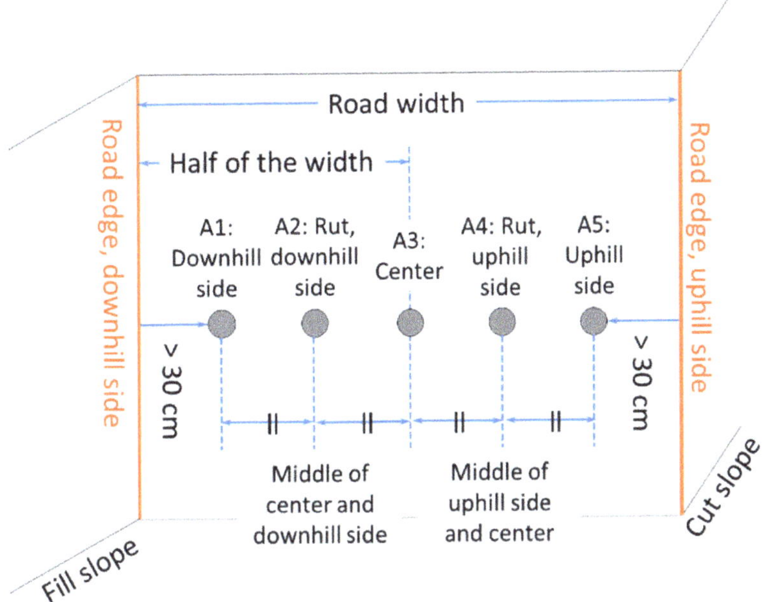

Figure 6. Measuring points on the spur road at each sample point of Site 1.

The number of repetitions of the measurement was set to five so that the distance between every two points would be the same. However, because of the slightly shorter section length of A3, the number of repetitions of A3 was four. Therefore, 24 points were investigated for Site 1 in total.

The obtained data were analyzed by two-way ANOVA [53,54]. The dependent variable was the N_d value up to a depth of 100 cm average weighted by Δh. Two independent variables were Factors A and B. However, in some points, the soil was too hard to reach a depth of 100 cm. When N_d exceeded 50 or more, the penetration was stopped, and the N_d value was averaged up to the maximum depth. One dummy data was added to the A3 level as the mean of four observations so that the number of repetitions is equal to the other levels of A, i.e., five. In the ANOVA table, the degree of freedom was adjusted for Factor A and the error concerning the dummy observations [53].

The second series of the investigation was carried out in Site 2 for the roadbed strength and bearing capacity. As for the roadbed strength, to compare the strength difference between natural soil and compacted soil as well as that between the downhill side and the uphill side, Factor A was set to the distance from the road center and the Factor B direction (Figure 7). The levels of Factor A were as follows: A1, 0.5 m from the center; A2, 1.5 m from the center; and A3, 2.5 m from the center. The points of A1 were nearly on the ruts. The distance between A1 and A2 was, in some cases, more than 1.0 m so that A2 was not on the fill slope. The distance between A2 and A3 was 1.0 m. The factor levels are B1, downhill side, and B2, uphill side.

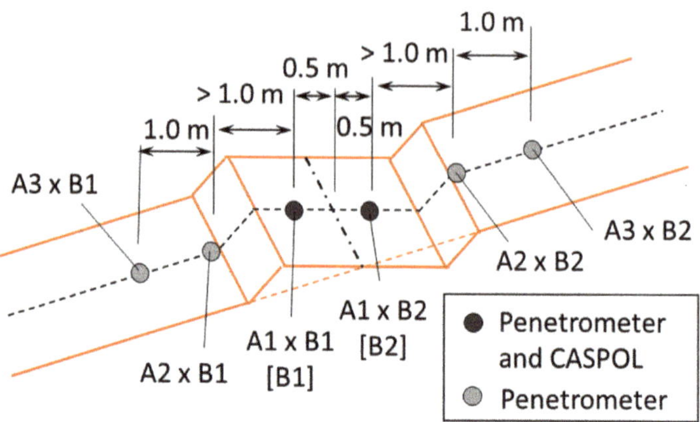

Figure 7. Measuring points on the spur road at each sample point of Site 2.

Factor C was set to the fiscal year of construction. The Site 2 spur road was constructed in 2019, 2020, and 2021. The section distance of 2021 was nearly twice that of the others. Therefore, section FY2021 was separated into two sections at the mid-point. To check the strength difference between the upper and lower parts, Factor D was set to the penetrating depth, i.e., the level D1 was 0–50 cm, and D2 was 50–100 cm. The N_d values were averaged for 0–50 cm and 50–100 cm, respectively. The number of repetitions was two. In Site 2, two points each were investigated at each section of Factor A, at a mid-point of the section and separated by 2 m, because the section length was not as long as that of Site 1. The total number of roadbed strength measurement points was $3 \times 2 \times 4 \times 2 \times 2 = 96$. The obtained data were analyzed via four-way ANVOVA [53,54]. The dependent variable was the N_d value averaged weighted by Δh. Four independent variables were Factors A, B, C and D.

As for the bearing capacity, in situ CBR measurement, three factors were set. The first two factors, Factors B and C, were set to be the same as those of the roadbed strength. The third factor, Factor R, was set to repetition, having two levels of R1 and R2. Although repetition was not treated as a factor in general, it was included to check for the homogeneity of the measurement. The total number of in situ CBR measurement data was $2 \times 4 \times 2 \times 5 = 80$ because five measurements each were required for the standard measurement of the in situ CBR tester. The obtained data were analyzed via three-way ANVOVA [53,54]. The dependent variable was the in situ value. The three independent variables were Factors B, C, and R.

The multi-way ANOVA was analyzed with SAS software using the super-computing service of the Institute for Information Management and Communication, Kyoto University. The two-way ANOVA and the following Tukey test for multiple comparisons [54] were carried out using Microsoft Excel software.

2.2.5. Date of Investigation

The investigation of Site 1 was carried out from November 2017 to January 2018. Surveying of the spur road route and road section measurement using a surveying pole with a precision of 0.1 m were conducted simultaneously.

The investigation of Site 2 was carried out from November 2021 to January 2022. The surveying of the spur road route and road section measurement with a surveying pole with 0.1 m precision were conducted simultaneously.

The soil samples were collected at the sections FY2013 and FY2015 of Site 1, the boundary of sections FY2019 and FY2020 (FY2019–2020), and the first half of FY2021 of Site 2. The FY2015 and the FY2019–2020 samples were collected in December 2020 and analyzed from December 2020 to February 2021. The FY2013 and FY2021 samples were collected in December 2021 and analyzed from December 2021 to January 2022.

3. Results and Discussion

3.1. Overview of the Sites

The total spur road lengths of the sites were 626 and 250 m for Sites 1 and 2, respectively (Tables 1 and 2). The route of Site 1 had a steep gradient in the section of FY2015, whereas the other sections were nearly flat or had a gentle gradient. In Site 2, the route extended upward; hence, its gradient was relatively high. The average road widths were nearly the same for both sites while the Site 1 route was slightly wider (3.0 m) than the Site 2 route (2.8 m).

Table 1. Descriptive statistics of Study Site 1: a spur road in the west block of Kochi University Forest.

Fiscal Year of Construction (Years Elapsed at Survey)		Length (m)	Average Gradient of the Road (%)	Num. of Measured Points	Average Road Width (m)	Average Cut Slope Height (m)	Average Cut Slope Gradient (deg.)	Average Fill Slope Gradient (deg.)	Average Slope of Adjacent Natural Ground (deg.)
2013-1	(4.2)	182.3	3.5	5	2.9	1.1	68.7	37.2	22.8
2013-2	(4)	98.7	−5.5	5	2.9	1.2	65.5	38.2	24.0
2014	(3)	87.6	−0.2	4	3.1	1.2	68.8	51.0	25.0
2015	(2)	100.2	13.4	5	3.2	1.6	62.1	45.8	28.1
2016	(1)	156.7	−4.5	5	3.1	1.4	76.8	43.8	29.6
Sum		625.5		24					
Total mean			1.2		3.0	1.3	68.4	42.9	26.0
STD					0.5	0.4	14.3	12.3	6.5
Max.					4.8	2.3	90.0	69.0	41.1
Min.					2.6	0.5	36.9	18.4	14.3

Fiscal year of survey: 2017 (November 2017–January 2018).

Table 2. Descriptive statistics of Study Site 2: a spur road in the east block of Kochi University Forest.

Fiscal Year of Construction (Years Elapsed at Survey)		Length (m)	Average Gradient of the Road (%)	Num. of Measured Points	Average Road Width (m)	Average Cut Slope Height (m)	Average Cut Slope Gradient (deg.)	Average Fill Slope Gradient (deg.)	Average Slope of Adjacent Natural Ground (deg.)
2019	(2)	41.8	11.0	2	2.8	1.3	52.4	41.2	23.5
2020	(1)	73.8	21.1	2	3.0	0.9	71.6	56.3	22.1
2021	(0.5)	134.6	15.4	4	2.7	0.6	62.9	45.4	20.5
Sum		250.2		8					
Total mean			16.3		2.8	0.9	62.5	47.1	21.7
STD					0.3	0.3	8.4	8.1	2.6
Max.					3.0	1.3	71.6	56.3	23.5
Min.					2.4	0.5	52.4	39.5	17.9

Fiscal year of survey: 2021 (November 2021–January 2022).

The slope gradients of adjacent natural ground were estimated by the connecting fill slope toes and top ends of cut slopes. The average slope gradients of the adjacent natural ground were $26.0 \pm$ STD $6.5°$ for Site 1 and $21.7 \pm$ STD $2.6°$ for Site 2. Such a steep natural ground slope of Site 1 resulted in a higher cut slope height ($1.3 \pm$ STD 0.4 m) compared to that of Site 2 ($0.9 \pm$ STD 0.3 m). This implies that the excavation volume as well as the dig over depth of Site 1 would be larger than those of Site 2 (Figure 2).

3.2. Soil Property

Figure 8 shows the grain size distributions of the obtained soil samples as a result of grain size analysis. All the samples were classified as GFS, fine-grained sandy gravel, of which the mechanical feature is classified as suitable for compaction [52]. Two samples, i.e., Site 1 FY2013 and Site 2 FY2021, were additionally classified to have a wide distribution, the uniformity coefficient of which is $U_c \geq 10$, and the coefficient of curvature U_c' is between 1 and 3.

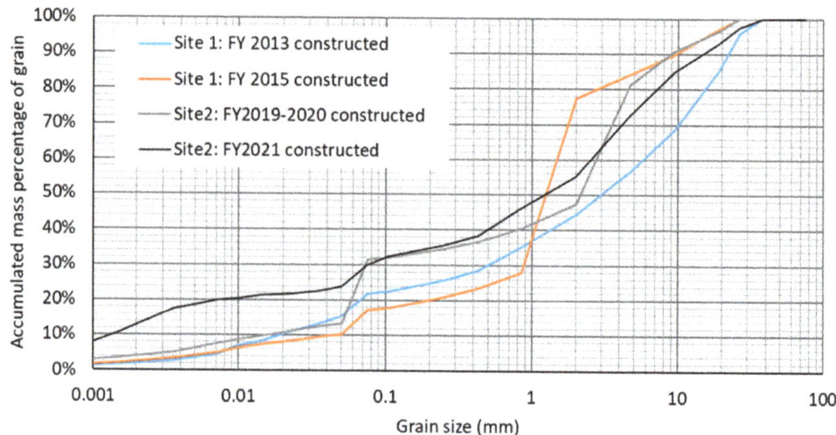

Figure 8. Grain size distribution of soil sampled at Sites 1 and 2.

The result of the grain size analysis indicated that the soil of both sites has a good property for compaction. It also can be said that there was no major difference between the soil properties of Site 1 and Site 2.

3.3. Site 1: Measurement of Roadbed Strength

The result of the two-way ANOVA on the N_d values in Site 1 indicated that two factors were significant, whereas their interaction was not: A, relative position within the road width, and B, fiscal year of construction (Table 3). In other words, the population means of all the levels were not the same in Factors A and B, and there was no interaction effect in the combination of Factor A levels and Factor B levels.

Table 3. ANOVA table of N_d in Site 1.

Factor	Degree of Freedom	Sum of Squares	Variance	F-Value	p-Value
A: Relative position within road width	4	5783.1	1445.8	6.74	0.00
B: Fiscal year of construction	4	6587.8	1647.0	7.68	0.00
Error (pooled)	111	23,807.1	214.5		
Total	119	36,614.2			

Total mean: 15.65.

From a brief inspection of the multiple comparison analysis of Factor A levels (Figure 9), it can be concluded that the roadbed soil strength was not even over the road width. In other words, it was stronger in the cut bunks and weaker in the fill bunks. Thus, the main objective of the Shimanto method was not achieved in the investigated spur road.

As for the effect of Factor B, level B3 (FY2014) had significantly greater N_d values than the other levels (Figure 10). From a visual inspection of the site, a possible reason would be that hard rock materials were more distributed in the FY2014 section than the others. The age-related change, such that older roadbeds gained a larger strength [48,49], is not observed in Figure 10.

Figure 11 shows all the observations of the N_d values in Site 1 accompanied with three thresholds: 1.4, 5.4, and 10.0. While no observation was lower than $N_d = 1.4$, 15 observations in B4 (FY2015) and B5 (FY2016) and 4 observations in B1 and B2 (FY2013) were lower than $N_d = 5.4$. In Figure 10, where the values were averaged, the age-related change was not clear. However, the lower limits of the values seem to have a tendency of age-related change and are hence lower for younger ages in Figure 11.

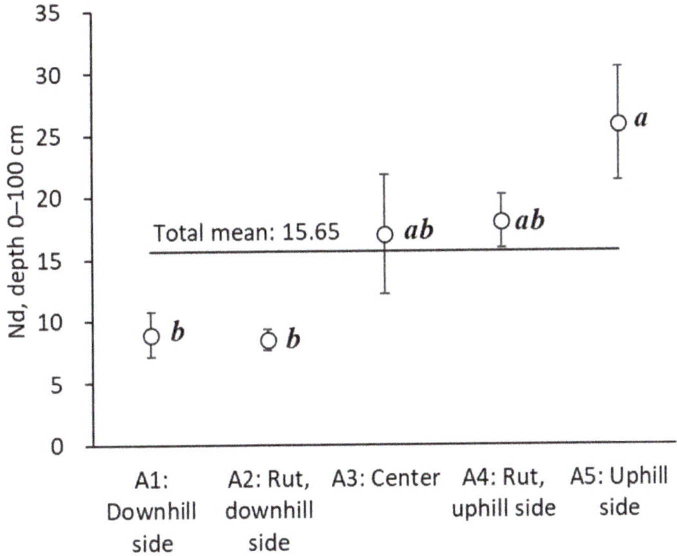

Figure 9. Main effect of Factor A, relative position within the road width, of Site 1 on the N_d -value. The error bars represent the standard error (SE). Different letters indicate significant differences between the population means. Tukey test, $p < 0.01$, $n = 24$.

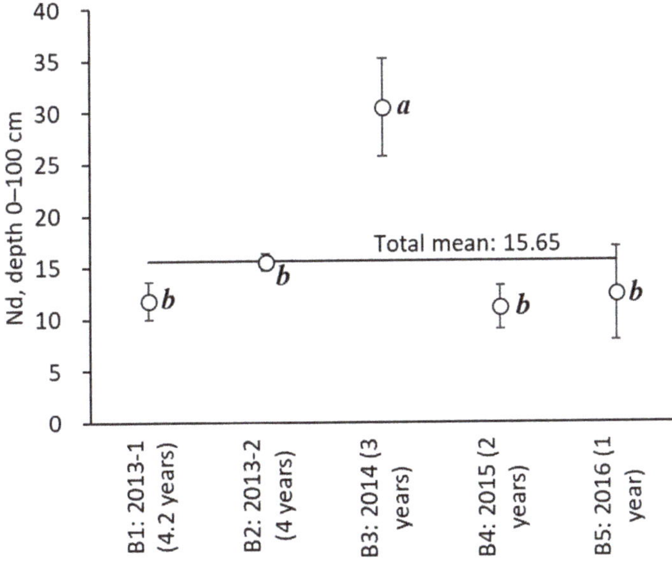

Figure 10. Main effect of Factor B, fiscal year of construction, of Site 1, on the N_d -value. The error bars and different letters have the same meaning as in Figure 9. $p < 0.01$, $n = 25$ except for B3 ($n = 20$).

3.4. Site 2

3.4.1. Measurement of Roadbed Strength

In the ANOVA table for the N_d values in Site 2, all the single factors and a few interactions were significant (Table 4). The insignificant interactions were pooled into the error.

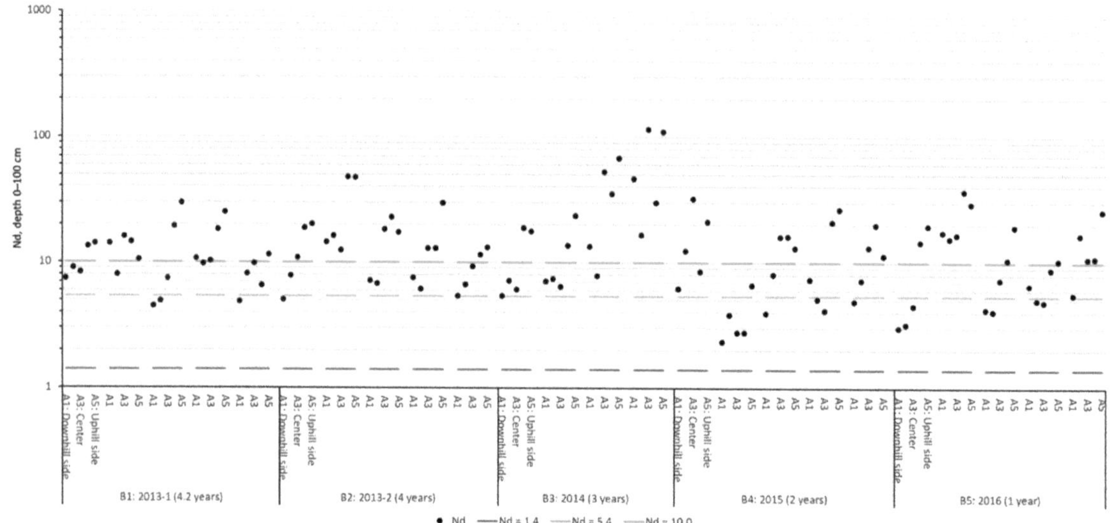

Figure 11. N_d—values of Site 1 accompanied with thresholds.

Table 4. ANOVA table of N_d in Site 2.

Factor	Degree of Freedom	Sum of Squares	Variance	F-Value	*p*-Value
A: Relative position	2	1028.0	514.0	4.31	0.02
B: Downhill side or uphill side	1	787.0	787.0	6.60	0.01
C: Fiscal year of construction	3	1783.8	594.6	4.99	0.00
D: Penetrationg depth	1	3098.4	3098.4	25.98	0.00
A × B	2	803.9	401.9	3.37	0.04
A × C	6	3140.2	523.4	4.39	0.00
B × C	3	1472.3	490.8	4.12	0.01
A × B × C	6	1563.4	260.6	2.19	0.05
B × C × D	3	1408.5	469.5	3.94	0.01
Error (pooled)	68	8108.4	119.2		
Total	95	23,193.9			

Total mean: 9.79.

As for Factor A (relative position in the road width), A1 (0.5 m from center, 14.4 ± SE (Standard Error) 4.4) was significantly greater than the others (A2, 1.5 m from center, 7.5 ± SE 1.2; A3, 2.5 m from center, 7.4 ± SE 1.3; Tukey test, $p < 0.05$, $n = 32$), implying that the soil strength increased from natural 7.4–7.5 to compacted 14.4. The effect of Factor B should be evaluated as the interaction A × B. The average N_d values of A1 × B2, the rut of the uphill side (21.1 ± SE 8.5), were significantly greater than the downhill side (7.6 ± SE 1.0), which was not significantly different from the other remaining interactions (Tukey test, $p < 0.05$, $n = 16$). The tendency is the same as that of Site 1 (Figure 9). Note that these N_d values were averaged through a depth of 0–100 cm (D1: 0–50 cm and D2: 50–100 cm). The average N_d values of D2 (15.5 ± SE 3.0) were significantly greater than those of D1 (4.1 ± SE 0.4; $p < 0.01$, $n = 48$). As for Factor C, the fiscal year of construction, the level C2 (FY2020, one year after construction, 17.2 ± SE 5.8) was significantly greater than the other levels (C1: FY2019, two years after construction, 6.6 ± SE 1.5; C3: FY2021-1, 0.5 year after construction, 8.4 ± SE 1.3; C4: FY2021-2, 0.5 year after construction, 7.0 ± SE 1.1; Tukey test, $p < 0.05$, $n = 24$).

The effect of Factor C can be more clearly observed as interactions between other factors. Figure 12 shows the interaction of Factors A and C. The average N_d values of the combined levels of A1 × C2 were significantly greater than the others, implying a possibility that there were harder subsoils or rocks under the investigated section of C2.

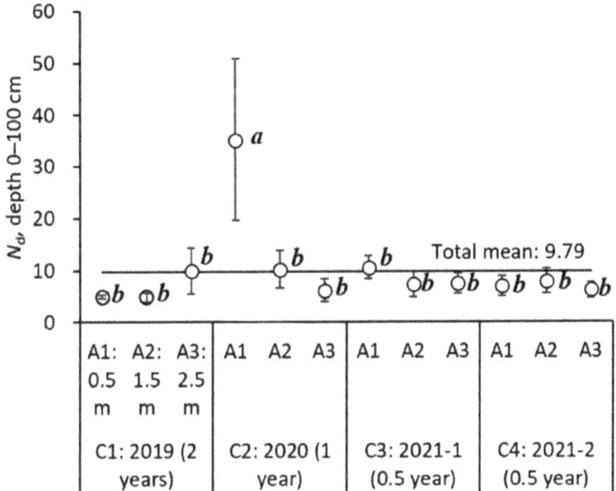

Figure 12. Interaction of Factor A (relative position) and Factor C (fiscal year of construction) of Site 2 on the N_d—value. The error bars and different letters have the same meaning as in Figure 9. $p < 0.01$, $n = 8$.

Figure 13 shows the interaction of Factors B, C, and D. From the figure, it can be observed that the average N_d values of the combined levels B1 × C2 × D2 were significantly greater than the others, also implying a possibility that there were harder subsoils or rocks under the rut of the uphill side at a depth of 50–100 cm on the investigated section of C2.

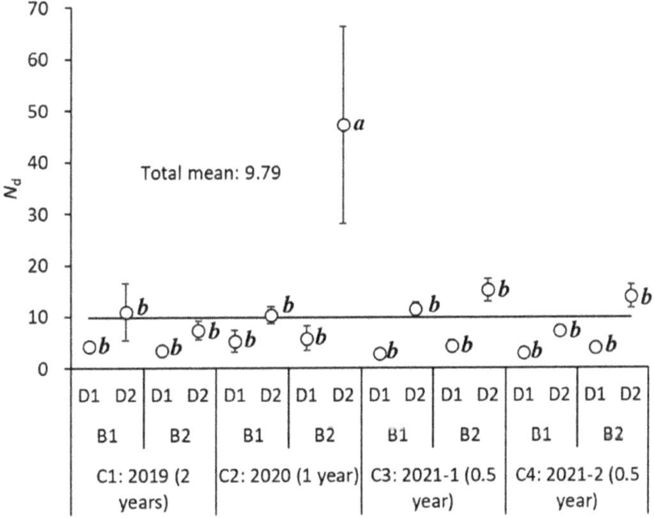

Figure 13. Interaction of Factor B (downhill side or uphill side), Factor C (fiscal year of construction), and Factor D (penetrating depth) of Site 2 on the N_d -value. The error bars and different letters have the same meaning as in Figure 9. $p < 0.01$, $n = 6$. B1: Downhill side, B2: Uphill side; D1: 0–50 cm, D2: 50–100 cm.

To provide an overview of all the aforementioned observations in Site 2, Figure 14 shows the all-observation average on Factor D, i.e., depth 0–100 cm for comparison with Figure 11 of Site 1. In the figure, the observations on the spur road are marked with filled circles, whereas those on natural ground are marked with open circles. From the figure, the N_d values on the spur road (closed circles) have a tendency of age-related change, except for C1 (FY2019). The N_d values on the natural ground (open circles) have a large variation, which probably reflects various situations of the natural ground.

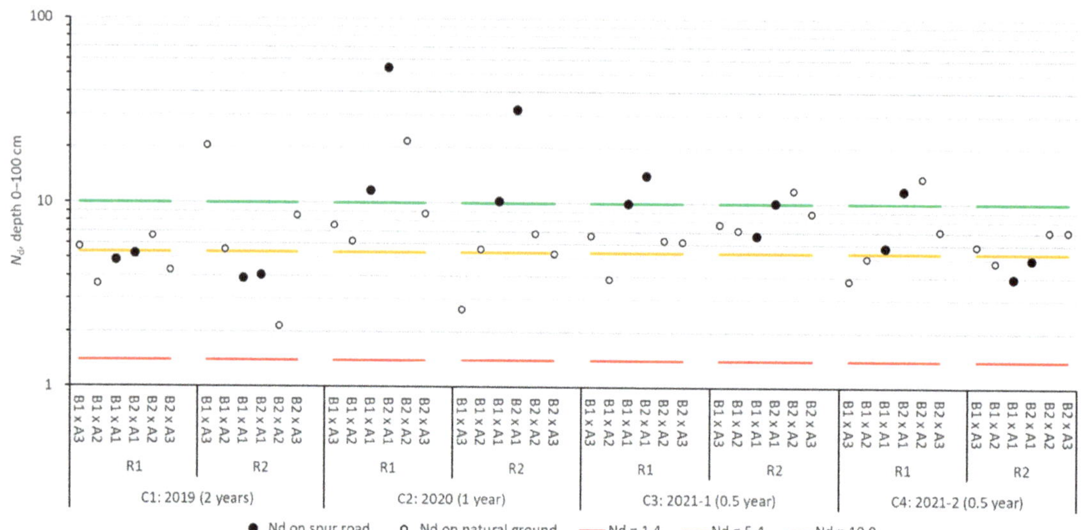

Figure 14. N_d—values of Site 2 accompanied with thresholds.

Compared with the threshold values for C1 (FY2019) and C4 (FY2021-2) × R2, the N_d values on the spur road were lower than $N_d = 5.4$. Thus, the compaction of the roadbed material soils, especially for FY2019, would have been insufficient to make the excavated soil sufficiently strong.

3.4.2. Measurement of Bearing Capacity

The result of the three-way ANOVA on the in situ CBR values in Site 2 is summarized in Table 5. Although Factor R was not significant, it was not pooled into the error because the interaction B × C × R, which includes Factor R, was significant [53].

Table 5. ANOVA table of in-situ CBR in Site 2.

Factor	Degree of Freedom	Sum of Squares	Variance	F-Value	*p*-Value
B: Downhill side or uphill side	1	315.2	315.2	23.00	0.00
C: Fiscal year of construction	3	433.0	144.3	10.53	0.00
R: Repetition	1	45.3	45.3	3.31	0.07
B × C × R	3	155.8	51.9	3.79	0.01
Error (pooled)	71	973.0	13.7		
Total	79	1922.3			

Total mean: 14.5%.

As for Factor B (downhill side or uphill side), B2 (uphill side, 16.5 ± SE 0.7%) was significantly greater than B1 (downhill side, 12.5 ± SE 0.7%; $p < 0.01$, n = 40), implying that the compaction of the road surface soil was better performed in the uphill side than in the downhill side. As for Factor C, the average in situ CBR values of C1 (FY2019, 17.1 ± SE 1.1) and C2 (FY2020, 16.6 ± SE 0.8) were significantly greater than those of C3 (FY2021-1, 12.5 ± SE 1.0) and C4 (FY2021-2, 12.0 ± SE 1.1; Tukey test, $p < 0.01$, $n = 20$). The tendency could be regarded as age-related change, i.e., natural compaction [49,50]. The effect of interaction among Factors B, C, and R is shown in Figure 15. In the combined levels of B1 × C1 × R2 and B2 × C4 × R2, the average values seem to be larger than the combined effects of Factors B and C. A reason would be the experimental error probably caused by cobbles or small rocks submerged in the road surface. However, according to the related studies that have evaluated the bearing capacities of the spur road surface using an in situ CBR tester [46–49], the total mean of 14.5% in Site 2 could be regarded as moderate, or not too low, for spur roads used for crawler-type machines.

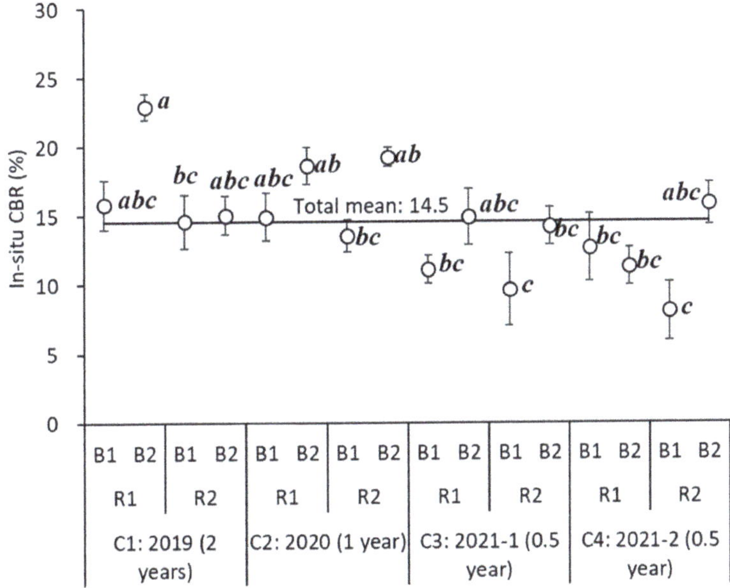

Figure 15. Interaction of Factor C (fiscal year of construction), Factor R (repetition), and Factor B (downhill side or uphill side) of Site 2 on in-situ CBR. The error bars and different letters have the same meaning as in Figure 9. $p < 0.05$, $n = 5$. B1: Downhill side, B2: Uphill side.

4. Conclusions

We investigated spur roads in Kochi University Forest constructed using the Shimanto method in order to verify whether sufficient roadbed strength and road surface bearing capacity were achieved. The main conclusions are as follows:

- Most parts of the investigated spur road sections had sufficient roadbed strength and surface bearing capacity;
- The roadbeds of the downhill sides tended to have lower strength because of insufficient compaction;
- The result indicated that the excavation over the entire road width and even compaction were not properly performed on parts with lower strength;
- Although it should be treated with caution, there appeared to be a tendency toward age-related change, i.e., natural compaction over time, for both the roadbed strength and the bearing capacity.

In the present study, the year of construction was treated as time elapsed since construction. However, in order to properly evaluate the strength development of a single pavement section, it is necessary to test the same pavement section at different ages. It would have been better to compare soil strength for three different depths, e.g., 0–30 cm, 30–60 cm, and 60–100 cm, to check the difference between fill and cut slopes and the effect of natural compaction. Although such analysis was not achieved in the present study because the resolution of N_d values was limited by the field situation, the authors would like to confirm the problem in the future study.

The lower strength of the roadbed on the downhill side of the spur roads or embankments was also observed in a related study on spur road construction. Furthermore, it was suggested that the soil under the road width should be excavated more widely toward the cut slope side before compaction to obtain the same roadbed strength on the downhill side as that on the cut slope side. Regarding the age-related change effect of the roadbed strength and bearing capacity, it was reported that in areas with low compaction ability soils, there is a spur road construction method that requires three years for completion. The method seems to effectively use the age-related change effect of the road strength. It is recommended that spur roads be left unused for a certain period of time before use in order to stabilize the materials of the constructed road body.

Author Contributions: Conceptualization, Y.S. and T.Y.; methodology, Y.S.; software, Y.S., S.H. and I.K.; validation, Y.S., S.H. and I.K.; formal analysis, Y.S., S.H. and I.K.; investigation, Y.S., S.H., H.A. and I.K.; resources, Y.S. and S.H.; data curation, Y.S. and S.H.; writing—original draft preparation, S.H. and I.K.; writing—review and editing, Y.S.; visualization, Y.S., S.H. and I.K.; supervision, Y.S.; project administration, Y.S.; funding acquisition, Y.S. and T.Y. All authors have read and agreed to the published version of the manuscript.

Funding: This study was supported by Japan Society for the Promotion of Science (JSPS) KAKENHI, grant numbers 21H03672, 22H03800, and 22K05747.

Data Availability Statement: The data used to support the findings of this study are available from the corresponding author upon request.

Acknowledgments: The authors are grateful to Ryohei Wada and Masa'aki Tateishi for supporting the field investigation in Site 2, and Naoyuki Hashimoto for supporting the GNSS survey in Site 2.

Conflicts of Interest: The authors declare no conflict of interest.

References

Note. The titles of Refs. [2,5,6,16–20,30,32,51–53] are tentative translations from the original Japanese by the authors.

1. Gotou, J. Current status and challenges for forest operations with high density forest road networks. *For. Econ.* **2011**, *63*, 19–22. (In Japanese) [CrossRef]
2. Japan Geotechnical Consultants Association. Geology and Geological Environment of the Japanese Islands—Geology of Japan and the West. Available online: https://www.zenchiren.or.jp/tikei/oubei.html (accessed on 31 December 2022). (In Japanese)
3. Lyons, C.K.; Borz, S.A.; Harvey, C.; Ramantswana, M.; Sakai, H.; Visser, R. Forest roads: Regional perspectives from around the world. *Int. J. For. Eng.* **2022**, 1–14. [CrossRef]
4. Sakai, H. Challenges in Road Construction and Timber Harvesting in Japan. *Croat. J. For. Eng.* **2017**, *38*, 187–195.
5. Forestry Agency. Development of Road Network Construction. Available online: https://www.rinya.maff.go.jp/j/seibi/sagyoudo/romousuisin.html (accessed on 30 December 2022). (In Japanese)
6. Suzuki, Y.; Aruga, K.; Iwaoka, M.; Saito, M.; Sakurai, R.; Suzuki, H.; Hasegawa, H.; Matsumoto, T.; Yabe, K.; Yoshioka, T.; et al. *Civil Engineering in Forestry*, 2nd ed.; Asakura Publishing Co., Ltd.: Tokyo, Japan, 2021; 200p. (In Japanese)
7. Suzuki, Y.; Setiawan, A.H.; Gotou, J. Timber Harvesting Systems in Japan's Aging Plantation Forests: Implications for Investment in Construction of Forest Road Networks. *J. Jpn. For. Soc.* **2015**, *97*, 191–202. (In Japanese with English summary) [CrossRef]
8. Suzuki, Y.; Yoshimura, T.; Hasegawa, H.; Aruga, K.; Saito, M.; Shirasawa, H.; Yamasaki, S. Perspective of road network improvement including public roads over forested area in Japan. *J. Jpn. For. Eng. Soc.* **2022**, *37*, 5–16. (In Japanese) [CrossRef]
9. Dietz, P.; Knigge, W.; Löffler, H. *Walderschließung*; Paul Parley: Hamburg, Germany; Berlin, Germany, 1984; 426p. (In German)
10. Heinimann, H.R. Forest Road Network and Transportation Engineering—State and Perspectives. *Croat. J. For. Eng.* **2017**, *38*, 173–188.

11. Keller, G.; Sherar, J. *Low-Volume Roads Engineering—Best Management Practices Field Guide*; USDA Forest Service/USAID: Virginia Polytechnic Institute and State University, Blacksburg, VA, USA, 2003; 158p. Available online: http://hdl.handle.net/10919/68420 (accessed on 30 December 2022).

12. Upfold, S. (Ed.) *South African Forest Road Handbook*; Forest Engineering Southern Africa and Institute for Commercial Forestry Research: Scottsville, South Africa, 2005; 214p. Available online: https://www.forestrysouthafrica.co.za/wp-content/uploads/2019/12/Roads-Handbook.pdf (accessed on 31 December 2022).

13. Forestry Agency. Annual Report on Forest and Forestry in Japan—Fiscal Year 2021 (Summary). Available online: https://www.maff.go.jp/e/data/publish/attach/pdf/index-96.pdf (accessed on 5 January 2023). (In Japanese)

14. Forestry Agency. Annual Report on Forest and Forestry for FY2021. Available online: https://www.rinya.maff.go.jp/j/kikaku/hakusyo/r3hakusyo/index.html (accessed on 5 January 2023). (In Japanese)

15. Kanzaki, K.; Ohasi, K.; Deki, T.; Miyake, T. On capillary path systems in steep mountain areas. *J. For. Eng.* **1990**, *2*, 17–21. [CrossRef]

16. Ohashi, K. *All of Road Construction*; Zenkoku-Ringyou-Kairyou-Fukyu-Kyokai: Tokyo, Japan, 2001; 159p. (In Japanese)

17. Ohashi, K. *Forestry Operation Road—Road Network Planning and Route Location*; Zenkoku-Ringyou-Kairyou-Fukyu-Kyokai: Tokyo, Japan, 2011; 121p. (In Japanese)

18. Ohashi, K. *How to Observe Mountain and Forest*; Zenkoku-Ringyou-Kairyou-Fukyu-Kyokai: Tokyo, Japan, 2012; 136p. (In Japanese)

19. Ohashi, K. *Inspection, Diagnosis, and Repair of Forest Road*; Zenkoku-Ringyou-Kairyou-Fukyu-Kyokai: Tokyo, Japan, 2015; 112p. (In Japanese)

20. Sakai, H. *Operation Road: Theory and Environmental Protection Function*; Zenkoku-Ringyou-Kairyou-Fukyu-Kyokai: Tokyo, Japan, 2004; 281p. (In Japanese)

21. Umeda, S.; Suzuki, H.; Yamaguchi, S. Considerations in the construction of a spur road network. *J. Jpn. For. Eng. Soc.* **2007**, *22*, 143–152, (In Japanese with English summary). [CrossRef]

22. Suzuki, Y.; Gotou, J.; Yamauchi, J.; Ohara, T. A case study on operational efficiency of downward felling, direct logging and winching operation on Hinoki plantation forest with dense road network and prospected effect of upward felling. *J. Jpn. For. Eng. Soc.* **2010**, *25*, 91–96, (In Japanese with English summary). [CrossRef]

23. Nakahata, C.; Aruga, K.; Uemura, R.; Saito, M.; Kanetsuki, K. Examining the optimal method to extract logging residues from small-scale forestry in the Nasunogahara area, Tochigi prefecture, Japan. *Small-Scale For.* **2014**, *13*, 251–266. [CrossRef]

24. Forest Survey. *Training Material 2010 Construction of Forest Operation Roads*; Forest Survey: Tokyo, Japan, 2010; 105p. (In Japanese)

25. Sawaguchi, I.; Sugawara, D.; Sasaki, K.; Tatsukawa, S. The number of optimum compaction times of skidding roads banked using hydraulic shovels. *J. Jpn. For. Eng. Soc.* **2012**, *27*, 225–232. (In Japanese with English summary) [CrossRef]

26. Watanabe, M.; Saito, M.; Shirasawa, H.; Ueki, T. The possibility of developing high-graded forest road networks by upgrading existing ones—A widening simulation-based investigation. *J. Jpn. For. Eng. Soc.* **2022**, *37*, 17–26. (In Japanese with English summary) [CrossRef]

27. Ikoma, N.; Saito, M.; Tatsukawa, S. Grasping the actual state of skidding road in clear cutting with a vehicle logging system. *J. Jpn. For. Eng. Soc.* **2022**, *37*, 39–46. [CrossRef]

28. Nagai, H.; Igarashi, S.; Takahashi, K.; Ichie, T. Evaluation of habitat selection by mammals in small vegetation patches of different forest physiognomies. *Jpn. J. For. Environ.* **2020**, *62*, 81–89. (In Japanese with English summary) [CrossRef]

29. Suzuki, Y.; Kusaka, H.; Yamaguchi, Y.; Aoki, H.; Hayata, Y.; Nagai, H. Productivity of a harvesting operation for a small clear-cut block by direct grappling using a processor in the Kochi University Forest. *Int. J. For. Eng.* **2019**, *30*, 195–202. [CrossRef]

30. Japan Meteorological Agency (JMA). Historical Weather Data Retrieval. Available online: https://www.data.jma.go.jp/obd/stats/etrn/index.php (accessed on 4 January 2023). (In Japanese)

31. BizStation Corp. DG-PRO1—Single Band GNSS Receiver. Available online: https://www.bizstation.jp/en/drogger/gps_index.html?tab=product (accessed on 4 January 2023).

32. Geological Survey of Japan, AIST. A Two-Hundred-Thousandth Seamless Geological Map of Japan. Available online: https://gbank.gsj.jp/seamless/ (accessed on 22 December 2022). (In Japanese)

33. Suzuki, Y.; Yamauchi, K. Practical investigation on affiliate structures for prevention of disaster and degradation on low-standard forest roads. *J. Jpn. For. Eng. Soc.* **2000**, *15*, 113–124, (In Japanese with English summary) [CrossRef]

34. Yamasaki, S.; Suzuki, Y.; Gotou, J.; Watanabe, N. Selection of terrain oriented appropriate logging system: A case study of the application on six watersheds in Kochi Prefecture, Japan. *J. Jpn. For. Eng. Soc.* **2021**, *36*, 13–20. (In Japanese with English summary) [CrossRef]

35. Umeda, S.; Yoshida, M. Development of penetration test equipment for mountain slopes and its practical application. *J. Jpn. For. Soc.* **1986**, *68*, 180–189. (In Japanese with English summary) [CrossRef]

36. Umeda, S.; Yoshida, M.; Takeshita, K. A standardization of the observed values obtained by various penetrometers on a mountain slope. *J. Jpn. For. Soc.* **1987**, *69*, 289–293, (In Japanese with English summary). [CrossRef]

37. Yoshinaga, S.; Ohnuki, Y. Estimation of Soil Physical Properties from a Handy Dynamic Cone Penetrometer Test. *J. Jpn. Soc. Eros. Control Eng.* **1995**, *48*, 22–28. (In Japanese with English summary) [CrossRef]

38. Gotou, J.; Yoshihara, O.; Mori, D.; Suzuki, Y. Considerations in the structure and the construction of spur roads for heavily loaded vehicles such as tower yarders. *J. Jpn. For. Eng. Soc.* **2014**, *29*, 5–12. (In Japanese with English summary) [CrossRef]

39. The Japan Geotechnical Society, Committee for Standard of Geotechnical Survey. *Japanese Standards and Explanations of Geotechnical and Geo-Environmental Investigation Methods*; Volume 1 and 2; The Japan Geotechnical Society: Tokyo, Japan, 2014; 1259p. (In Japanese)

40. Uchida, T.; Yahata, T.; Inagaki, M.; Kore'eda, K. Determination of Ground Bearing Capacity of Wide-Area Embankment Soils Using a Simple Penetration Tester. *Proc. 15th Ann. Meet. Soil Mech. Geotech. Eng.* **1980**, 57–60. (In Japanese)

41. Okimura, T.; Tanaka, S. Researches on Soil Horizon of Weathered Granite Mountain Slope and Failured Surface Depth in a Test Field. *J. Jpn. Soc. Eros. Control Eng.* **1980**, *33*, 7–16. (In Japanese with English summary) [CrossRef]

42. Sugiyama, T.; Okada, K.; Noguchi, T.; Muraishi, H. Spatial Distribution Characteristics of Soil Strength in Embankment Surface. *Doboku Gakkai Ronbunshu* **1992**, *1992*, 33–40. (In Japanese with English Summary) [CrossRef] [PubMed]

43. Ogawa, K. Study of Profile Patterns and Moisture Variations in Surface Soil Layers on Slopes of a Mountain Range. *Res. BUll. Hokkaido Univ. For.* **1997**, *54*, 87–141. Available online: http://hdl.handle.net/2115/21411 (accessed on 5 January 2023). (In Japanese with English Summary)

44. Hiramatsu, S.; Bitoh, K. Study on easy set up method of input data to slope failure predicting model from a handy dynamic cone penetrometer. *J. Jpn. Soc. Eros. Control Eng.* **2001**, *54*, 12–21. (In Japanese with English summary) [CrossRef]

45. Koyama, K. A new simple inspection method to check collapse at shoulder of spur road. *J. Jpn. Soc. Eros. Control Eng.* **2011**, *64*, 15–23. (In Japanese with English summary) [CrossRef]

46. Kobayashi, H.; Fukuda, M. Some factors influencing bearing capacity of forest road surfaces. *J. Jpn. For. Soc.* **1975**, *57*, 135–143. (In Japanese with English summary) [CrossRef]

47. Kobayashi, H.; Nitami, T.; Aruga, K.; Sakurai, R. On the Structure and Bearing Capacity of Forest Road. *Bull. Univ. Tokyo For.* **2005**, *113*, 241–256. Available online: http://hdl.handle.net/2261/22567 (accessed on 5 January 2023). (In Japanese with English summary)

48. Suzuki, H.; Yamaguchi, S.; Muneoka, H.; Tanaka, Y.; Shimizu, N.; Kariya, Y. The influence of soil property on strength of strip road bed and road surface. *J. Jpn. For. Eng. Soc.* **2014**, *29*, 21–29. (In Japanese with English summary) [CrossRef]

49. Sawaguchi, I.; Takahashi, Y.; Tatsukawa, S.; Takahashi, T. Aged change of bearing capacity of new construction skidding road. *J. Jpn. For. Eng. Soc.* **2011**, *26*, 97–104. (In Japanese with English summary) [CrossRef]

50. Marui & Co., Ltd. Portable Bearing Tester "CASPFOL". Available online: https://www.marui-group.co.jp/en/products/items2_4/item2_4_3/ (accessed on 3 January 2023).

51. Ministry of Land, Infrastructure and Tourism, Kinki Regional Development Bureau. Simplified Bearing Capacity Measuring Instrument (CASPFOL) User's Guide. Available online: https://www.kkr.mlit.go.jp/kingi/kensetsu/gijutusien/bcu0ke0000002vrs-att/caspol_guide.pdf (accessed on 31 December 2022).

52. Japanese Geotechnical Society. Soil Testing Practice Manual. In *Soil Testing—Basics and Guidance*, 3rd ed.; Japanese Geotechnical Society: Tokyo, Japan, 2001; 251p. (In Japanese)

53. Taguchi, G. *The Design of Experiments*, 3rd ed.; Vols. I and II.; Maruzen: Tokyo, Japan, 1977; 1095p. (In Japanese)

54. Zar, J.H. *Biostatistical Analysis*, 4th ed.; Prentice-Hall Inc.: Upper Saddle River, NJ, USA, 1999; 663p.

Article

Possibilities of Using UAV for Estimating Earthwork Volumes during Process of Repairing a Small-Scale Forest Road, Case Study from Kyoto Prefecture, Japan

Hisashi Hasegawa [1,*], Azwar Azmillah Sujaswara [2], Taisei Kanemoto [2] and Kazuya Tsubota [2]

[1] Field Science Education and Research Center, Kyoto University, Kyoto 6068502, Japan
[2] Graduate School of Agriculture, Kyoto University, Kyoto 6068502, Japan
*   Correspondence: hasegawa.hisashi.6x@kyoto-u.ac.jp

Abstract: Although forest road networks are an important infrastructure for forestry, recreation, and sustainable forest management, they have a considerable effect on the environment. Therefore, a detailed analysis of the various benefits and associated costs of road network construction is needed. The cost of earthwork in road construction can be estimated based on the change in topography before and after construction. However, accurate estimation of the earthwork volume may not be possible on steep terrain where soil placement is limited. In this study, an unmanned aerial vehicle was flown under the tree canopy six times during a road repair work to measure the changes in topography using structure from motion analysis. Comparing the obtained 3D model with the measurement results from the total station, the average vertical error and root mean square error were -0.146 m and 0.098 m, respectively, suggesting its good accuracy for measuring an earthwork volume. Compared to the amount of earthwork estimated from the topographic changes before and after the repair work, the actual earthwork volume was 3.5 times greater for cutting and 1.9 times greater for filling. This method can be used to calculate the earthwork volume accurately for designing forest road networks on steep terrain.

Keywords: small-scale forest road; UAV; Structure from Motion (SfM); earthwork volume; under tree canopy

Citation: Hasegawa, H.; Sujaswara, A.A.; Kanemoto, T.; Tsubota, K. Possibilities of Using UAV for Estimating Earthwork Volumes during Process of Repairing a Small-Scale Forest Road, Case Study from Kyoto Prefecture, Japan. *Forests* **2023**, *14*, 677. https://doi.org/10.3390/f14040677

Academic Editor: Stefano Grigolato

Received: 19 January 2023
Revised: 13 March 2023
Accepted: 21 March 2023
Published: 24 March 2023

1. Introduction

Forest road networks can be used for timber production, recreational activities, and preventing the spread of pests and fires [1]. However, constructing a road network on sloping terrain increases the risk of slope failure due to heavy rainfall and the fragmentation of wildlife habitat [2]. For designing an optimal road network layout, it is necessary to comprehensively evaluate candidate routes from the economic and environmental perspectives by setting objective variables, such as the construction and maintenance cost, effect of incurring a minimal cost for timber harvesting and transportation, social benefits related to recreation and disaster prevention, and the risk of damaging environmental health.

The cost–benefit evaluation in designing a forest road network requires a complex analysis, as the benefits of road construction are diverse; additionally, the costs should be calculated while considering the environmental risks [3]. Several methods, such as the analytic hierarchy process, have been proposed to evaluate the optimal route planning to maximize the benefits and minimize the costs of various construction objectives [4–8]. Recently, the precise microtopographic data obtained using an airborne laser scanner (ALS) have been used for forest road construction, because the construction cost is related to the amount of earthwork. Many studies have proposed designing forest road networks with minimal earthwork using high-resolution digital elevation model (DEM), because the routes with less earthwork are economical and discharge less sediment into the environment [9–14]. Aruga et al. [15] comprehensively studied a route selection method that combines the prediction of surface erosion and discharge of the related sediment into streams with the amount of earthwork. Ghajara et al. [16] developed a model in which the total cost of road construction may be estimated with an accuracy range of ±6.5%, assuming that the total construction cost is determined by six factors: clearing operations, embankments, pavements, gradings, culverts, and ditches. These studies have enabled us to estimate the cost of road construction precisely based on the difference in shape change between the landform before and after the forest road construction. Particularly, earthwork volume accounts for a large portion of the cost of small-scale road construction on sloping terrain [17,18], so these models have already been put to practical use as a route selection method to reduce road construction costs [19].

On the other hand, small-scale road construction operations, mainly earthwork operations, often occur on slopes, where there needs to be more space for temporary soil storage. It is necessary to repeatedly excavate, temporarily place, and fill soil in a limited area. Therefore, it is difficult to predict and quantify the amount of earthwork performed since it is difficult to ascertain all of the earthwork performed simply by comparing the shape of the soil before construction with the shape of the completed ground. It is necessary to conduct surveys during construction to grasp the precise amount of earthwork that changes with time. However, accurate quantification has been impractical.

Recently, however, with the development of unmanned aerial vehicles (UAVs), various sensors, and analysis technology, it has become possible to perform high-frequency surveying from the sky without stopping construction work [20]. The two surveying methods using UAVs are as follows: (1) obtaining a 3D model of the ground surface via structure from motion (SfM) analysis [21] based on images taken using an onboard camera and (2) measuring the distance to the ground surface using an onboard laser scanner. As the first method requires inexpensive sensors, it has recently been widely used for surveying.

UAVs are indispensable for earthwork estimation in the construction industry owing to their speed and efficiency [22]. Akgul et al. [23] compared the results of earthwork volume estimation of small-scale roads using DEMs created by fixed-wing UAV and Network real time kinematic global navigation satellite system (NRTK-GNSS). It was revealed that the UAV-based method was more cost-effective and accurate than NRTK-GNSS. Buğday [24] used UAVs and SfM analysis to create 5 cm mesh data after road construction in a forest area and showed that post-evaluation of earthwork volumes is possible. Hrůza et al. [25] operated a UAV at a speed of 1 m/s and 1 frame/s from an altitude of 4–6 m to create a point cloud with an average of 3.2 points/cm^2 and accuracy with a root mean square error (RMSE) of 0.0198 m for use in inspection work on asphalt-paved forest roads. Hrůza et al. [26] also reported that SfM analysis was performed on asphalt paved road surfaces using photographs taken from a height of 1.5 m above the ground and that the accuracy was higher than various LiDAR methods. Many other studies have been conducted for use in the calculation of earthwork quantities at construction sites [27–29].

Although SfM analysis using UAVs is used for constructing forest road network, previous studies have used data obtained from locations higher than the tree canopy. In the SfM analysis, identifying the same object in images taken from multiple locations is necessary. The ground surface in forests cannot be photographed due to the obstruction caused by standing trees. Therefore, flying the UAV under the tree canopy is necessary for performing a detailed analysis of the changes in the amount of earthwork during construction. An automatic flight is difficult under the tree canopy, as stable global navigation satellite system (GNSS) reception and positioning accuracy cannot be ensured. Consequently, manual operation with advanced flight techniques is required in such conditions. Therefore, few studies have been conducted by flying UAVs under the canopy to obtain detailed information of forests [30,31].

In this study, we attempted to quantify the changes in topography by taking six UAV mapping and conducting an SfM analysis at a small-scale forest road repair site, where it is relatively easy to fly UAVs even under the tree canopy because earthwork is performed in a narrow area with few standing trees. At the same time, we compared the accuracy of the topography measurement with that of a total station. The obtained data were used to verify whether it was possible to estimate the amount of earthwork at each stage of the repair work. A forest road repair site was chosen for this study, instead of a forest road construction site, because the earthwork is concentrated in a small area in forest road repair sites; therefore, it is convenient to collect data and obtain accurate evaluation.

2. Materials and Methods

The measurement test was conducted at a 2.5 m wide road repair site in Kizugawa City, located in the southern part of Kyoto Prefecture, Japan. The length of the road is 706.3 m, and it was constructed in 2020 as a branch line from a larger forest road. This road was constructed using the Ohashi-type road construction method, which assumed that 3–5-ton class excavators and 2–3-ton trucks will be used in the construction [32]. The logs used to make the retaining wall with log structure were used for the frame to distribute the loads [33].

Repair work was conducted from 22–25 August 2022 to repair a defective embankment near the middle of this line caused by heavy rainfall in May 2021. According to official rainfall data in the neighboring city of Kyotanabe, the total rainfall from 16–22 May was 204.5 mm, 160 mm of which was concentrated over two days. The road was carefully constructed with logs on the upper part of the embankment and the roadbed, but the bedrock was exposed in the affected area, making it challenging to form the embankment.

The repair consisted of first creating a simple road to access the lower part of the missing embankment and then reshaping the embankment with logs from the valley flowing under the lower part of the road embankment to the upper part of the road. A rectangular plot with a length of about 30 m and a width of about 16 m was set along the work road in the repair work area, and this was used as the analysis range.

UAV photography was conducted six times: once before the repair work (morning of 22 August), four times during the repair work (afternoon of 22 August, morning of 23 August, afternoon of 23 August, and morning of 24 August), and once after the repair work (13 September). A DJI Mavic3 drone was used for photography. We used PiX4Dmapper (Pix4D S.A., Prilly, Switzerland) for the SfM analysis. ArcGIS 10.8.1 (ESRI Inc., Redlands, CA, USA) and QGIS 3.16.7 (QGIS Development Team, 2021) were used for earthwork volume estimation.

Four aerial markers were set up as ground control points (GCPs) to measure absolute coordinates and eliminate distortion of the obtained model. In addition, surveying was conducted at two other open locations in the sky using a survey GNSS receiver (Spectra Precision SP80). Absolute coordinates of the GNSS points were precisely measured using electronic reference points of the Geospatial Information Authority of Japan, and absolute coordinates of the GCPs were calculated using a total station (TS) TAJIMA TT-N45 (TJM Design Co., Tokyo, Japan) from GNSS points. According to the TS specifications, the ranging accuracy, in this case, is "(2 + 2 ppm × D) mm (for measuring distances of 4 m or more)," with an error of 0.01 mm at a distance of 10 m. The angular accuracy is 5 and the error at a distance of 10 m is 0.24 mm.

The ground height was measured a total of four times using TS during the lunch break on each day when repair work was suspended. The measurement results were compared with the 3D model obtained by SfM analysis from the images taken by the UAV every morning. As the UAV imaging on the 2nd and 3rd days was conducted during the repair work, there was a slight time lag with the TS survey and the topography may have slightly changed. We used 41, 76, 91, and 88 TS survey points for comparison on days 1 (before the repair work), 2, 3, and 4 (after the repair work), respectively. The accuracy of the ground height obtained by the UAV was calculated by the TS measurement points as the validation points, and the results of the TS measurements were true values.

The UAV was operated at two altitudes: close up (3–5 m) and low altitude (7–10 m). In the SfM analysis, 3D models were created using two altitude images at the same time. It captured the area where the repair work was being done, with a flight duration of about 30 min. The UAV was operated manually along the area, and the image was automatically captured at 2 s intervals to ensure a high overlap between images. We also captured images from various directions in the flyable area.

DEMs of the ground surface were created from the 3D model obtained by SfM analysis using a UAV, and the amount of cut and fill at each stage was calculated using the Cut Fill tool in ArcGIS.

3. Results and Discussion

3.1. Generation of 3D Models for Each Phase of the Small-Scale Forest Road Repair Project

The 3D models of each phase were successfully constructed using SfM analysis. The resolution of the generated models ranged from 1.1 cm to 1.4 cm. For example, the 3D models generated from the data taken before, during, and after the work was completed and viewed from various angles are shown in Figures 1–3.

In this test, the flight route of the UAV was not fixed, and the appropriate course for photography was determined manually during the operation. Because the site was located in a valley, which made GNSS reception difficult, and the flight altitude was lower than the height of the surrounding trees, there was a risk of colliding with branches or tree trunks. Under these conditions, GNSS control is insufficient, and UAV flight control technology for real-time obstacle detection using various sensors and SLAM technology is needed.

Figure 1. A 3D model generated by SfM analysis before the repair work started.

Figure 2. A 3D model generated by SfM analysis during the repair work.

Figure 3. A 3D model generated by SfM analysis after the repair work completed.

Because the shooting area differed depending on the shooting session, the area where the 3D model was obtained also differed in each shooting session. There were some differences between the modeled trees and other land objects. However, no inconsistencies in their positions or shapes were observed. Even for the complex and finely patterned subsoil vegetation and soil and stone surfaces, no significant distortions were observed, and it was found that 3D models with clean shapes could be obtained.

3.2. Verification of Height Accuracy of Surface Data Obtained by SfM Analysis by UAV

The elevation values obtained at each point of the 3D model were compared with those measured by the TS. However, measurement by the TS takes time, which causes a significant time lag between drone photography and TS surveying. Therefore, verification of measurement accuracy was limited to four times: the morning of 22 August before the repair work, the morning of 23 August and 24 August during the repair work, and 13 September after the repair work. The results are shown in Table 1 and Figure 4.

Table 1. Height measurement error (m) in SfM analysis compared with the total station survey.

	Mean	Max	Min	RMSE
Before the repair work	−0.102	0.136	−1.333	0.105
During the repair work 1	−0.259	0.290	−0.910	0.162
During the repair work 2	−0.198	0.059	−1.218	0.121
After the repair work	−0.017	0.113	−0.494	0.017
Overall	−0.146			0.098

The overall measurement error averaged −0.146 m, leading to a general tendency toward underestimation. The RMSE was 0.098 m, indicating that elevation values could be obtained with an accuracy of about 10 cm. However, the average error varied greatly from one measurement session to the next. It was −0.102 m before the repair work, and −0.259 m and 0.198 m during the repair work, whereas it was −0.017 m after the repair work, a significant improvement. Lee and Lee [28] reported an RMSE error of 0.11 m in the vertical direction for the SfM analysis using vertical and 45-degree photographs from 50 m altitude at a construction earthwork site. The error in this study was similar, but the accuracy in this study was slightly less at closer distances to the ground surface and in forest areas.

The final survey was measured after the repair work had been completed and the area had been compacted. However, before and during the repair work surveys may not have measured the complete ground surface during TS surveying because of the softness of the soil on the ground.

Additionally, there was a considerable time lag between drone photography and TS surveying during the measurements before and during the repair work. This was because the survey was conducted safely so as not to interfere with the repair work as much as possible. In some locations, earthwork, or log construction may have been undertaken after drone flight and before the TS survey.

With this in mind, we discuss the accuracy of the measurements at each stage. The absolute value of the error at the TS measurement points before the repair work is shown in Figure 5a. It is shown in the figure that the error is small on slopes where the soil is visible, and slightly larger errors are observed in vegetated areas and on the shoulders of embankment slopes.

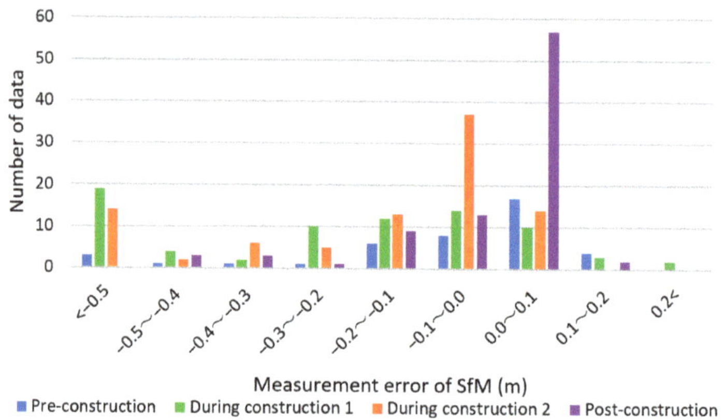

Figure 4. Frequency distributions of height measurement error in SfM analysis.

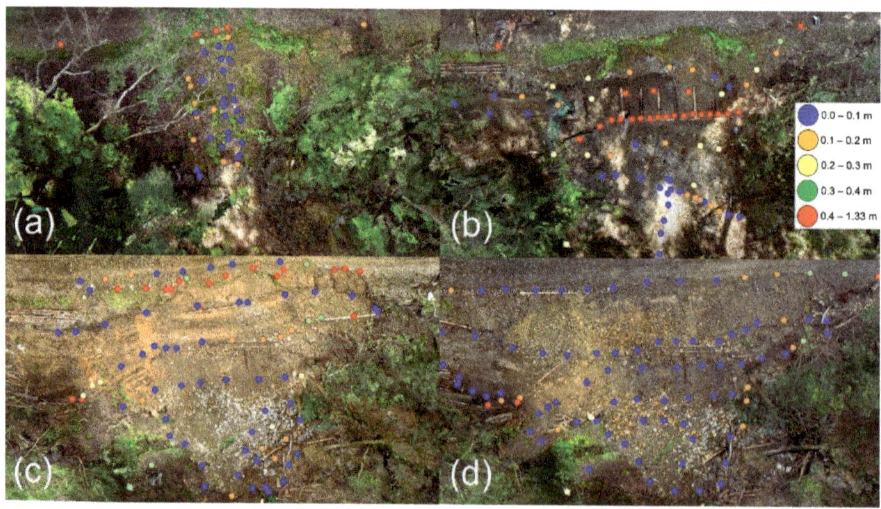

Figure 5. Absolute values of the error at each process of the repair work. (**a**) before the repair work, (**b**,**c**) during the repair work, (**d**) after the repair work.

Similarly, the situation during the repair work is shown in Figure 5b,c, and the situation after the repair work is shown in Figure 5d. It is shown in Figure 5b,c that errors were minor, even during the repair work, in areas where the soil was visible. However, substantial errors occurred in areas where work was being done, such as in areas where wood framing was being installed. A significant error at the shoulder of the embankment is shown in Figure 5c. Still, further verification was needed to determine whether the accuracy of edge detection in the SfM analysis was reduced by this problem or whether the soil tended to move easily at the shoulder of the embankment.

On the other hand, Figure 5d shows that the accuracy was generally good after the repair work, including the shoulders of the embankment. Even on the shoulders of the embankment, most of the errors were lower than 0.1 m. Slightly larger errors were observed in the upper right and lower left parts of the figure, which are close to the vegetation. Together with the edge mentioned above, a more precise analysis of the accuracy in areas easily shaded by vegetation is considered necessary.

3.3. Estimation of Earthwork Volume and Earthwork Allocation Plan

It was suggested by these results that SfM analysis is sufficiently accurate for soil volume calculations in units of cubic meters. However, there was room for further study on the accuracy of measurements under subsoil vegetation and at sharp slope conversion points. Therefore, we calculated the amount of earthwork for cut and fill at each stage. The results are shown in Figure 6.

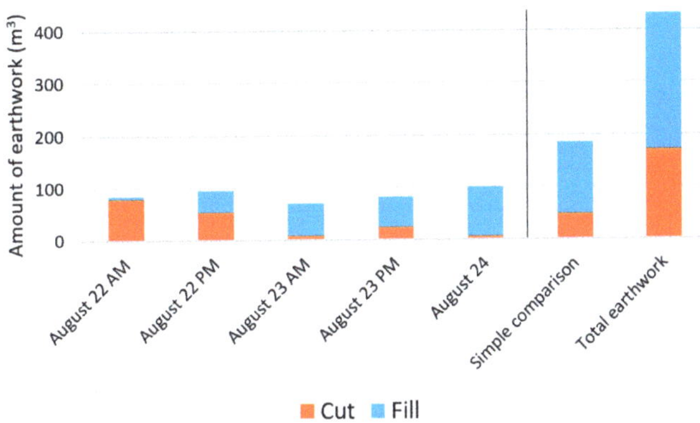

Figure 6. Amount of earthwork for cut and fill at each stage.

It was estimated that 78.6 m^3 of cut soil was generated on the morning of the first day of work (22 August), and another 5.49 m^3 was used as fill. In the early stage of the repair work, it was assumed that much of the cut soil was generated to shape collapsed areas and construct an access road to the lower part of the embankment. In the afternoon of 22 August, the cut, and fill volumes were 54.1 m^3 and 41.4 m^3, respectively. This was consistent with the fact that much soil was used during the formation of the log structure at the base of the fill. On the morning of 23 August, the cut decreased to 8.8 m^3 and the fill increased to 60.9 m^3 when the log structure continued to be installed. The cut soil increased again to 23.6 m^3 and the fill to 58.9 m^3 when the upper portion of the embankment was re-excavated simultaneously with the formation of the fill in the afternoon of 23 August. On 24 August, the last day of the repair work, the cut was 6.2 m^3 and the fill was 93.9 m^3.

A simple comparison between before and after the repair work had a difference of 48.6 m^3 for cut and 137.5 m^3 for fill, but when the amounts of earthwork at each stage were added up, 171.3 m^3 for cut and 260.7 m^3 for fill were performed. In the case of civil construction sites, the difference in the amount of earthwork before and after construction and the total amount of earthwork at each stage was not considered to be significantly different. However, it is necessary to temporarily bring in soil from another location or temporarily place cut soil in another location in forest work. The amount of cut and fill soil in this study was about 3.5 times and 1.9 times larger than that measured before and after the repair work. This is a significant error factor in estimating earth work volume and maintenance costs.

The cumulative earthwork volume at each stage is shown in Figure 7. The rate of change in the volume of loosened soil, the change in the volume of soil in the excavation of the ground soil volume (L), was calculated as 1.2 in calculating the amount of earthwork. The rate of change in the volume of compacted soil and the difference in soil volume in forming the ground soil volume into fill (C) was 0.9. The volume of the log structure embedded in the fill was not considered.

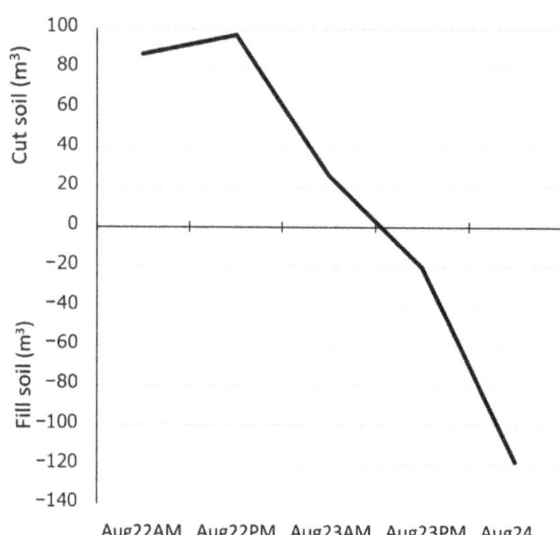

Figure 7. Cumulative amount of earthwork at each stage.

According to this figure, it can be seen that in the early stage of the repair work, when there was much cut soil, there was a maximum of about 100 m³ of cut soil. Securing a temporary storage place for this cut soil around the work site is necessary. On the other hand, the cut soil forms the embankment, and at some point, the soil for the embankment becomes insufficient. Therefore, it is necessary to secure soil from another location. This amount of fill was considered to be about 120 m³.

Figure 8 shows a 3D model of the earthwork sections at each stage. The cut and fill areas are shown in blue and red, respectively. In Figure 8f, showing the overall earthwork volume in the lower right corner, almost the entire missing fill area was covered with fill. The cut and fill areas were indicated in each step, making the procedure easy to understand. The entire fill was shown in blue, except for some parts, in Figure 8e. This is because the photographs after the repair work were not taken until three weeks after the completion of the repair work, and it is presumed that the entire fill has tightened and settled.

These results indicate that it may be difficult to accurately estimate the actual volume of earthwork in road construction on steep terrain where there are few temporary soil storage areas based only on the topography change between before and after construction. This study was conducted at a road repair site, and the procedure differed from that of earthwork during road construction. However, while constructing small-scale roads that do not require much artificial structures, the method of excavating the ground once and piling up topsoil blocks may be used to strengthen the road structure [32,33]. In addition to the topographical conditions, the capacity of the temporary soil storage area varies with the construction method. In steep terrain, it is necessary to determine the optimal construction procedures and earthwork volume in each step according to topographic conditions, road standards, and construction methods.

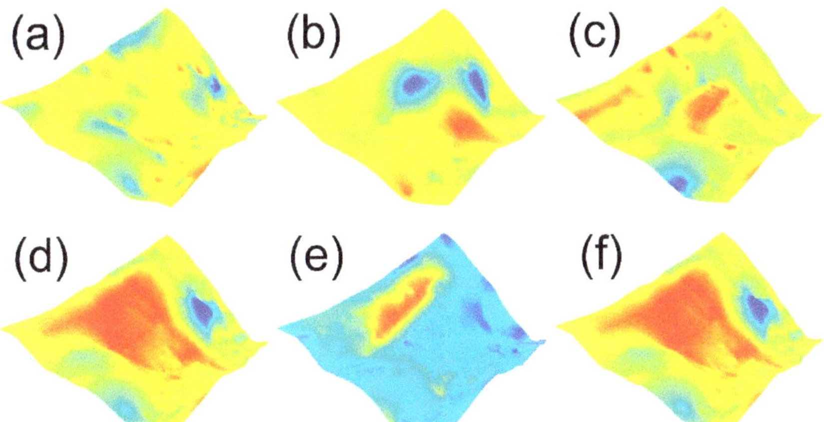

Figure 8. 3D model of the earthwork sections at each stage. (**a**) constructing an access road to the lower part of the embankment and the formation of the base, (**b**) forming the embankment base by earthwork and logs, (**c**) forming the lower part of the embankment by earthwork and logs, (**d**) forming the upper part of the embankment by earthwork and logs, (**e**) shaping the entire embankment, (**f**) overall earthwork.

4. Conclusions

It is shown in this study that SfM analysis using a UAV can be used to quantify the amount of earthwork and record the procedures for repair work on forest roads. The SfM analysis with UAV was an accurate means of measuring landforms comparable to TS surveying, creating detailed 3D models, calculating precise earthwork quantities, and recording detailed area-specific work procedures at each construction stage. Since the earthwork performed at each stage of the work was closely related to the cost of construction, it should be possible to accurately estimate the cost of construction, plan the work, and predict the number of days required for the work, which are deeply related to the cost–benefit analysis of forest roads.

Studying a stable flying method under the tree canopy was necessary as a future issue. However, the drone flight plan must be thoroughly examined when the work area changes by several tens of meters daily. For example, in conditions where a small-scale forest road is being constructed. The method of capturing images must also be verified to enable highly accurate 3D modeling. The drone flight plan must be thoroughly considered, and the method of capturing images must be verified for accurate 3D modeling.

Various measurement conditions, such as light environment, and ground surface conditions, will significantly impact the measurement evaluation, and an assessment of these shooting conditions is also required. The evaluation of soil movement during the construction process will also need to be considered.

This study evaluates only one case of small-scale forest road repair work. It is expected that completely different methods or procedures will be selected for different soil types, topographies, damage conditions, and soil placement locations. Particularly in forests, UAV flight has many obstacles, such as tree trunks and branches. In addition, the GNSS reception environment is poor, making automated flights difficult. In recent years, artificial intelligence drones equipped with sensors to avoid obstacles in real time have appeared [34]. Thus, the automatic flight into forests will be possible soon. In the future, it will be possible to collect a large amount of data on various tasks at various construction sites. It will be possible to formulate optimal road network establishment plans for complex local conditions by converting this information into big data. By collecting a large amount of data under various conditions of small-scale forest road construction and repair works using the methodology of this study, it is possible to estimate precise earthwork volume in road construction and repair works.

The disadvantage of SfM analysis is that it is computationally expensive and takes time to analyze. Still, a characteristic of SfM analysis is that it provides a point cloud with color information. Using color information for object identification is a significant advantage over other methods, such as LiDAR measurements, in forests with many different objects. SfM analysis in forests using UAVs can be used to evaluate the impact of human activities on the environment, such as vegetation change, and sediment transport.

Author Contributions: Conceptualization, H.H.; methodology, H.H. and A.A.S.; validation, H.H.; formal analysis, H.H. and A.A.S.; investigation, H.H., A.A.S., T.K. and K.T.; resources, H.H.; data curation, H.H. and A.A.S.; writing—original draft preparation, H.H.; writing—review and editing, A.A.S., T.K. and K.T.; visualization, H.H. and A.A.S.; supervision, H.H.; project administration, H.H.; funding acquisition, H.H. All authors have read and agreed to the published version of the manuscript.

Funding: This research was funded by JSPS KAKENHI Grant Numbers JP21H03672 and JP19K06125.

Data Availability Statement: Not applicable.

Acknowledgments: We would like to express our sincere appreciation to Suntory Holdings Limited, Albero Cuore Limited, Yamashiro-cho Forest Cooperative, Kitaharima Forest Cooperative, Forest Media Works Inc., Sumitomo Forestry Co., Ltd. for their great cooperation in conducting this study.

Conflicts of Interest: The funders had no role in the design of the study; in the collection, analyses, or interpretation of data; in the writing of the manuscript, or in the decision to publish the results.

References

1. Picchio, R.; Pignatti, G.; Marchi, E.; Latterini, F.; Benanchi, M.; Foderi, C.; Venanzi, R.; Verani, S. The Application of Two Approaches Using GIS Technology Implementation in Forest Road Network Planning in an Italian Mountain Setting. *Forests* **2018**, *9*, 277. [CrossRef]
2. Colchero, D.; Conde, A.; Manterola, C.; Chávez, C.; Rivera, A.; Ceballos, G. Jaguars on the move: Modeling movement to mitigate fragmentation from road expansion in the Mayan Forest. *Anim. Conserv.* **2011**, *14*, 158–166. [CrossRef]
3. Gumus, S.; Acar, H.H.; Toksoy, D. Functional forest road network planning by consideration of environmental impact assessment for wood harvesting. *Environ. Monit. Assess.* **2008**, *142*, 109–116. [CrossRef] [PubMed]
4. Shiba, M. Analytic Hierarchy Process (AHP)-Based Multi-Attribute Benefit Structure Analysis of Road Network Systems in Mountainous Rural Areas of Japan. *Int. J. For. Eng.* **1995**, *7*, 41–50. [CrossRef]
5. Abdi, E.; Majnounian, B.; Darvishsefat, A.; Mashayekhi, Z.; Sessions, J. A GIS-MCE based model for forest road planning. *J. For. Sci.* **2009**, *55*, 171–176. [CrossRef]
6. Pellegrini, M.; Grigolato, S.; Cavalli, R. Spatial Multi-Criteria Decision Process to Define Maintenance Priorities of Forest Road Network: An Application in the Italian Alpine Region. *Croat. J. For. Eng.* **2013**, *34*, 31–42.
7. Dragoi, M.; Palaghianu, C.; Miron-Onciul, M. Benefit, cost and risk analysis on extending the forest roads network: A case study in Crasna Valley (Romania). *Ann. For. Res.* **2015**, *58*, 333–345. [CrossRef]
8. Laschi, A.; Neri, F.; Montorselli, N.B.; Marchi, E. A Methodological Approach Exploiting Modern Techniques for Forest Road Network Planning. *Croat. J. For. Eng.* **2016**, *37*, 319–331.
9. Coulter, E.D.; Chung, W.; Akay, A.; Sessions, J. Forest road earthwork calculations for linear road segments using a high resolution digital terrain model generated from LIDAR data. In Proceedings of the First Precision Forestry Symposium, Seattle, WA, USA, 17–20 June 2001; pp. 125–129.
10. Aruga, K. Tabu search optimization of horizontal and vertical alignments of forest roads. *J. For. Res.* **2005**, *10*, 275–284. [CrossRef]
11. Aruga, K.; Sessions, J.; Akay, A.E. Application of an airborne laser scanner to forest road design with accurate earthwork volumes. *J. For. Res.* **2005**, *10*, 113–123. [CrossRef]
12. Akay, A.E.; Sessions, J. Applying the Decision Support System, TRACER, to Forest Road Design. *West. J. Appl. For.* **2005**, *20*, 184–191. [CrossRef]
13. Contreras, M.; Aracena, P.; Chung, W. Improving Accuracy in Earthwork Volume Estimation for Proposed Forest Roads Using a High-Resolution Digital Elevation Model. *Croat. J. For. Eng.* **2012**, *33*, 125–142.
14. Saito, M.; Goshima, M.; Aruga, K.; Matsue, K.; Shuin, Y.; Tasaka, T. Study of Automatic Forest Road Design Model Considering Shallow Landslides with LiDAR Data of Funyu Experimental Forest. *Croat. J. For. Eng.* **2013**, *34*, 1–15.
15. Aruga, K.; Sessions, J.; Miyata, E.S. Forest road design with soil sediment evaluation using a high-resolution DEM. *J. For. Res.* **2005**, *10*, 471–479. [CrossRef]
16. Ghajara, I.; Najafi, A.; Ali, S.; Karimimajd, A.M.; Boston, K.; Torabi, S.A. A program for cost estimation of forest road construction using engineer's method. *For. Sci. Technol.* **2013**, *9*, 111–117. [CrossRef]

17. Hare, W.L.; Koch, N.R.; Lucet, Y. Models and algorithms to improve earthwork operations in road design using mixed integer linear programming. *Eur. J. Oper. Res.* **2011**, *215*, 470–480. [CrossRef]
18. Papa, I.; Picchio, R.; Lovrinčević, M.; Janeš, D.; Pentek, T.; Validžić, D.; Venanzi, R.; Đuka, A. Factors Affecting Earthwork Volume in Forest Road Construction on Steep Terrain. *Land* **2023**, *12*, 400. [CrossRef]
19. Brown, K.; Visser, R. Adoption of emergent technology for forest road management in New Zealand. *NZ J. For.* **2018**, *63*, 23–29.
20. Siebert, S.; Teizer, J. Mobile 3D mapping for surveying earthwork projects using an Unmanned Aerial Vehicle (UAV) system. *Autom. Constr.* **2014**, *41*, 1–14. [CrossRef]
21. Hudzietz, B.P.; Saripalli, S. An experimental evaluation of 3D terrain mapping with an autonomous helicopter. *Int. Arch. Photogramm. Remote Sens. Spat. Inf. Sci.* **2011**, *XXXVIII-1/C22*, 137–142. [CrossRef]
22. Park, H.C.; Rachmawati, T.S.N.; Kim, S. UAV-Based High-Rise Buildings Earthwork Monitoring—A Case Study. *Sustainability* **2022**, *14*, 10179. [CrossRef]
23. Akgul, M.; Yurtseven, H.; Gulci, S.; Akay, A.E. Evaluation of UAV- and GNSS-Based DEMs for Earthwork Volume. *Arab. J. Sci. Eng.* **2018**, *43*, 1893–1909. [CrossRef]
24. Buğday, E. Capabilities of using UAVs in Forest Road Construction Activities. *Eur. J. For. Eng.* **2018**, *4*, 56–62. [CrossRef]
25. Hrůza, P.; Mikita, T.; Janata, P. Monitoring of Forest Hauling Roads Wearing Course Damage Using Unmanned Aerial Systems. *Acta Univ. Agric. Et Silvic. Mendel. Brun.* **2016**, *64*, 1537–1546. [CrossRef]
26. Hrůza, P.; Mikita, T.; Tyagur, N.; Krejza, Z.; Cibulka, M.; Procházková, A.; Patočka, Z. Detecting Forest Road Wearing Course Damage Using Different Methods of Remote Sensing. *Remote Sens.* **2018**, *10*, 492. [CrossRef]
27. Kim, Y.H.; Shin, S.S.; Lee, H.K.; Park, E.S. Field Applicability of Earthwork Volume Calculations Using Unmanned Aerial Vehicle. *Sustainability* **2022**, *14*, 9331. [CrossRef]
28. Lee, K.; Lee, W.H. Earthwork Volume Calculation, 3D Model Generation, and Comparative Evaluation Using Vertical and High-Oblique Images Acquired by Unmanned Aerial Vehicles. *Aerospace* **2022**, *9*, 606. [CrossRef]
29. Rachmawati, T.S.N.; Park, H.C.; Kim, S. A Scenario-Based Simulation Model for Earthwork Cost Management Using Unmanned Aerial Vehicle Technology. *Sustainability* **2023**, *15*, 503. [CrossRef]
30. Kuželka, K.; Surový, P. Mapping Forest Structure Using UAS inside Flight Capabilities. *Sensors* **2018**, *18*, 2245. [CrossRef]
31. Krisanski, S.; Taskhiri, M.S.; Turner, P. Enhancing Methods for Under-Canopy Unmanned Aircraft System Based Photogrammetry in Complex Forests for Tree Diameter Measurement. *Remote Sens.* **2020**, *12*, 1652. [CrossRef]
32. Suzuki, Y.; Hashimoto, S.; Aoki, H.; Katayama, I.; Yoshimura, T. Verification of Structural Strength of Spur Roads Constructed Using a Locally Developed Method for Mountainous Areas: A Case Study in Kochi University Forest, Japan. *Forests* **2023**, *14*, 380. [CrossRef]
33. Sakai, H. Challenges in Road Construction and Timber Harvesting in Japan. *Croat. J. For. Eng.* **2017**, *38*, 187–195.
34. Taki, S.; Aoki, M.; Koujimaru, M.; Inada, J. Aerial photography for in-forest using an artificial intelligence-enabled drone, and the construction of an in-forest three-dimensional model. *J. Jpn. For. Eng. Soc.* **2021**, *36*, 151–160.

 *forests*

Article

Assessing the Productivity of Forest Harvesting Systems Using a Combination of Forestry Machines in Steep Terrain

Tetsuhiko Yoshimura [1,*], Yasushi Suzuki [2] and Noriko Sato [3]

1   Faculty of Life and Environmental Sciences, Shimane University, Matsue 690-8504, Japan
2   Faculty of Agriculture and Marine Science, Kochi University, Nankoku 783-8502, Japan
3   Department of Agro-Environmental Sciences, Faculty of Agriculture, Kyushu University, Fukuoka 819-0395, Japan
*   Correspondence: t_yoshimura@life.shimane-u.ac.jp

Abstract: Despite similarly steep terrain, the productivity of forest harvesting operations in Japan is lower than in Central Europe. Harvesting systems in Japan are typically characterized by the four production processes of felling, yarding, processing, and forwarding, whereas in Central Europe they have mostly been reduced to just two through the use of a PTY (Processor Tower Yarder). This study investigated the number of production processes as a reason for the relatively lower productivity of forest harvesting in Japan using the Combined Machine Productivity (CMP) and Combined Labor Productivity (CLP) indices. The CMP and CLP were 1.81 m^3/h and 0.45 m^3/worker/h, respectively, for a parallel production model based on a typical Japanese forest harvesting system in Japan. The CMP and CLP values were improved to 2.51 m^3/h and 0.63 m^3/worker/h, respectively, when the forwarding process was removed from the model. The CMP and CLP values were further improved to 3.04 m^3/h and 0.76 m^3/worker/h, respectively, when yarding and processing were integrated into a single process. Reducing the number of the production processes can therefore improve the productivity of forest harvesting operations in Japan.

Keywords: productivity; harvesting system; processor tower yarder; combined machine productivity; combined labor productivity

Citation: Yoshimura, T.; Suzuki, Y.; Sato, N. Assessing the Productivity of Forest Harvesting Systems Using a Combination of Forestry Machines in Steep Terrain. *Forests* **2023**, *14*, 1430. https://doi.org/10.3390/f14071430

Academic Editor: Gianni Picchi

Received: 1 June 2023
Revised: 4 July 2023
Accepted: 10 July 2023
Published: 12 July 2023

1. Introduction

Productivity is defined as the rate of product output per time unit for a given production system [1]. In forestry, productivity in harvesting operations is a central concern in achieving an optimum balance between profitability and sustainability. When the slope gradient exceeds 40%, ground-based harvesting technology cannot provide good results [2]. Therefore, forest management in steep terrain depends on cable yarding as the primary extraction technology, which is usually deployed on difficult sites [3]. Cable yarding is a well-established practice for timber extraction in mountainous regions of the world where fully mechanized harvesting systems such as harvester–forwarder combinations cannot operate due to the steep terrain [4]. Analysis of 12 years of cable logging studies, from 2000 to 2011, showed that cable system efficiency was the most frequent keyword, with 78 out of 172 scientific references [5]. In the current study, the productivity of mechanized forest harvesting systems in steep terrain was discussed in relation to further improving profitability and sustainability.

Today, most forest harvesting operations are carried out with modern forestry machines, and their efficiency depends not only on the performance of each forestry machine, but also on how they are used in combination and on various site characteristics. In Japan, chainsaws, swing yarders, processors, and forwarders are typically used for felling, yarding, processing, and forwarding, respectively, on steep slopes where the most efficient harvester–forwarder system suitable for gentler slopes cannot be used. Cable yarding has been used for many years for timber extraction in the mountainous forests of Central

European countries [6] and is an efficient and effective harvesting system in steep terrain [7]. Regarding technical development, by the end of the 1990s, tower yarders, combined with a processor (PTY, Processor Tower Yarder) equipped with a radio-controlled carriage that automatically moves and stops, had become the standard for all manufacturers [8]. In Central Europe, system integration through the mounting of a tower yarder, crane and processing unit on a single carrier has been the main innovation path towards maintaining low harvesting costs and improving the productivity of cable yarding operations [9]. As a result, the productivity of forest harvesting operations in Japan is relatively low compared to Central Europe, as represented by Austria, where there have been significant technological innovations in cable-based harvesting machines and systems.

Many previous studies have determined the productivity of variations of the above-mentioned forest harvesting operations in Japan. According to the authors of [10], the labor productivity was 1.2 to 2.6 m^3/worker/day when a chainsaw, swing yarder, and processor were used for felling, yarding, and processing, respectively. The study described in [11] found that the labor productivity was 3.62 m^3/worker/day when a chainsaw, swing yarder, processor, and forwarder were used for felling, yarding, processing, and forwarding, respectively, and identified the use of a swing yarder as the reason for the relatively low productivity. The productivity of forest harvesting operations on moderate slopes was compared for different sizes of forestry machines and different log lengths when a chainsaw, harvester, and forwarder were used for felling, processing, and forwarding, respectively, and ranged from 5.3 to 7.5 m^3/worker/h [12]. Labor productivity was 6.64 m^3/worker/day when a chainsaw, grapple loader with a small winch, harvester, and forwarder were used for felling, wood extraction, processing, and forwarding, respectively [13]. According to the authors of [14], the overall average productivity of forest harvesting operations in Japan is 7.14 m^3/worker/day and 4.17 m^3/worker/day for final cutting and thinning, respectively.

While m^3/h or m^3/day is often used in Japan as a unit of productivity, Productive Machine Hour (PMH) is commonly used as the unit of productivity for mechanized forest harvesting operations. PMH represents the time during which the machine actually performs work, and this excludes time lost due to both mechanical and non-mechanical delays from Scheduled Machine Hours (SMH) that includes all time the machine is scheduled to work [15]. On the other hand, the study in [16] pointed out that Productive System Hour (PSH) must be used for systems consisting of several machines, while PMH has been widely used for systems consisting of a single machine and an operator. PSH is similar to PMH, but PSH includes two or more machines or sequential operations necessary to complete the task [17]. While PMH is commonly used as the unit of productivity for mechanized forest harvesting operations, PSH has been used by many studies, especially in Central Europe [3,4,8,18–27]. In addition, PMH_0 and PSH_0 do not include delays while PMH_{15} and PSH_{15} include delays of up to 15 min.

The PTY systems developed in Central Europe, such as Syncrofalke (MM Forsttechnik, Frohnleiten, Austria), as shown in Figure 1, can efficiently yard and process trees in a single production process. Productivity with the Syncrofalke for uphill and downhill yarding was found to be 11.54 m^3/PSH_0 and 8.25 m^3/PSH_0, respectively [6]. The average productivity of Syncrofalke in Central Bulgaria was calculated to be 15.20 m^3/PMH [28]. In Romania, the production rate of a PTY system consisting of a Mounty 4100 tower yarder and a Woody 60 processor (Konrad Forsttechnik) was 11.89 m^3/h, including delays [29]. In all cases, the productivity of forest harvesting operations with PTY systems was much higher than the average productivity in Japan.

According to the authors of [30,31], the low productivity of forestry operations in Japan is mainly due to the structural characteristics of small-scale forest ownership. We investigated additional reasons for the relatively low productivity of forest harvesting operations in Japan in terms of the way forestry machinery is used in combination. In this study, we evaluated the productivity of Japanese harvesting systems using two indices of the Combined Machine Productivity (CMP) and Combined Labor Productivity (CLP) and compared them with those observed in Central Europe.

Figure 1. Syncrofalke, one of the most typical PTY systems developed in Austria.

2. Materials and Methods

2.1. Classification of Production Models

2.1.1. Serial Production System

Forest harvesting systems were classified into two types, i.e., the serial and parallel production systems [32]. In the serial system the production processes are performed one after another in sequence (Figure 2). For example, after all the trees are felled by chainsaw, the yarding process starts. Conversely, in the parallel system the production processes are performed at the same time, either partially or completely (Figure 3). In this system, for example, trees are felled with a chainsaw while the yarding process is in progress. As shown in Figure 2, the total production time for the serial model (T_0) is expressed by the following equation:

$$T_0 = T_A + T_B \tag{1}$$

where T_0 is the total production time for the serial model, T_A and T_B are the production times for processes A and B, respectively. Then, the production rate of the serial production model (R_0) is expressed by the following equation:

$$R_0 = \frac{V}{T_0} = \frac{V}{T_A + T_B} = \frac{V}{\frac{V_A}{P_A} + \frac{V_B}{P_B}} = \frac{V}{\frac{V}{P_A} + \frac{V}{P_B}} = \frac{1}{\frac{1}{P_A} + \frac{1}{P_B}} \tag{2}$$

where V is the total production volume, V_A and V_B are the processed volume in the production processes A and B, respectively, and P_A and P_B are the production rate in the production processes A and B, respectively. In Equation (2), the following equation holds:

$$V = V_A = V_B \tag{3}$$

2.1.2. Parallel Production System

The total production time can be shortened by using the parallel production model as shown in Figure 3, and the time-saving rate (s) or the ratio of the total production time for the parallel model (T_s) to the total production time for the serial model (T_0) is expressed by the following equation:

$$s = \frac{T_s}{T_0}$$ (4)

where Ts ($<T_0$) is the total production time for the parallel model and s (<1) is the time-saving rate. Here, s is equal to 1 for the serial production model in Equation (4). In Figure 3, T_A and T_B are expressed by the following equations:

$$T_A = r_A \times T_s$$ (5)

$$T_B = r_B \times T_s$$ (6)

where r_A and r_B are the ratios of the production time for the processes A and B, respectively, to the total production time for the parallel model (T_s). It should be noted that the basic idea of r_A and r_B correspond to the utilization rate, which is calculated by dividing PMH by SMH and then multiplying by 100 to obtain a percentage [33]. Then, the following equation holds under the above conditions:

$$s = \frac{T_s}{T_0} = \frac{T_s}{T_A + T_B} = \frac{T_s}{r_A T_s + r_B T_s} = \frac{1}{r_A + r_B}$$ (7)

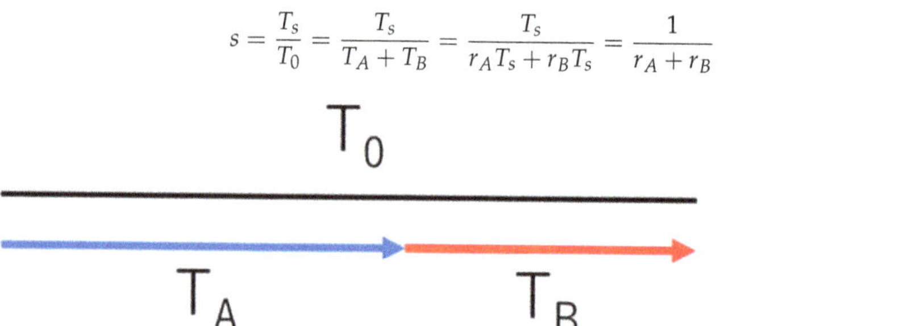

Figure 2. Serial production model. Note: T_0 = total production time for the serial model, T_A = production time for the process A, T_B = production time for the process B.

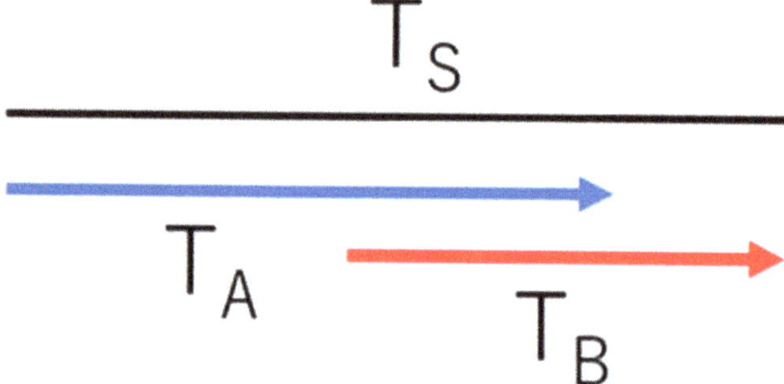

Figure 3. Parallel production model. Note: T_s = total production time for the parallel model, T_A = production time for the process A, T_B = production time for the process B.

The production rate of the parallel production model (R_s) is expressed by the following equation:

$$R_s = \frac{V}{T_s} = \frac{V}{s \times T_0} = \frac{V(r_A + r_B)}{T_0} = \frac{V}{\frac{V}{P_A} + \frac{V}{P_B}} \times (r_A + r_B) = \frac{1}{\frac{1}{P_A} + \frac{1}{P_B}} \times (r_A + r_B)$$ (8)

In Equation (8), $r_A + r_B$ is equal to 1 for the serial production model, and the following equation holds:

$$\frac{r_A}{r_B} = \frac{T_A/T_s}{T_B/T_s} = \frac{T_A}{T_B} = \frac{T_A/V}{T_B/V} = \frac{1/P_A}{1/P_B} = \frac{P_B}{P_A} \qquad (9)$$

2.2. Combined Machine Productivity

The equation to calculate the production rates for the serial and parallel models (R) can be generalized for two or more production processes as follows:

$$R = \left[\frac{1}{\left\{ \sum_{i=1}^{n} \left(\frac{1}{P_i} \right) \right\}} \right] \times \sum_{i=1}^{n} r_i \qquad (10)$$

where P_i is the production rate in the production process, i ($i = 1, n$; where n is the total number of production processes), and r_i is the ratio of the production time of the process i to the total production time. In Equation (8), it should be noted that $\sum_{i=1}^{n} r_i$ is equal to 1 for the serial production model, and it is more than 1 for the parallel production model.

The *CMP* is the total or overall productivity when two or more forestry machines work in combination, and it is calculated based on Equation (11) as follows:

$$CMP = \left[\frac{1}{\left\{ \sum_{i=1}^{n} \left(\frac{1}{P_i} \right) \right\}} \right] \times \sum_{i=1}^{n} r_i \qquad (11)$$

where *CMP* is the combined machine productivity (m^3/h), P_i is the production rate (m^3/h) in the production process, i ($i = 1, n$; where n is the total number of production processes), and r_i is the ratio of the production time of the process i to the total production time. The *CMP* is based on the idea proposed by the authors of [34] and further developed as a measure of system productivity by the authors of [32], which has been widely used in Japan to estimate productivity based on the performance of each forestry machine. The PMH or PSH can be used as a unit of productivity instead of using the unit of m^3/h.

2.3. Combined labor Productivity

We also used a labor productivity model for when two or more forestry machines are used in combination. The *CLP* is the total labor productivity, and is calculated as follows:

$$CLP = \frac{1}{\sum_{i=1}^{n} \left(\frac{N_i}{P_i} \right)} \qquad (12)$$

where *CLP* indicates the combined labor productivity (m^3/worker/h), P_i is the production rate (m^3/h) in the production process, i ($i = 1, n$; where n is the total number of production processes), and N_i is the number of workers in the production process, i. Equation (12) can be applied to both serial and parallel models. In Japan the CLP is referred to as system labor productivity and it has been used in many studies there [10,12,34–44]. As with the CMP, the PMH or PSH can be used as a unit of productivity instead of using the unit of m^3/h.

2.4. Productivity Assessment

We evaluated the productivity of several production models representing forest harvesting systems in mountainous conditions by calculating the CMP and CLP. Table 1 shows the performance values of the forestry machines used for the productivity calculation in Japan and Central Europe. The performance values of forestry machines depend on the operating conditions such as slope, terrain, yarding distances, tree volumes, machine types, operator experience, and so on, and they vary widely from one site to another. In addition, the purpose of this study is not to estimate the exact productivity but to identify the disadvantages of forest harvesting systems in Japan by comparing them with those in

Central Europe. Therefore, we determined the approximate performance values of forestry machines based on the published literature.

Table 1. Performance of forestry machines used in Japan and Central Europe for productivity calculation.

Type of Operation	Type of Machine	Country/Region	Hourly Processed Volume (m^3/h)	Number of Worker(s)
Felling	Chainsaw ($\times$2)	Japan	8.30 (4.15 $\times$ 2)	2
Yarding	Swing yarder	Japan	3.24	2
Processing	Processor	Japan	7.32	1
Forwarding	Forwarder	Japan	5.89	1
Felling	Chainsaw ($\times$2)	Central Europe	15.00 (7.50 $\times$ 2)	2
Yarding/Processing	PTY	Central Europe	11.54	2

In Japan, the labor productivity of chainsaw felling operations with a chainsaw at three research sites was reported to be 7.73, 2.97, and 1.75 m^3/worker/h [42], and we determined the hourly processed volume to be 4.15 m^3/h by averaging them. We also determined the hourly processed volume of yarding operations with a Japanese swing yarder to be 3.24 m^3/h by averaging the production rates of 3.29 m^3/h [44] and 3.18 m^3/h [45] for uphill yarding operations. The hourly processed volume of processing operations with a processor in Japan was determined to be 7.32 m^3/h by following [36]. The hourly processed volume of forwarding operations with a Japanese forwarder for two different-skilled operators was reported to be 3.10 and 8.67 m^3/h [46], and the average was determined to be 5.89 m^3/h.

In Germany, the labor productivity of chainsaw felling operations with a chainsaw was reported to be 7.50 m^3/worker/h [47]. The hourly processed volume of yarding/processing operations with PTY in Austria was determined to be 11.54 m^3/h based on the production rate for uphill yarding with a Syncrofalke tower yarder [6].

Figure 4 shows Model A1 for a Japanese harvesting system, where the production processes are connected in series and carried out one after another. There are four production processes in Model A1, namely felling, yarding, processing, and forwarding, and the forestry machines used for these production processes are shown in Figure 5. This type of production system is the most efficient in terms of labor productivity or CLP in this study, but it requires the longest total production time.

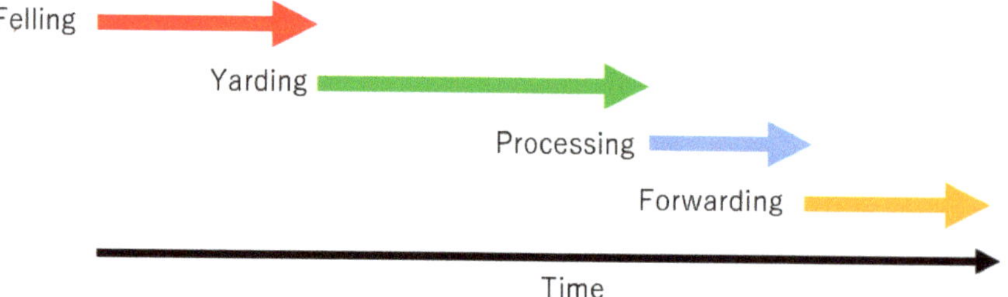

Figure 4. Production chart for Model A1.

Figure 6 shows Model A2 for a Japanese harvesting system, which omits the forwarding process from Model A1; there are thus three production processes in the model. In Model A3, the processing is merged with yarding and the timber extraction is carried out in just two processes as shown in Figure 7. Considering the poor forest road infrastructure in Japan, it may be almost impossible to use a PTY system due to its size and weight. However, it is possible to perform the same type of yarding and processing operations as a PTY system by using a tower yarder and a processor in combination, as shown in Figure 8.

Figure 5. Typical forestry machines used in Japan: (**a**) chainsaw; (**b**) swing yarder; (**c**) harvester; (**d**) forwarder.

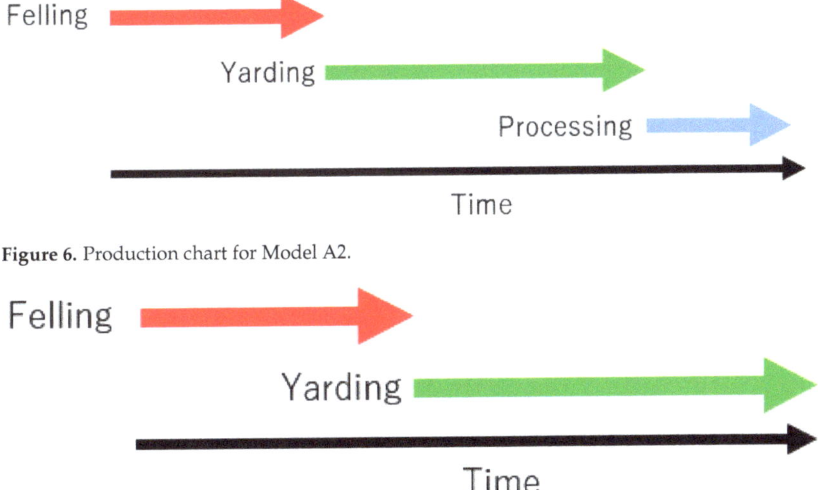

Figure 6. Production chart for Model A2.

Figure 7. Production chart for Model A3.

In Model B1 (Figure 9), all machines and crew members work together to make the total production time shorter than in Model A1, and the production processes are connected in parallel. The time-saving rate of this model is set to 0.75. This type of production model is the most typical in Japan, where teamwork has traditionally been highly valued in the workplace. It should be noted that this production model requires four crew members, even if the chainsaw operators also operate the processor and forwarder. The critical disadvantages of this harvesting system are that all crew members have to remain at the harvesting site from the beginning to the end of the operations, and are expected to perform additional tasks such as cleaning or organizing even when they are not involved in operating the machines. Such situations can result in a significant loss of time and skilled manpower and, ultimately, lead to a reduction in overall productivity.

Figure 8. A truck-mounted tower yarder, radio-controlled carriage, and excavator-based harvester used in combination in Shimane Prefecture, Japan.

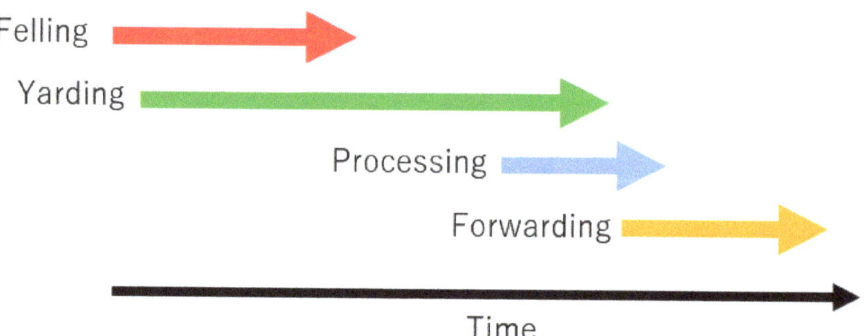

Figure 9. Production chart for Model B1.

Model B2 omits the forwarding process from Model B1, meaning there are three production processes involved (Figure 10). The time-saving rate is 0.71 in this model. In addition, Model B3 merges the processing and yarding processes, and timber extraction is therefore performed in two processes (Figure 11). The time-saving rate is 0.77 in this model. As in Model B1, Models B2 and B3 require that all crew members work together at the harvesting site.

Figure 12 shows a production model (Model C), which represents forest harvesting operations using PTY systems in Central Europe as a benchmark to evaluate the productivity of forest harvesting systems in Japan. The advantages of a PTY system are that it yards logs while processing them without requiring additional time or workforce, and there are only two production processes, i.e., felling and yarding/processing, as shown in Model C.

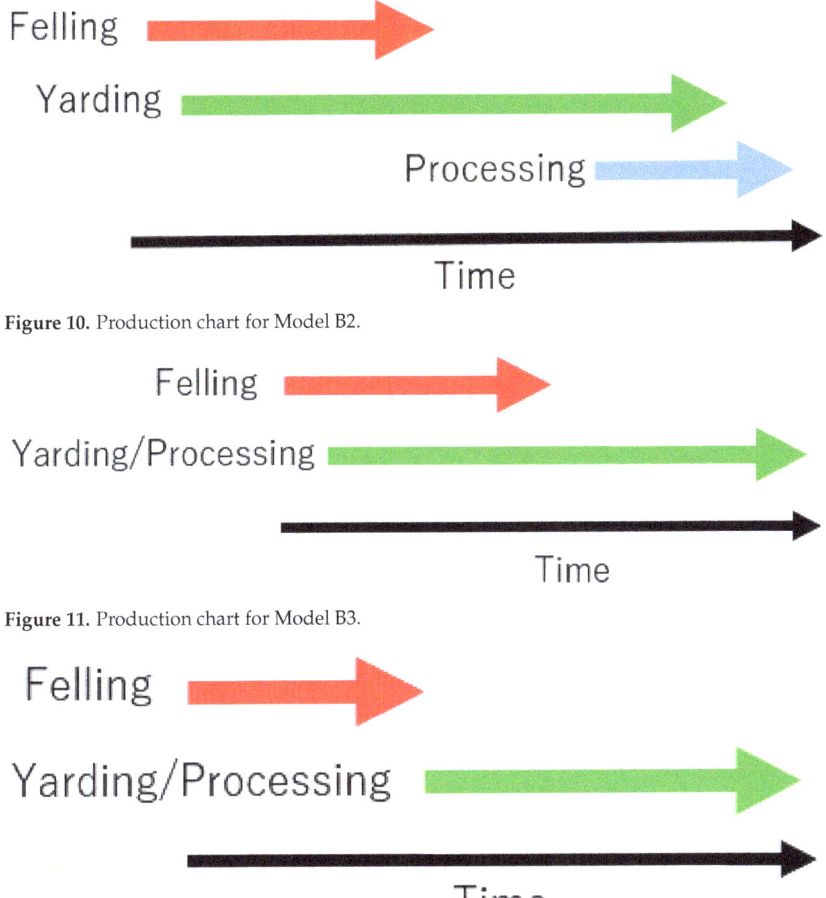

Figure 10. Production chart for Model B2.

Figure 11. Production chart for Model B3.

Figure 12. Production chart for Model C.

3. Results

Figures 13 and 14 show the results of the CMP and CLP calculations, respectively, for all production models. The CMP and CLP are the highest for Model C for Central Europe. The most typical production model in Japan is Model B1: it requires less production time than that of Model A1, but the labor productivity is also lower. In addition, process management in Model B1 is much harder than in Model A1 because it requires an efficient combination of machines and workforce. Although Model A1 is better than Model B1 in terms of process management, it has a critical practical disadvantage in that it is often difficult to store full trees on the forest road or landing until the next process starts.

The CMP and CLP were 1.81 m^3/h and 0.45 m^3/worker/h, respectively, for Model B1 based on a typical Japanese forest harvesting system in Japan. The CMP and CLP values were improved to 2.51 m^3/h and 0.63 m^3/worker/h, respectively, for Model B2 when the forwarding process was removed from Model B1. The CMP and CLP values were further improved to 3.04 m^3/h and 0.76 m^3/worker/h, respectively, for Model B3 when yarding and processing were integrated into a single process.

Parallel production models are supposed to be desirable because the forest harvesting operation can be completed in a shorter time than in serial models [32,48]. In fact, the respective CMP of Models B1, B2, and B3 are higher than those of Models A1, A2, and A3 (Figure 13). That is why parallel harvesting systems have been widely utilized in Japanese forestry. However, as shown in Figure 14, parallel models are less productive than serial models in terms of the CLP. A comparison of CMP and CLP in these models highlights the disadvantages of typical Japanese harvesting systems, which require more forestry machines and workforce at the same time, and process management is highly complicated and difficult to carry out efficiently. Forestry machines and workforce are sometimes idle at the harvesting site, as shown in Models B1, B2, and B3, introducing further inefficiency.

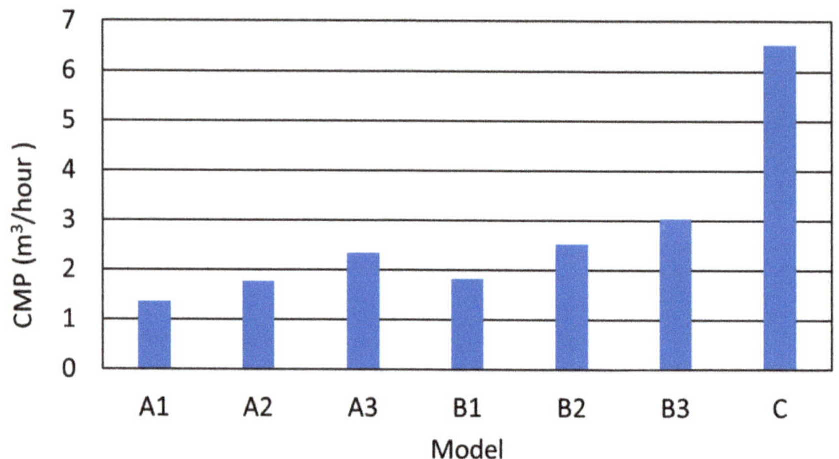

Figure 13. Comparison of the CMP among models for productivity evaluation.

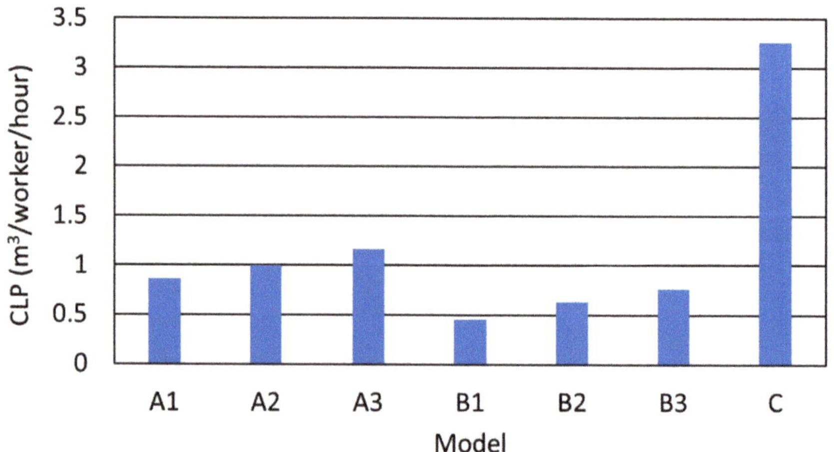

Figure 14. Comparison of the CLP among models for productivity evaluation.

Figure 15 shows the variation of the ratios of the production time for each process to the total production time for the parallel models B1, B2, and B3. As shown in this figure, the ratios become higher as the number of production processes is reduced. The ratio of idle time for Model B1 is the highest among the three parallel models, and the ratio of idle time for felling is the highest among the four production processes in Model B1. As a result, the imbalance of production rates between processes is a fundamental problem in the parallel production systems typical in Japan.

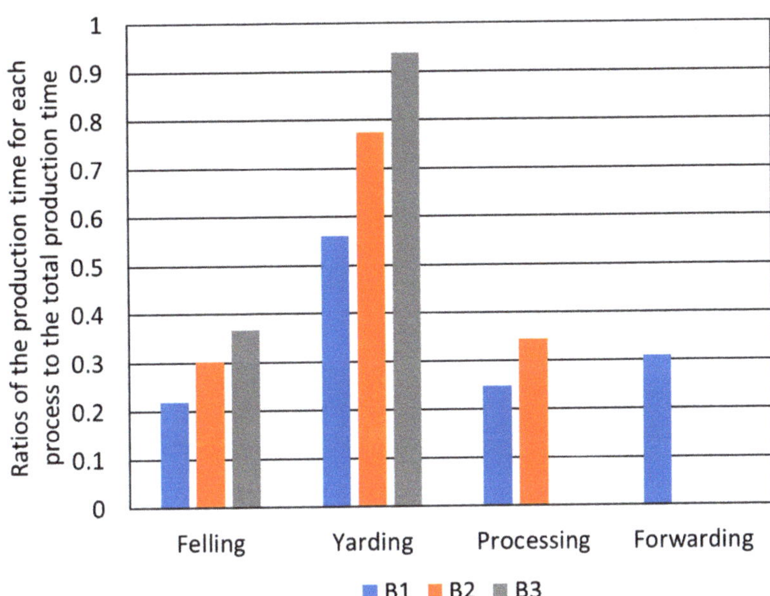

Figure 15. Ratios of the production time for each production process to the total production time for the parallel models.

Figure 16 shows the variation in the number of crew members for each production model according to the elapsed time. In this figure, it is assumed that 300 m^3 of timber is harvested and that the total production time is 220.7 h for Model A1. The number of crew members required for Models B1, B2, and B3 is higher than for Models A1, A2, A3, and C. The number of crew members for Models A1, A2, A3, and C varies in the same way. It should be noted that production models with PTY systems used in Central Europe refer to the serial models A1, A2, and A3 and not to the parallel models B1, B2, and B3, which are more common in Japan.

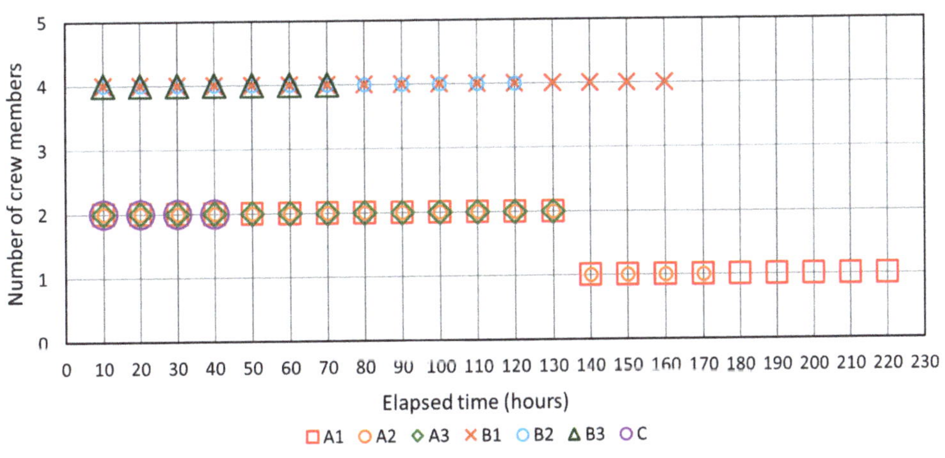

Figure 16. Variation in the number of crew members for each production model according to the elapsed time.

Figure 17 shows the variation in the volume of yarded full trees for each production model as a function of time. Cable yarding systems require well-coordinated transport to available storage space at the landing, otherwise the operation will run out of storage space and be shut down until serviced by a log truck [9].

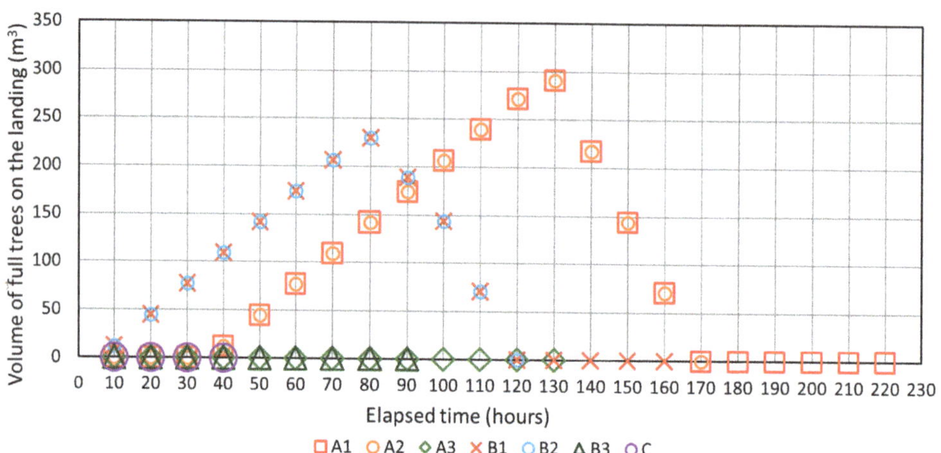

Figure 17. Variation in the volume of yarded full trees on the landing for each production model according to the elapsed time.

For Models A1 and A2, the volume of yarded full trees on the landing reaches 300 m^3 after 128.7 h. For Models B1 and B2, the volume reaches the maximum (235.2 m^3) after 78.7 h. As a result, these production models can be applied to harvesting sites that have a large area available for full tree storage. On the other hand, such a large area for full tree storage is not necessary for Models A3, B3, and C, where PTY or an equivalent system is used for the combined processes of yarding and processing. In a typical Japanese harvesting system such as Model B1, the imbalance of productivity between yarding and processing often causes extra waiting time for a processor, and the use of a PTY system can be an appropriate solution to this problem.

4. Discussion

In general, the productivity of forest harvesting operations is influenced by many factors. The main factors influencing the productivity of mechanized harvesting are the environment such as tree and terrain characteristics and climate, machine characteristics including bucking instructions, and the operator's mental and physical capabilities and work technique [1]. The productivity of the cable crane application is strongly influenced by the log volume, length of skyline, silvicultural prescription (harvesting intensity), and lateral extraction distance, and, in addition, the terrain slope, stand density, and direction of the yarding (uphill/downhill) have an influence on the extracted volume per time [47,49]. The productivity of harvesters depends on several factors, such as the tree size, stand type and productivity class, species, harvesting intensity and type, ground conditions, machine class, and operator skill [50]. A review of 70 studies of cable yarding operations found that different yarding conditions result in different effective factors, and that factors that have a large effect on one system studied may not be present in another system, or at least may not have a significant effect on performance [4]. The study in [4] also noted that weather is likely to have a significant effect on performance. It is very likely that heavy rain and wet conditions in the stand will decrease performance because the choker setter's movement will be more cautious and consequently slower and, in addition, the workers' motivation may be lower in rainy working conditions [4].

Due to the large number of factors that need to be considered as independent variables, there have been no practical models to accurately estimate the productivity by using all of the above factors. In this study, the CMP and CLP were calculated using a limited number of examples from the published literature, and this is one of the major limitations of this study. Nevertheless, the productivity of each Japanese machine shown in Table 1 is close to the standard productivity data published by the Forestry Agency of Japan [51], where 3 m³/h for chainsaw felling, 10 m³/h for processors, 8.3 m³/h for forwarders, and 3.3 m³/h for swing yarders was recorded. It should also be noted that the purpose of this study is to clarify the reasons for the relatively low productivity of forest harvesting operations in Japan in terms of the way forestry machinery is used in combination, and that it is not an accurate estimate of productivity.

The results showed the strong advantages of forest harvesting systems in Central Europe compared to those in Japan. It was suggested that the number of production processes is key to productive harvesting when two or more machines are used in combination. In Europe, the most successful forest harvesting systems in terms of productivity are based on two combined production processes. As an example of two-process production on gentle slopes, the harvester–forwarder system is widely used, where trees are felled and processed by a harvester, and then logs are collected and transported by a forwarder. The study in [52] calculated the system productivity of a small-scale combination harvester–forwarder in industrial plantation first thinning operations and pointed out that the system productivity is limited by the least productive component. On steep slopes, trees have been felled with a chainsaw, yarded, and then processed by a PTY and, nowadays, the harvester–forwarder system is feasible even on steep slopes by using harvesters [53,54] and forwarders [55], or by using a winch-assisted system [56–58]. All these systems are two-production processes and, furthermore, there are also potential possibilities of single process production, as seen in the development of harwarders [59–61]. The harwarder has shown the greatest potential to compete with the two-machine system in final fellings with relatively small stand volumes and short extraction distances [62]. The two-in-one harvester/forwarder [52] is another innovative technology, and the system productivity using this machine can be calculated as a serial production system.

Despite the importance of the number of production processes, the typical Japanese harvesting system consists of four production processes, with the addition of a forwarder to transport logs along the spur/secondary road, whereas harvesting systems on steep slopes in Japan used to consist of the three processes of felling with a chainsaw, yarding with a tower yarder, and processing with a processor in the 1990s [63]. In addition, tower yarders have been replaced by swing yarders or grapple loaders, which have a shorter reach for extracting trees than tower yarders. As a result, Japanese harvesting systems require high-density spur roads, especially when grapple loaders are used for tree extraction. The study in [64] raised concerns about the sustainability of this type of practice because it is usually carried out on steep slopes in high rainfall areas with little or no planning, drainage, or stabilization. Today, this type of practice often attracts public criticism in Japan because it can cause landslides during periods of torrential rain. In fact, on 4 July 2020, record-breaking rainfall hit the southern Japanese island of Kyushu, and many landslide disasters occurred in clear-cut areas, where spur roads had not been maintained after their construction. It is necessary to improve the productivity of forest harvesting operations on steep slopes in Japan, especially with a tower yarder and swing yarder, not only for increasing profitability but also for environmental sustainability.

5. Conclusions

This study showed that the productivity of forest harvesting systems in Japan is typically lower than in Central Europe in terms of the CMP and CLP. The reasons for this are not only the lower performance of forestry machines and the lower standard of forest roads in Japan, but also the difference in harvesting systems, especially the number of production processes employed in timber extraction. It was also found that the parallel production models are less productive than the serial production models in terms of the CLP, and that serial production models achieve the best productivity when processing is merged with the yarding process, and the forwarding process is eliminated. Forest harvesting systems in the mountainous countries of Central Europe, which have steep slope conditions similar to those in Japan, suggest the direction that Japanese forestry mechanization should follow. Finally, it is recommended that Japanese forestry should reduce the number of production processes to two by introducing the cable yarding system used in Central Europe, rather than making any further efforts by maintaining the four-process production system.

Future research is needed to flexibly adjust the hourly processed volume of each machine based on variables such as site slope, tree volume, road networks, silvicultural methods, and yarding and hauling distances by using computer simulation. This will allow us to generalize the production models in this study and take the next step in accurately determining the productivity according to different operating conditions.

Author Contributions: Conceptualization, T.Y., Y.S. and N.S.; methodology, T.Y., Y.S. and N.S.; validation, T.Y., Y.S. and N.S.; formal analysis, T.Y.; investigation, T.Y. and N.S.; resources, T.Y. and N.S.; data curation, T.Y. and N.S.; writing—original draft preparation, T.Y.; writing—review and editing, T.Y. and Y.S.; visualization, T.Y.; supervision, Y.S.; project administration, T.Y.; funding acquisition, T.Y. and Y.S. All authors have read and agreed to the published version of the manuscript.

Funding: This study was supported by the MEXT/JSPS KAKENHI Grant Numbers 21H03672 and 22K05747.

Data Availability Statement: All data used in this study is shown in Table 1. Anybody can get the same results as ours by using this data.

Acknowledgments: The authors are grateful to Masahiro Iwaoka, Tokyo University of Agriculture and Technology and Kazuya Sugimoto, Gifu Academy of Forest Science and Culture for their valuable advice. We also thank the Faculty of Life and Environmental Sciences at Shimane University for their financial support in publishing this report.

Conflicts of Interest: The authors declare no conflict of interest.

References

1. Liski, E.; Jounela, P.; Korpunen, H.; Sosa, A.; Lindroos, O.; Jylhä, P. Modeling the productivity of mechanized CTL harvesting with statistical machine learning methods. *Int. J. For. Eng.* **2020**, *31*, 253–262. [CrossRef]
2. Spinelli, R.; Magagnotti, N.; Visser, R. Productivity models for cable yarding in Alpine forests. *Eur. J. For. Eng.* **2015**, *1*, 9–14. Available online: https://dergipark.org.tr/en/pub/ejfe/issue/5186/70454 (accessed on 10 February 2023).
3. Varch, T.; Erber, G.; Spinelli, R.; Magagnotti, N.; Stampfer, K. Productivity, fuel consumption and cost in whole tree cable yarding: Conventional diesel carriage versus electrical energy-recuperating carriage. *Int. J. For. Eng.* **2021**, *32*, 20–30. [CrossRef]
4. Böhm, S.; Kanzian, C. A review on cable yarding operation performance and its assessment. *Int. J. For. Eng.* **2022**, *34*, 229–253. [CrossRef]
5. Cavalli, R. Prospects of research on cable logging in forest engineering community. *Croat. J. For. Eng.* **2012**, *33*, 339–356. Available online: https://hrcak.srce.hr/en/116850 (accessed on 7 February 2023).
6. Ghaffariyan, M.R.; Stampfer, K.; Sessions, J. Production equations for tower yarders in Austria. *Int. J. For. Eng.* **2009**, *20*, 17–21. [CrossRef]
7. Stampfer, K.; Visser, R.; Kanzian, C. Cable corridor installation times for European yarders. *Int. J. For. Eng.* **2006**, *17*, 71–77. [CrossRef]
8. Heinimann, H.R.; Stampfer, K.; Loschek, J.; Caminada, L. Perspectives on Central European cable yarding systems. In Proceedings of the International Mountain Logging 11th Pacific Northwest Skyline Symposium, Seattle, WA, USA, 10–12 December 2001. Available online: https://depts.washington.edu/sky2001/proceedings/papers/Heinimann.pdf (accessed on 10 February 2023).
9. Holzfeind, T.; Kanzian, C.; Gronalt, M. Challenging agent-based simulation for forest operations to optimize the European cable yarding and transport supply chain. *Int. J. For. Eng.* **2021**, *32*, 77–90. [CrossRef]
10. Hara, Y.; Sakagoshi, H. Survey and analysis of exploitation thinning in Shimane Prefecture. *Bull. Shimane Pref. Mount. Reg. Res. Ctr.* **2007**, *3*, 21–26. Available online: https://www.pref.shimane.lg.jp/admin/region/kikan/chusankan/syoseki/research/No3_kenkyuhoukoku.data/kenkyuhoukokudai3gou04.pdf (accessed on 9 February 2023). (In Japanese with English Summary)
11. Yoshimura, T.; Noba, T. Productivity analysis of thinning operations using a swing yarder on steep slopes in western Japan. In Proceedings of the IUFRO Unit 3.06 International Conference Forest Operations in Mountainous Conditions, Honne, Norway, 2–5 June 2013.
12. Nakazawa, M.; Yoshida, C.; Sasaki, T.; Taki, S.; Uemura, T.; Ito, T.; Yamaguchi, H.; Mozuna, M.; Usui, K.; Inomata, Y.; et al. Productivity of logging large diameter logs and long logs during final cutting in a mountain forest in Japan. *Int. J. For. Eng.* **2019**, *30*, 203–209. [CrossRef]

13. Kameyama, S.; Suzuki, R.; Yoshioka, T.; Inoue, K. Productivity and economic balance in the plantation forest behind thinning: A case study of Mishima city, Shizuoka. *J. Jpn. For. Eng. Soc.* **2017**, *32*, 137–142. (In Japanese) [CrossRef]

14. Forestry Agency. *Annual report on forest and forestry in Japan (Fiscal Year 2021)*; Forestry Agency: Tokyo, Japan, 2022; p. 207. Available online: https://www.rinya.maff.go.jp/j/kikaku/hakusyo/r3hakusyo/attach/pdf/zenbun-34.pdf (accessed on 9 February 2023). (In Japanese)

15. Picchio, R.; Venanzi, R.; Di Marzio, N.; Tocci, D.; Tavankar, F. A comparative analysis of two cable yarder technologies performing thinning operations on a 33 year old pine plantation: A potential source of wood for energy. *Energies* **2020**, *13*, 5376. [CrossRef]

16. Heinimann, H. *Operational Productivity Studies in Forestry Based on Statistical Models—A Tutorial*; Inst. Terrestrial Ecosystems Forest Engineering: Zurich, Switzerland, 2021; p. 56. [CrossRef]

17. Talbot, B.; Stampfer, K.; Visser, R. Machine function integration and its effect on the performance of a timber yarding and processing operation. *Biosyst. Eng.* **2015**, *135*, 10–20. [CrossRef]

18. Heinimann, H.R.; Visser, R.; Stampfer, K. Harvester-cable yarder system evaluation on slopes: A Central European study in thinning operations. In Proceedings of the Harvesting Logistics: From Woods to Markets, Portland, OR, USA, 20–23 July 1998. Available online: https://cofe.org/wp-content/uploads/2013/08/COFE_1998.pdf (accessed on 10 February 2023).

19. Stampfer, K. (1999) Influence of terrain conditions and thinning regimes on productivity of a track-based steep slope harvester. In Proceedings of the International Mountain Logging 10th Pacific Northwest Skyline Symposium, Corvallis, OR, USA, 28 March–1 April 1999.

20. Stampfer, K.; Steinmüller, T. A new approach to derive a productivity model for the harvester "Valmet 911 Snake". In Proceedings of the International Mountain Logging 11th Pacific Northwest Skyline Symposium, Seattle, WA, USA, 10–12 December 2001. Available online: https://depts.washington.edu/sky2001/proceedings/papers/Stampfer.pdf (accessed on 10 February 2023).

21. Bogo, A.; Cavalli, R. Productivity of a Tracked Excavator-Based Processor in the North-Eastern Italian Alps. Austro 2003: High Tech Forest Operations for Mountainous Terrain 2003. Available online: https://www.formec.org/images/proceedings/2003/4 5_bogo_cavalli.pdf (accessed on 7 February 2023).

22. Affenzeller, G.; Stampfer, K. Comparison of integrated with conventional harvester-forwarder-concepts in thinning operations. In Proceedings of the International Mountain Logging 13th Pacific Northwest Skyline Symposium, Corvallis, OR, USA, 1–6 April 2007. Available online: https://www.iufro.org/download/file/4702/4504/30000-mountain-logging-symp07_p_pdf/ (accessed on 10 February 2023).

23. Rottensteiner, C.; Affenzeller, G.; Stampfer, K. Evaluation of the feller-buncher Moipu 400E for energy wood harvesting. *Croat. J. For. Eng.* **2008**, *29*, 117–128. Available online: https://hrcak.srce.hr/en/32188 (accessed on 10 February 2023).

24. Erber, G.; Holzleitner, F.; Kastner, M.; Stampfer, K. Effect of multi-tree handling and tree-size on harvester performance in small-diameter hardwood thinnings. *Silva Fenn.* **2016**, *50*, 1428. [CrossRef]

25. Huber, C.; Kroisleitner, H.; Stampfer, K. Performance of a mobile star screen to improve woodchip quality of forest residues. *Forests* **2017**, *8*, 171. [CrossRef]

26. Holzfeind, T.; Stampfer, K.; Holzleitner, F. Productivity, setup time and costs of a winch-assisted forwarder. *J. For. Res.* **2018**, *23*, 196–203. [CrossRef]

27. Cadei, A.; Mologni, O.; Roser, D.; Cavalli, R.; Grigolato, S. Forwarder productivity in salvage logging operations in difficult terrain. *Forests* **2020**, *11*, 341. [CrossRef]

28. Papandrea, S.F.; Stoilov, S.; Angelov, G.; Panicharova, T.; Mederski, P.S.; Proto, A.R. Modeling productivity and estimating costs of processor tower yarder in shelterwood cutting of pine stand. *Forests* **2023**, *14*, 195. [CrossRef]

29. Borz, S.A.; Bîrda, M.; Ignea, G.H.; Popa, B.; Câmpu, V.R.; Iordache, E.; Derczeni, R.A. Efficiency of a Woody 60 processor attached to a Mounty 4100 tower yarder when processing coniferous timber from thinning operations. *Ann. For. Res.* **2014**, *57*, 333–345. [CrossRef]

30. Aruga, K.; Yamada, T.; Yamamoto, T. Comparative analyses of the cycle time, productivity, and cost between 62- and 107-year-old Japanese cypress clear-cutting operations using a small-scale cable logging system. *Small-Scale For.* **2019**, *18*, 279–289. [CrossRef]

31. Nakahata, C.; Aruga, K.; Saito, M.; Hayashi, U. Productivity and cost of clear-cutting and regeneration operations with small and medium-sized forestry machines in Utsunomiya city, Tochigi Prefecture, Japan. *Small-Scale For.* **2020**, *19*, 275–289. [CrossRef]

32. Oka, M. Selection of harvesting systems. In *Management of Mechanization for How to Utilize High-Efficient Forestry Machines to Increase Local Profitability*; Zenrinkyo, Ed.; Zenrinkyo: Tokyo, Japan, 2001; pp. 105–123. (In Japanese)

33. Sessions, J.; Berry, M.; Han, H.-S. Machine rate estimates and equipment utilization—A modified approach. *Croat. J. For. Eng.* **2021**, *42*, 437–443. [CrossRef]

34. Kanzaki, K. Think much of the labor productivity on the logging system with the help of high efficient forestry machines. *J. For. Mechanization Soc.* **1999**, *547*, 6–11. (In Japanese)

35. Sawaguchi, I.; Sasaki, S.; Tsuchiya, Y.; Tatsukawa, S. Labor productivity of line thinning with a mini swing-yarder. *Bull. Iwate Univ. For.* **2004**, *35*, 47–57. (In Japanese with English Summary)

36. Goto, M. Systematic productivity of thinning timber harvesting in Tokushima Prefecture. *Appl. For. Sci.* **2005**, 101–105. (In Japanese)

37. Nakazawa, M.; Imatomi, Y.; Oka, M.; Tanaka, Y.; Yoshida, C.; Uemura, T.; Yamaguchi, H.; Suzuki, H.; Umeda, S.; Takahashi, M.; et al. Development of thinning system using excavator with long-reach grapple-Productivity and cost of the system. *J. Jpn. For. Eng. Soc.* **2011**, *26*, 173–180. (In Japanese with English Summary) [CrossRef]

38. Nakazawa, M.; Yoshida, C.; Sasaki, T.; Uemura, T.; Suzuki, H.; Jinkawa, M.; Kondo, M.; Oya, S.; Toda, K.; Takano, T. Development of a thinning system with a small wheel mounted harvester and forwarder-Productivities of ordinary and line thinning. *J. Jpn. For. Eng. Soc.* **2013**, *28*, 187–192. (In Japanese with English Summary) [CrossRef]

39. Oya, S.; Saito, M.; Shirota, T.; Otsuka, D.; Miyazaki, T.; Yanagisawa, N.; Kobayashi, N. The productivity of one continuous clearcutting and reforestation operation with the vehicle logging system on moderate slopes in Nagano Prefecture. *J. Jpn. For. Soc.* **2016**, *98*, 233–240. (In Japanese with English Summary) [CrossRef]

40. Katagiri, T. Comparative study of production cost for clear cutting between the vehicle logging system and cable logging system in Okayama Prefecture. *J. Jpn. For. Eng. Soc.* **2018**, *33*, 37–45. (In Japanese) [CrossRef]

41. Yamasaki, S.; Yamasaki, T.; Suzuki, Y.; Mitani, Y.; Morimoto, M.; Nagasawa, Y. A verification study of a spur road improvement project on harvesting operational efficiency and total cost reduction. *J. Jpn. For. Eng. Soc.* **2018**, *33*, 25–35. (In Japanese with English Summary) [CrossRef]

42. Hirayama, K.; Nozue, N.; Hakamata, T.; Nakazawa, M.; Sasaki, T.; Yoshida, C. Labor productivity of full tree and tree length logging system using a self-propelled carriage introduced from Europe. *J. Jpn. For. Eng. Soc.* **2019**, *34*, 141–144. (In Japanese with English Summary) [CrossRef]

43. Kimura, S.; Sasaki, S.; Hoshikawa, T. Work efficiency at the clear cutting site using daily work reports. *Chubu For. Res.* **2020**, *68*, 59–60. (In Japanese) [CrossRef]

44. Watai, J.; Kondo, K. The labor productivity of a thinning operation system according to forest-road density-Study of an operation system using a swing yarder and processor. *Bull. Shizuoka Res. Inst. Agric. For.* **2012**, *5*, 53–58. Available online: https://agriknowledge.affrc.go.jp/RN/2010831114 (accessed on 20 March 2023). (In Japanese with English Summary)

45. Katagiri, T. Productivity of the felling-yarding alternate system using a swing-yarder-Combination of the felling-yarding alternate system and bucking, improvement of the felling-yarding alternate system. *Bull. Okayama Pref. Tech. Ctr. Agric. For. Fish. Res. Inst. For. For. Prod.* **2016**, *31*, 1–7. Available online: https://agriknowledge.affrc.go.jp/RN/2010921501 (accessed on 20 March 2023). (In Japanese)

46. Miyazaki, T.; Imai, M.; Shiraishi, R. Development of forest operation systems using high-efficient forestry machines. *Bull. Nagano Pref. For. Res. Ctr.* **2011**, *25*, 1–7. Available online: https://agriknowledge.affrc.go.jp/RN/2010900102 (accessed on 20 March 2023). (In Japanese)

47. Schweier, J.; Ludowicy, C. Comparison of a cable-based and a ground-based system in flat and soil-sensitive area: A case study from southern Baden in Germany. *Forests* **2020**, *11*, 611. [CrossRef]

48. Sakai, H. *Seminar on Forestry Production Techniques for Practical Management-Logging, Road Networks and Supply Chain*; Zenrinkyo: Tokyo, Japan, 2012; p. 349. (In Japanese)

49. Schweier, J.; Klein, M.-L.; Kirsten, H.; Jaeger, D.; Brieger, F.; Sauter, U.H. Productivity and cost analysis of tower yarder systems using the Koller 507 and the Valentini 400 in southwest Germany. *Int. J. For. Eng.* **2020**, *31*, 172–183. [CrossRef]

50. Borz, S.A.; Secelean, V.-N.; Iacob, L.-M.; Kaakkurivaara, N. Bucking at landing by a single-grip harvester: Fuel consumption, productivity, cost and recovery rate. *Forests* **2023**, *14*, 465. [CrossRef]

51. Committee for improving productivity in forestry enterprises. In *Guide to Productivity Improvement for Establishing a High Productivity Forestry Industry*; Forestry Agency: Tokyo, Japan, 2022; p. 108. Available online: https://www.rinya.maff.go.jp/j/gyoumu/gijutu/attach/pdf/torikumi-11.pdf (accessed on 3 July 2023). (In Japanese)

52. Ackerman, S.A.; McDermid, H.; Ackerman, P.; Terblanche, M. The effectiveness of a small-scale combination harvester/forwarder in industrial plantation first thinning operations. *Int. J. For. Eng.* **2022**, *33*, 56–65. [CrossRef]

53. Ackerman, P.; Martin, C.; Brewer, J.; Ackerman, S. Effect of slope on productivity and cost of Eucalyptus pulpwood harvesting using single-grip purpose-built and excavator-based harvesters. *Int. J. For. Eng.* **2018**, *29*, 74–82. [CrossRef]

54. Strandgard, M.; Alam, M.; Mitchell, R. Impact of slope on productivity of a self-levelling processor. *Croat. J. For. Eng.* **2014**, *35*, 193–200. Available online: https://hrcak.srce.hr/en/clanak/187578 (accessed on 12 February 2023).

55. Borz, S.A.; Marcu, M.V.; Cataldo, M.F. Evaluation of an HSM 208F 14tone HVT-R2 forwarder prototype under conditions of steep-terrain low-access forests. *Croat. J. For. Eng.* **2021**, *42*, 185–200. [CrossRef]

56. Berendt, F.; Tolosana, E.; Hoffmann, S.; Alonso, P.; Schweier, J. Harvester productivity in inclined terrain with extended machine operating trail intervals: A German case study comparison of standing and bunched trees. *Sustainability* **2020**, *12*, 9168. [CrossRef]

57. Holzfeind, T.; Visser, R.; Chung, W.; Holzleitner, F.; Erber, G. Development and benefits of winch-assist harvesting. *Curr. For. Rep.* **2020**, *6*, 201–209. [CrossRef]

58. Visser, R.M.; Spinelli, R. Benefits and limitations of winch-assist technology for skidding operations. *Forests* **2023**, *14*, 296. [CrossRef]

59. Wester, F.; Eliasson, L. Productivity in final felling and thinning for a combined harvester-forwarder (harwarder). *Int. J. For. Eng.* **2003**, *14*, 45–51. [CrossRef]

60. Laitila, J.; Asikainen, A. Energy wood logging from early thinnings by harwarder method. *Baltic For.* **2006**, *12*, 94–102. Available online: https://balticforestry.lammc.lt/bf/PDF_Articles/2006-12[1]/94_102%20Juha%20Laitila%20&%20Asikainen.pdf (accessed on 22 February 2023).

61. Karha, K.; Poikela, A.; Palander, T. Productivity and costs of harwarder systems in industrial roundwood thinnings. *Croat. J. For. Eng.* **2018**, *39*, 23–33. Available online: https://hrcak.srce.hr/en/clanak/285567 (accessed on 11 February 2023).

62. Jonsson, R.; Rönnqvist, M.; Flisberg, P.; Jönsson, P.; Lindroos, O. Comparison of modeling approaches for evaluation of machine fleets in central Sweden forest operations. *Int. J. For. Eng.* **2023**, *34*, 42–53. [CrossRef]
63. Inoue, M. The present situation and further research of the high productive mechanization system in logging operation. *Res. J. Food Agric.* **1994**, *17*, 6–12. Available online: https://agriknowledge.affrc.go.jp/RN/2030520036.pdf (accessed on 10 February 2023). (In Japanese)
64. Bjerketvedt, J.; Talbot, B.; Kindernay, D.; Ottaviani Aalmo, G.; Clarke, N. Evaluation of site impact after harvesting in steep terrain with excavator assisted ground based systems. In Proceedings of the IUFRO Unit 3.06 International Conference Forest Operations in Mountainous Conditions, Honne, Norway, 2–5 June 2013.

Article

Examination of Social Factors Affecting Private Forest Owners' Future Intentions for Forest Management in Miyazaki Prefecture: A Comparison of Regional Characteristics by Forest Ownership Size

Nariaki Onda [1,*], Shunsuke Ochi [2] and Nobuyuki Tsuzuki [3]

[1] Tohoku Research Center, Forestry and Forest Products Research Institute, Iwate 0200123, Japan
[2] Sugisachi Forestry Consulting Firm, Miyazaki 8892152, Japan
[3] Forestry and Forest Products Research Institute, Ibaraki 3058687, Japan
* Correspondence: ondanariaki@ffpri.affrc.go.jp

Abstract: Although Japan's planted forest resources are mature, efficient timber production and reforest postharvest are hindered by the small-scale forest ownership and private forest owners' (PFOs') low willingness to engage in forest management. A New Scheme of Forest Management (NSFM) has been established under which Japan's municipalities can aggregate forest management rights which PFOs with low future intentions for forest management. Therefore, this study explores the socioeconomic factors that determine PFOs' future intentions for forest management and examines NSFM challenges. PFOs were surveyed via questionnaires in two regions of Miyazaki Prefecture with different forest ownership sizes. The results showed that forest size and the presence of successors affect PFOs' future intention for forest management. In addition, PFOs with low future intentions were less aware of their forests, and their forests were the source of reforest abandonment. Although aggregating forest management rights of PFOs with low future intention by the municipalities may contribute to sustainable forest management, the increased workload on municipalities is a challenge. Overall, accessibility to sufficient decision-making information is a prerequisite for evaluating PFOs' future intention to manage their forests.

Keywords: New Scheme of Forest Management; Forest Management Law; private forest; small-scale forestry; typology of forest owners; Japan

Citation: Onda, N.; Ochi, S.; Tsuzuki, N. Examination of Social Factors Affecting Private Forest Owners' Future Intentions for Forest Management in Miyazaki Prefecture: A Comparison of Regional Characteristics by Forest Ownership Size. *Forests* **2023**, *14*, 309. https://doi.org/10.3390/f14020309

Academic Editor: Andrej Ficko

Received: 23 December 2022
Revised: 29 January 2023
Accepted: 30 January 2023
Published: 4 February 2023

1. Introduction

In Japan, planted forests cover 10 million ha or approximately 40% of the 25 million ha of the forest area, of which 65% are privately owned [1]. The planted forest resources, it is mainly converted from primarily broadleaf forests to conifer forests Japanese Cedar (*Cryptomeria japonica*) and Japanese Cypress (*Chamaecyparis obtusa*), which were established after World War II have matured; half of them are over 50 years old and are now in their utilization period. Notably, the amount of timber produced increased from 15.1 million m^3 in 2002 to 31.2 million m^3 in 2020 [1]. However, the Forestry Agency has found that the log supply from clearcutting is approximately 40% of the growing volume of planted forests accumulated, indicating that the resource must be utilized more efficiently.

The Japanese government has promoted forestry promotion policies, including the revision of the Forestry Basic Law in 1964, in response to the declining profitability of the domestic forestry industry against the backdrop of the strong yen and increasing timber imports [2]. Aggregating small-scale forest stands have been promoted as a policy measure to improve operational efficiency and reduce timber production cost in Japan, where the majority of forest owners are small-scale private forest owners (PFOs) [3]. However, as planted forests matured and entered the harvesting stage, further aggregation of forest management was required. In 2019, the Forestry Agency has started a "New Scheme of Forest Management" (NSFM) to solve the following problems: the low willingness of PFOs to manage their forests, the disparity between the intentions of PFOs and forestry managers to expand the scale of their operations, the slow introduction of road network maintenance and high-performance machinery, and low productivity. The small-scale ownership of forest conditions is a major obstacle to efficient timber production, as 74% of forest areas owned by PFOs are less than 5 ha [4]. The NSFM ensures the following: (1) A growth industry is compatible with proper forest management, (2) aggregation of forest management rights, not including ownership, to highly motivated and sustainable forestry enterprises, and (3) improvement conditions for aggregation of forest management [4]. As part of this effort, in 2018, the government enacted a Forest Management Law to achieve effective timber production and appropriate forest management by aggregating forest management rights. Specifically, this law stipulates that PFOs must manage their own forests and are responsible for harvesting, silviculture, and nursery at the appropriate times. It also allows PFOs to entrust forest management to the municipality in accordance with PFOs' future intentions. Forests suitable for forestry management are re-entrusted to forestry enterprises from the municipality according to their economic values, whereas the municipality manages those that are unsuitable. Thus, the NSFM is based on PFOs' willingness to manage their forests in the future, and municipalities survey PFOs' intentions to ascertain their willingness to do so. The survey selection criteria included planted forest owners without a forest management plan and forest management that had not been implemented for the past decade.

The Forest Environment Transfer Tax (FETT) allocation began in 2019 in Japan as a financial resource for municipalities and prefectures that will be directly responsible for NSFM. The FETT amount allocated to municipalities is calculated as follows: 50%, 20%, and 30% of the municipality allocations are based on the area of private forest plantations, forestry worker population, and municipality population, respectively. The area of privately planted forest is corrected according to the forest area ratio, with 1.5 for municipalities with a forest area ratio of 85% or more and 1.3 for municipalities with a forest area ratio of 75% or more but less than 85%. Notably, the criteria for determining the allocation amount deviate from the system's intention [5]. As FETT has only been operational for a limited time, system evaluation is a future concern. However, the actual situation regarding tax utilization is being assessed for prefectures [6] and large cities [7], prefectural support for municipalities [8,9], urban–rural partnerships [10]. In addition, Ishizaki et al. [11] mentioned the increased workload of municipal officials on the NSFM and the FETT administration.

The Forest Management Law, which directs the future management of the PFOs' forest according to the PFOs' willingness, may promote immediate timber production and forest improvement. However, because the transfer of rights related to the property rights of PFOs must be done cautiously, the development of PFOs willing to manage their forests must be balanced with the smooth aggregating forest management rights from PFOs unwilling to manage their forests. Clarifying the factors that influence future management intentions is thus necessary when considering the maintenance or enhancement of PFOs' willingness to forest management.

Effective forest policy implementation requires identifying the determinants of PFOs' decisions [12] and developing forest policies that can influence PFOs' behaviors [13]. Forest owner typologies are being utilized to develop a method for identifying forest owner values [14]. The typology studies are mostly based on ownership objectives [15]. Previous studies noted that forest owners could be divided into five types: "economist," "multiobjective owner," "recreationist," "self-employed," and "passive owner," based on the purpose of their forest ownership [12,13]. Boon et al. [12] classified the Danish PFOs into three types: "classic owner," "hobby owner," and "indifferent farmer" based on a survey of PFOs' interest in forests. Ingemarson et al. [13] classified the Swedish PFOs into five types: "traditionalist," "economist," "conservationist," "passive," and "multiobjective," according to the purpose of ownership, and showed differences in the forest ownership size, frequency of visits to their own forest, and presence of successors. Ficko and Boncina [15] classified PFOs as "materialists" and "nonmaterialists". On the other hand, some studies categorized PFOs focused on their forest management behaviors. The willingness of landowners to harvest woody biomass as a characteristic of woody biomass suppliers has been noted as a factor of ownership purpose, owned forest size, tree species structure and composition, and demographics in the southern United States [16]. In contrast to these previous studies, an approach that categorizes PFOs according to their expressed future intention for their forest management and who identifies the underlying factors that can contribute to their decision-making process is required to clarify the issues involved in the NSFM, Japan.

The declining willingness of PFOs to forest management has been identified as a problem [17], with multiple factors influencing PFOs' management behaviors in Japan. Considering the PFOs' forest status, these factors included forest ownership size [18], especially planted forest size, and the distance between the residence and owned forest [19,20]. For PFOs' perceptions and management behaviors, PFOs' awareness of forests as property [21], the awareness of owned forest boundaries [22], perception of planted forest locations [19,20], and registration status [19,22] are noted. PFOs' attributes were mentioned in terms of age [23], occupation [24], and the existence of successors [24]. In addition, social relationships in local communities [25], membership in a forest owners' cooperative (FOC) [22], deteriorated functioning of FOC's regional organizations [26], and residence or absence in the village have been identified as factors influencing the owner–local community relationship [20]. Since the late 1990s, neglecting reforest postharvest has emerged as a problem resulting from PFOs' poor forest management practices [27]. Low prices of standing timber as economic factors [28] and the failure to continue the management of forest divisions upon contract expiration as institutional factors contribute to reforest abandonment [29]. Indicated by these results are the factors that define PFOs' management behaviors and their perception of forests. However, studies on future management intentions are limited. Hayashi et al. [24] mentioned regional differences in the factors that influence PFOs' willingness to sell, as well as occupation and successors. Kushiro and Ito [30] described that many PFOs, notably absentee village owners, want to disengage from forest management despite acknowledging the necessity of continuing forest management for reasons including uncertainty of inheritance, loss of boundary, and economic evaluation.

This study aimed to identify the socioeconomic factors affecting the future intention for forest management by classifying the PFOs' future intentions.

2. Methodology

2.1. Study Site

The study survey was conducted in Miyazaki Prefecture, which ranks third in Japan for timber production (1,879 thousand m^3) and first for cedar production (1739 thousand m^3) [31]. Small-scale PFOs dominate southern Miyazaki Prefecture, whereas large-scale PFOs dominate northern Miyazaki Prefecture, indicating regional variations. Southern PFOs in this prefecture have a low willingness to manage their forests [32]. In addition, the identifying PFOs and their confirming their rights is an obstacle to timber production in the southern region [33,34].

In this study, considering the difference in the forest ownership scale [35], Kunitomi Town (hereafter Kunitomi) was selected as the study site from the southern part (primarily small-scale PFOs), and Kitakata district in Nobeoka City (hereafter Kitakata) (Figure 1) from the northern part (mainly large-scale PFOs) (Figure 2).

Kunitomi is a suburban area adjacent to Miyazaki City, the capital city of Miyazaki Prefecture, with a population of 18,027 [35], an area of 130.6 km^2, and a forest area of 7736 ha (59.2% forest area) [36]. Ownership of less than 5 ha accounts for 98.8% [37]. Former Kitakata Town merged with Nobeoka City in 2006 and became a part of Nobeoka City. The population is 3321 [38], with an area of 200.1 km^2, forest areas of 17,770 ha, and a forest area ratio of 88.4% [36].

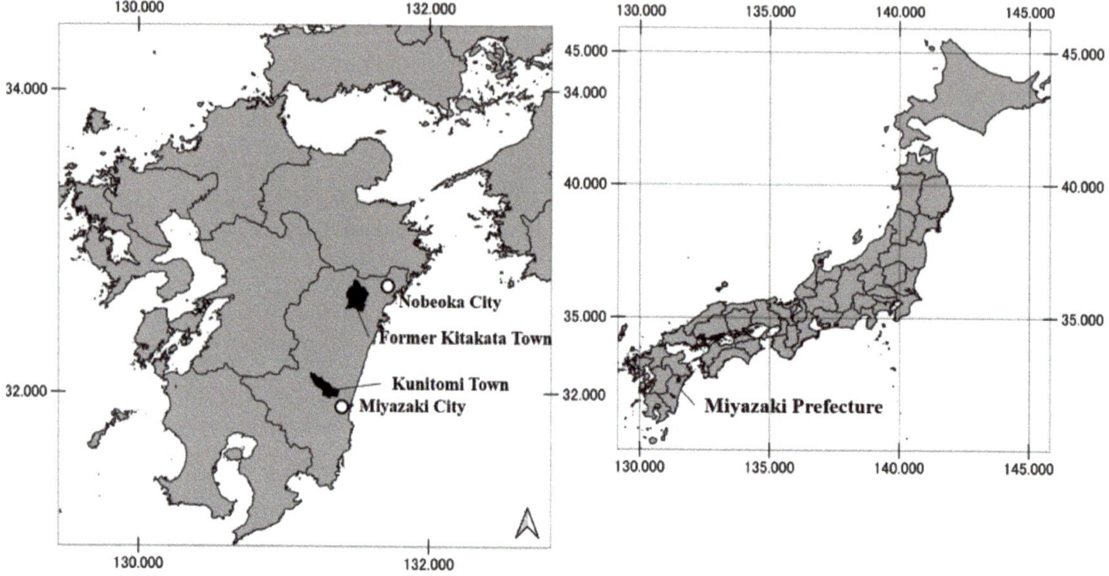

Figure 1. Study site.

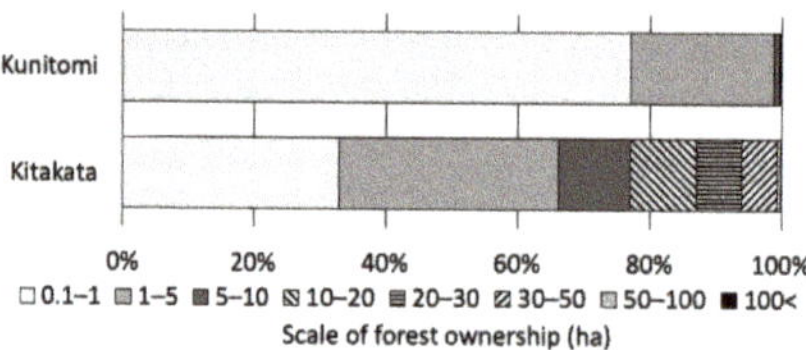

Figure 2. Proportion of forest ownership scale [34].

2.2. Method and Data Collection

All PFOs who belonged to FOC in both regions were surveyed using questionnaires to collect data. The questionnaires included questions on the following (1) forest conditions, (2) PFOs' characteristics, and (3) PFOs' management behaviors and attitudes; these factors were considered to influence the differences in the future intention of forest management. Although relatives of PFOs may have responded to the questionnaire, they were treated as PFOs in this study. In Kunitomi, 502 questionnaires were mailed out, and of the 367 questionnaires sent out, excluding 135 that were unaddressed, 166 were returned (response rate: 45.2%), and the number of valid responses was 162 (valid response rate: 44.1%). In Kitakata, out of 613 letters sent by mail (625 questionnaires were mailed and 12 were unaddressed), 299 questionnaires (response rate: 48.8%) were received, and the number of valid responses was 298 (48.6%). The questionnaire surveys were conducted from October to November 2020 in Kunitomi, and from December 2020 to January 2021 in Kitakata.

Based on the results obtained, the PFOs were classified into two groups according to their future intentions for the forest management scale: "expansion/maintenance PFOs" (hereafter EM group) who want to expand or maintain the management scale, and "decrease PFOs" (hereafter D group) who want to decrease the forest management scale. We then compared the situation in the two regions and examined the effects of regional differences in forest ownership size on PFOs' future intentions to manage their forests. Then, we compared the forest conditions factors and PFOs' characteristics. After that, we compare PFOs' behaviors and attitudes to forest management in each region, elucidating the factors that influence future intentions to forest management and regional differences. The chi-square test was used to make comparisons at a 0.05 significance level. Based on this analysis, we discuss the chances and challenges of NSFM (Figure 3).

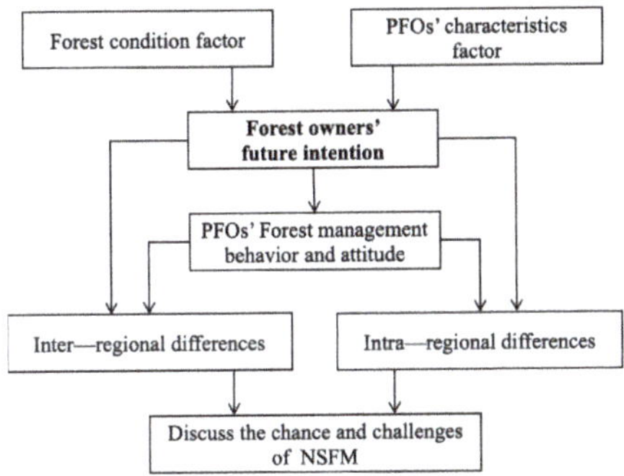

Figure 3. Research structure.

3. Results

3.1. Regional Comparisons

3.1.1. PFOs' Future Intentions

Differences were observed between the two regions' PFOs' intentions regarding the future forest management scale ($p = 0.000$). In Kunitomi, 47.3% (70 PFOs) belonged to the EM group, whereas 52.7% (78 HHs) belonged to the D group. In Kitakata, 70.9% (185 PFOs) belonged to the EM group, whereas 29.1% (76 PFOs) belonged to the D group.

Differences between the two regions were also seen in forest management intentions ($p = 0.000$). In Kunitomi, most PFOs (29.7%, 46 PFOs) wanted to "sell and transfer" their land, whereas 25.8% (40 PFOs) were "undecided," meaning that they were not thinking about or were considering the future management method. However, in Kitakata, "undecided" was the most common response (30.8%, 88 PFOs), followed by "entrustment" (28.7%, 82 PFOs). PFOs who answered "sell or transfer" accounted for 13.6% (39 PFOs). Overall, these results indicate that PFOs' future management intentions were low in Kunitomi and that many PFOs were willing to relinquish their land (Tables 1 and A1).

Table 1. Comparison of future intention to manage forests between two regions.

	Kunitomi		Kitakata		*p*-Value
	n	**%**	**n**	**%**	
Future intention of management scale *					0.000
Increase and maintain	70	47.3	185	70.9	
Decrease	78	52.7	76	29.1	
Total	148	100.0	261	100.0	
Future intention of forest management *					0.000
Independent	27	17.4	70	24.5	
Entrustment	32	20.6	82	28.7	
Sell or transfer	46	29.7	39	13.6	
Suspense	40	25.8	88	30.8	
Others	10	6.5	7	2.4	
Total	155	100.0	286	100.0	

Note: * *p*-Value < 0.05.

3.1.2. Forest Ownership Size and Forest Conditions

Forest ownership size between the two regions differed, and Kunitomi tended to have smaller forest ownership size than Kitakata ($p = 0.000$). The most prevalent response in both regions was that PFOs were unaware of their forest sizes, with 29.2% (45 PFOs) in Kunitomi and 38.7% (106 PFOs) in Kitakata. Concerning the trends by size of PFOs who were aware of their forest areas, Kunitomi had the highest percentage of PFOs with a forest ownership size of 1–3 ha (34.4%, 53 PFOs), followed by PFOs with a forest ownership size of less than 1 ha. In Kitakata, the highest percentage of PFOs owned 10–30 ha (16.1%, 44 PFOs), followed by those who owned 5–10 ha (11.7%, 32 PFOs). Similarly, the planted forest size was unknown, with the highest percentage of PFOs in Kunitomi (40.1%, 59 PFOs) and Kitakata (41.4%, 110 PFOs). Kunitomi tended to have smaller PFOs than Kitakata ($p = 0.000$) based on the planted forest size known.

Regarding the degree of maturity of planted forests, 46.7% (70 PFOs) of the PFOs in Kunitomi and 60.0% (168 PFOs) in Kitakata indicated that their planted forests were "mature," whereas 0.7% (1 PFOs) in Kunitomi and 2.5% (7 PFOs) in Kitakata said they were "partially" at the harvest stage. The northern region tended to have a greater proportion of mature forests ($p = 0.014$). In addition, 20.7% (31 PFOs) of the respondents in Kunitomi and 12.9% (36 PFOs) in Kunitomi answered that they were unsure. No significant difference was observed between the two regions in the status of the cadastral survey, with 60.9% (95 PFOs) completed in Kunitomi and 58.5% (172 PFOs) in Kunitomi ($p = 0.161$).

Regarding forest registration methods, Kunitomi tended to favor single-title registration, whereas Kitakata favored shared-title registration ($p = 0.002$).

These results indicate that Kunitomi has smaller forest and planted forest areas than Kitakata and that Kunitomi's forest ownership size is smaller than that of Kitakata. In both regions, most PFOs were unaware of their own forest areas and planted forests (Table 2).

Table 2. Comparison of forest conditions between the two regions.

	Kunitomi		Kitakata		*p*-Value		Kunitomi		Kitakata		*p*-Value
	n	%	n	%			n	%	n	%	
Forest ownership size *					0.000		**Condition of plantation forest ***				0.014
<1 ha	29	18.8	14	5.1		Maturity	70	46.7	168	60.0	
1–3 ha	53	34.4	26	9.5		Immature	48	32.0	69	24.6	
3–5 ha	8	5.2	18	6.6		Both	1	0.7	7	2.5	
5–10 ha	9	5.8	32	11.7		Unknown	31	20.7	36	12.9	
10–30 ha	6	3.9	44	16.1		Total	159	100.0	283	100.0	
30–50 ha	1	0.6	21	7.7		**Condition of cadastral survey**				0.161
50 ha<	3	1.9	13	4.7		Completion	95	60.9	172	58.5	
Unknown	45	29.2	106	38.7		Partially	19	12.2	28	9.5	
Total	154	100.0	274	100.0		Not-yet	27	17.3	75	25.5	
Plantation forest size *					0.000		Unknown	15	9.6	19	6.5	
0 ha	5	3.4	6	2.3		Total	156	100.0	294	100.0	
<1 ha	29	19.7	20	7.5		**Registration type ***				0.002
1–3 ha	39	26.5	40	15.0		Sole	145	91.2	240	84.8	
3–5 ha	6	4.1	18	6.8		Joint	10	6.3	10	3.5	
5–10 ha	6	4.1	25	9.4		Both	4	2.5	33	11.7	
10–30 ha	1	0.7	41	15.4		Total	159	100.0	283	100.0	
30–50 ha	0	0.0	4	1.5							
50 ha<	2	1.4	2	0.8							
Unknown	59	40.1	110	41.4							
Total	147	100.0	266	100.0							

Note: * *p*-Value < 0.05.

3.1.3. Demographic Characteristics of PFOs

The largest proportion of PFOs in both regions were in their 60s (Kunitomi: 31.0%; Kitakata: 37.9%), followed by those in their 70s (Kunitomi: 29.0%; Kitakata: 29.1%) ($p = 0.371$). Gender was predominantly male (Kunitomi: 86.5%; Northern: 87.8%) ($p = 0.764$). The relationship between the PFO and FOC was as follows: in Kunitomi, 50.7% (74 PFOs) were cooperative members, 15.1% (22 PFOs) were heirs of cooperative members, and 34.2% (50 PFOs) were unknown. In Kitakata, 83.4% (226 PFOs) were cooperative members, 3.7%

(10 PFOs) were heirs of cooperative members, and 12.9% (35 PFOs) were unknown, indicating that more PFOs in Kitakata were cooperative members than in Kunitomi, whereas more PFOs in Kunitomi were unaware of their relationship with FOC than Kitakata ($p = 0.000$). No differences existed in the two regions regarding the PFOs' primary source of income. 59.1% (94 PFOs) of PFOs in Kunitomi and 63.5% (183 PFOs) in Kitakata reported having a successor ($p = 0.362$).

3.1.4. Awareness of Forest Ownership and Forest Management Behaviors

The proportion of respondents who were registered PFOs was 78.2% (122 PFOs) in Kunitomi and 72.1% (202 PFOs) in Kitakata, with no significant difference between the two regions ($p = 0.097$). The forest was primarily managed by its PFOs in both regions (Kunitomi: 55.0%, Kitakata: 64.0%). In comparison, 40.0% (64 PFOs) of the PFOs in Kunitomi and 28.1% (82 PFOs) in Kitakata indicated that they did not manage the forest ($p = 0.190$). The proportion of PFOs recognizing their forest locations was 75.6% (121 PFOs) in Kunitomi and 73.1% (209 PFOs) in Kitakata, whereas the proportion of PFOs not recognizing the location was 10.6% (17 PFOs) in Kunitomi and 10.8% (31 PFOs) in Kitakata, showing a similar trend ($p = 0.573$). Furthermore, in Kunitomi, 60.9% (98 PFOs) were aware of the PFO's boundaries, whereas 21.1% (34 PFOs) were unaware, and in Kitakata, 67.8% (192 PFOs) were aware, whereas 15.2% (43 PFOs) were unaware. No differences were observed between the two regions ($p = 0.149$). The PFOs' frequency of visits to the forests did not differ between the two regions ($p = 0.650$). The most common response in Kunitomi was "rarely" (23.8%, 38 PFOs), followed by "once a year" (21.9%, 35 PFOs); 3.8% (6 PFOs) of PFOs visited monthly, 18.1% (29 PFOs) visited several times a year; and 9.4% (15 PFOs) of the PFOs never visited. The most common response in Kitakata was "once every few years" (25.7%; 76 PFOs), followed by "almost never" (22.0%; 65 PFOs); 3.0% of PFOs visited monthly (9 PFOs), 19.6% (58 PFOs) visited several times a year; and 6.4% (19 PFOs) of the PFOs never visited (Table A1).

Notably, 73 PFOs (46.8%) in Kunitomi and 125 PFOs (44.8%) in Kitakata ($p = 0.690$) had harvesting experience within the previous five years. Of these, 72 PFOs (excluding 1 PFO in Kunitomi due to no response) and 125 PFOs in Kitakata were compared. The "suitable age of harvesting" was answered by 26.4% (19 PFOs) of the respondents in Kunitomi and 45.6% (57 PFOs) in Kitakata ($p = 0.010$); 6.9% (5 PFOs) of the respondents in Kunitomi and 16.8% (21 PFOs) in Kitakata responded "to earn extra income" ($p = 0.052$). "Recommended from FOC" was selected by 1.4% (1 PFO) in Kunitomi and 6.4% (8 PFOs) in Kitakata ($p = 0.159$). "Recommended from the private logging company" was 51.4% (37 PFOs) in Kunitomi and 25.6% (32 PFOs) in Kitakata ($p = 0.000$). An "increase in timber prices" was unobserved in Kunitomi but was 0.8% (1 PFO) in Kitakata ($p = 1.000$). The "expiration of sharing contract" was unobserved in Kunitomi but was 6.4% (8 PFOs) in Kitakata ($p = 0.028$). "Wind damage, disease, and insect damage" accounted for 11.1% (8 PFOs) in Kunitomi and 3.2% (4 PFOs) in Kitakata ($p = 0.033$). "Others" accounted for 9.7% (7 PFOs) in Kunitomi and 10.4% (13 PFOs) in Kitakata ($p = 1.000$).

We then determined whether these PFOs had reforested postharvest. The results showed that 29.6% (21 PFOs) of the PFOs in Kunitomi and 55.2% (64 PFOs) in Kitakata had reforested, whereas 57.7% (41 PFOs) in Kunitomi and 37.1% (43 PFOs) in Kitakata had not reforested, indicating that more PFOs in Kitakata had reforested ($p = 0.026$). Regarding the intention to conduct harvest and reforest in the future, 20.6% (32 PFOs) of the PFOs in Kunitomi and 31.3% (89 PFOs) in Kitakata wanted to conduct both harvesting and reforestation. In comparison, 10.3% (16 PFOs) in Kunitomi and 8.5% (24 PFOs) in Kitakata wanted only to harvest and not reforest (24 PFOs) in Kitakata. The PFOs who answered that they had no plans to do so were the most numerous in both areas, with 66.5% (103 PFOs) in Kunitomi and 57.7% (164 PFOs) in Kitakata.

When comparing the factors that would be important in the decision-making process for harvesting, the most common factors that differed between the two regions were "reasonable profit" (Kunitomi: 40.4%; Kitakata: 54.6%, $p = 0.007$) and "sell with the land"

(Kunitomi: 66.7%; Kitakata: 12.6%, $p = 0.000$). No differences were indicated for "trust of buyer" ($p = 0.098$), "reforestation postharvest" ($p = 0.211$), "only harvesting" ($p = 1.000$), and "the adjacent landowner also harvesting" ($p = 0.800$) (Table 3).

Table 3. Comparison of reasons for harvesting/reforestation and harvesting decisions between the two regions.

	Kunitomi		Kitakata		*p*-Value
	n	**%**	**n**	**%**	
Logging experience in the past 5 years					0.690
Yes	73	46.8	125	44.8	
No	83	53.2	154	55.2	
Total	156	100.0	279	100.0	
Reason for harvesting (multiple answer)					
Suitable age of harvesting *	19	26.4	57	45.6	0.010
To earn extra income	5	6.9	21	16.8	0.052
Recommended by FOC	1	1.4	8	6.4	0.159
Recommended by private logging company *	37	51.4	32	25.6	0.000
Increase in timber prices	0	0.0	1	0.8	1.000
Expiration of sharing contract *	0	0.0	8	6.4	0.028
Wind damage, disease, and insect damage *	8	11.1	4	3.2	0.033
Others	7	9.7	13	10.4	1.000
Total	72	100.0	125	100.0	
Reforestation postharvest *					0.026
Totally	21	29.6	64	55.2	
Partially	7	9.9	8	6.9	
Not planted	41	57.7	43	37.1	
Others	2	2.8	1	0.9	
Total	71	100.0	116	100.0	
Intention to harvesting and reforestation					0.122
Harvesting and reforestation	32	20.6	89	31.3	
Only harvesting	16	10.3	24	8.5	
Undecided	103	66.5	164	57.7	
No place to harvesting	4	2.6	7	2.5	
Total	155	100.0	284	100.0	
Factors to consider in the decision to harvesting (multiple answer)					
Reasonable benefits *	57	40.4	147	54.6	0.007
Sell with the land *	47	33.3	34	12.6	0.000
Trust of buyer	54	38.3	81	30.1	0.098
Reforestation postharvest	36	25.5	86	32.0	0.211
Not reforestation postharvest	9	6.4	18	6.7	1.000
The adjacent landowner also harvesting	7	5.0	11	4.1	0.800
Others	8	5.7	17	6.3	1.000
Total	161	100.0	283	100.0	

Note: * *p*-Value < 0.05.

3.2. In the Case of Kunitomi

3.2.1. Size of Ownership and Forest Conditions in Kunitomi

The EM group owned most (34.8%) of the forest management size (1–3 ha), whereas the D group had most (36.8%) of the PFOs who did not know the size, but no difference was indicated in the trend of the forest size ($p = 0.112$). Similarly, the size of plantations did not differ ($p = 0.094$), but the EM group had the largest number of PFOs with 1–3 ha (34.8%), whereas the D group had the largest number of PFOs who did not know the size of their plantations (48.8%). The maturity of planted forests was most frequently answered by PFOs of both the EM group (47.0%) and the D group (60.0%) as mature, whereas the D group had the largest proportion of PFOs who did not know the status ($p = 0.518$). Land cadastral surveys ($p = 0.316$) and land registration titles were most often under a single title, with no differences indicated ($p = 0.159$) (Table 4).

Table 4. Comparison of forest conditions between the two groups in Kunitomi.

	EM Group		D Group		Total		*p*-Value
	n	**%**	**n**	**%**	**n**	**%**	
Forest owned area							0.112
<1 ha	12	17.6	16	21.1	28	19.4	
1–3 ha	30	44.1	20	26.3	50	34.7	
3–5 ha	4	5.9	4	5.3	8	5.6	
5–10 ha	4	5.9	5	6.6	9	6.3	
10–30 ha	3	4.4	3	3.9	6	4.2	
30–50 ha	1	1.5	0	0.0	1	0.7	
50 ha<	2	2.9	0	0.0	2	1.4	
Unknown	12	17.6	28	36.8	40	27.8	
Total	68	100.0	76	100.0	144	100.0	
Plantation forest area							0.094
0 ha	1	1.5	4	5.4	5	3.6	
<1 ha	15	22.7	14	18.9	29	20.7	
1–3 ha	23	34.8	14	18.9	37	26.4	
3–5 ha	4	6.1	2	2.7	6	4.3	
5–10 ha	2	3.0	4	5.4	6	4.3	
10–30 ha	1	1.5	0	0.0	1	0.7	
30–50 ha	0	0.0	0	0.0	0	0.0	
50 ha<	1	1.5	0	0.0	1	0.7	
Unknown	19	28.8	36	48.8	55	39.3	
Total	66	100.0	74	100.0	140	100.0	
Age of plantation forest							0.518
Maturity	31	47.0	35	60.0	66	47.5	
Immature	23	34.8	21	24.6	44	31.7	
Both	1	1.5	0	2.5	1	0.7	
Unknown	11	16.7	17	12.9	28	20.1	
Total	66	100.0	73	100.0	139	100.0	
Status of land cadastral survey							0.316
Completion	43	63.2	41	55.4	84	59.2	
Partially	11	16.2	8	10.8	19	13.4	
Not-yet	8	11.8	16	21.6	24	16.9	
Unknown	6	8.8	9	12.2	15	10.6	
Total	68	100.0	74	100.0	142	100.0	
Registration type							0.159
Sole	68	97.1	69	89.6	137	93.2	
Joint	2	2.9	6	7.8	8	5.4	
Both	0	0.0	2	2.6	2	1.4	
Total	70	100.0	77	100.0	147	100	

3.2.2. Demographic Characteristics of PFOs

The age distribution of PFOs (*p* = 0.108), gender (*p* = 0.430), relationship with the FOC (*p* = 0.960), and PFOs' primary income source showed no differences between the two groups. Significant differences were observed in the presence or absence of successors, with the D group tending to have no successors (*p* = 0.019) (Table 5).

Table 5. Comparison of PFOs' characteristics between the two groups in Kunitomi.

	EM Group		D Group		Total		*p*-Value
	n	%	n	%	n	%	
Relationship with FOC							0.960
Member	38	60.3	31	42.5	69	50.7	
Member successor	9	14.3	12	16.4	21	15.4	
Unknown	16	25.4	30	41.1	46	33.8	
Total	63	100.0	73	100.0	136	100.0	
Successors in forest management *							0.019
Existence	49	70.0	39	50.6	88	59.9	
Absence	21	30.0	38	49.4	59	40.1	
Total	70	100.0	77	100.0	147	100.0	
Main income (multiple answer)							
Agriculture	27	38.6	20	27.0	47	32.6	0.158
Forestry	3	4.3	0	0.0	3	2.1	0.112
Independent business	4	5.7	0	0.0	4	2.8	0.053
Salary and wages	16	22.9	18	24.3	34	23.6	0.847
Pension	30	42.9	42	56.8	72	50.0	0.133
Real estate	3	4.3	0	0.0	3	2.1	0.112
Other	0	0.0	2	2.7	2	1.4	0.497
Total	70	100.0	74	100.0	144	100.0	

Note: * *p*-Value < 0.05.

3.2.3. PFOs' Forest Management Behaviors and Attitudes

In both regions, the registered person was typically the principal owner ($p = 0.521$). Most primary managers were owners themselves (74.3%) in the EM group, whereas the majority were not managing (50.6%) in the D group ($p = 0.002$). A difference existed in the number of respondents who knew the location of their forest, with the EM group tending to have a higher percentage of respondents who knew the location of their forest ($p = 0.010$). The same was true for the boundaries, with the EM group tending to have more PFOs who knew the boundaries in person ($p = 0.020$). Comparing the frequency of forest visits revealed that the EM group tended to visit the forest more frequently ($p = 0.002$) (Table 6).

Table 6. Comparison of forest management behaviors between the two groups in Kunitomi.

	EM Group		D Group		Total		*p*-Value
	n	%	n	%	n	%	
Registration name							0.521
Owner	57	82.6	57	74.0	114	78.1	
Previous generation	11	15.9	18	23.4	29	19.9	
Varies depending on site	1	1.4	1	1.3	2	1.4	
Unknown	0	0.0	1	1.3	1	0.7	
Total	69	100.0	77	100.0	146	100.0	
Managing person *							0.002
Owner	52	74.3	32	41.6	84	57.1	
Relatives	0	0.0	3	3.9	3	2.0	
FOC	1	1.4	2	2.6	3	2.0	
Private company	0	0.0	0	0.0	0	0.0	
Not managing	16	22.9	39	50.6	55	37.4	
Varies depending on site	0	0.0	0	0.0	0	0.0	
Others	1	1.4	1	1.3	2	1.4	
Total	70	100.0	77	100.0	292	100.0	
Person or organization recognizes location *							0.010
Owner	61	88.4	50	64.1	111	75.5	
Relatives	0	0.0	3	3.8	3	2.0	
FOC	0	0.0	2	2.6	2	1.4	
Municipality	1	1.4	8	10.3	9	6.1	
Not recognized	4	5.8	12	15.4	16	10.9	
Others	3	4.3	3	3.8	6	4.1	
Total	69	100.0	78	100.0	147	100.0	

Table 6. *Cont.*

	EM Group		D Group		Total		*p*-Value
	n	%	n	%	n	%	
Person or organization recognizes boundary *							0.020
Owner	52	74.3	38	49.4	90	61.2	
Relatives	1	1.4	4	5.2	5	3.4	
FOC	0	0.0	2	2.6	2	1.4	
Municipality	3	4.3	12	15.6	15	10.2	
Not recognized	11	15.7	19	24.7	30	20.4	
Others	3	4.3	2	2.6	5	3.4	
Total	70	100.0	77	100.0	147	100.0	
Frequency of owned forest visits by owner *							0.002
Every month	5	7.1	1	1.3	6	4.1	
Several times a year	21	30.0	7	9.1	28	19.0	
Once a year	17	24.3	16	20.8	33	22.4	
Once every few years	10	14.3	19	24.7	29	19.7	
Almost never	9	12.9	24	31.2	33	22.4	
Never	4	5.7	8	10.4	12	8.2	
Varies depending on site	4	5.7	2	2.6	6	4.1	
Total	70	100.0	77	100.0	147	100.0	

Note: * *p*-Value < 0.05.

No differences between regions were indicated in the percentage of PFOs who harvested in the last five years ($p = 0.742$), and the reasons that led to harvesting were similar. Regarding reforestation postharvest, 46.4% of the PFOs in the EM group reforested, whereas 17.1% in the D group did, indicating that PFOs with low motivation for forest management tended not to reforest ($p = 0.037$). The EM group was likelier to prioritize the following factors in their decision to log: the prospect of substantial profit ($p = 0.015$) and reforestation postharvest ($p = 0.030$). However, the D group demonstrated a greater likelihood of selling stumpage with the land ($p = 0.000$) (Table 7).

Table 7. Comparison of reasons for harvesting/reforestation and harvesting decisions between the two groups in Kunitomi.

	EM Groups		D Group		Total		*p*-Value
	n	%	n	%	n	%	
Harvesting experience in the past 5 years							0.742
Yes	31	45.6	38	48.7	69	47.3	
No	37	54.4	40	51.3	77	52.7	
Total	68	100.0	78	100.0	146	100.0	
Reason for logging (multiple answer)							
Suitable age of logging	10	34.5	7	20.0	17	26.6	0.258
To earn extra income	2	6.9	2	5.7	4	6.3	1.000
Recommended by FOC	1	3.4	0	0.0	1	1.6	0.453
Recommended by private company	13	44.8	22	62.9	35	54.7	0.208
Wind damage, disease, and insect damage	4	13.8	3	8.6	7	10.9	0.692
Others	2	6.9	4	11.4	6	9.4	0.681
Total	29	100.0	35	100.0	64	100.0	
Reforest postharvest *							0.037
Totally	13	46.4	7	17.1	19	30.2	
Partially	4	14.3	4	8.6	7	11.1	
Not planted	10	35.7	27	71.4	35	55.6	
Others	1	3.6	0	2.9	2	3.2	
Total	28	100.0	38	100.0	63	100.0	
Intention to harvesting and reforestation *							0.000
Harvesting and reforestation	29	42.0	3	3.9	32	2.1	
Only harvesting	3	4.3	13	17.1	16	11.0	
Undecided	35	50.7	59	77.6	94	64.8	
No place to harvesting	2	2.9	1	1.3	3	2.1	
Total	69	100.0	76	100.0	145	100.0	

Table 7. *Cont.*

	EM Groups		D Group		Total		*p*-Value
	n	%	n	%	n	%	
Factors to consider in the decision to harvesting (multiple answer)							
Reasonable benefits*	33	52.4	23	31.5	56	41.2	0.015
Sell with land *	4	6.3	40	54.8	44	32.4	0.000
Trust of buyer	28	44.4	24	32.9	52	38.2	0.216
Reforest postharvest *	22	34.9	13	17.8	35	25.7	0.030
Not reforest postharvest	4	6.3	5	6.8	9	6.6	1.000
Neighboring owner also logs	2	3.2	5	6.8	7	5.1	0.450
Others	5	7.9	2	2.7	7	5.1	0.249
Total	63	100.0	73	100.0	136	100.0	

Note: * *p*-Value < 0.05.

3.3. In the Case of Kitakata

3.3.1. Future Intentions of PFOs

Most PFOs in the EM and D groups reported that they were unsure of their forest size (EM group: 32.4%; D group: 48.5%). Most respondents in the EM group (18.8%) owned 10–30 ha, whereas most respondents in the D group (13.2%) owned 1–3 ha, indicating that the EM group tended to own a larger forest. In contrast, a higher percentage of respondents in the D group did not know their owned forest size ($p = 0.007$). Similarly, the plantation forest size of the EM group was the most common (32.0% in the EM group; 57.6% in the reduced size group). As with the forest size, the EM group tended to have a larger planted forest size and to know the size ($p = 0.000$). The maturity of planted forests was most frequently reported as mature by both the EM (63.8%) and D groups (57.5%), although a higher percentage of PFOs in the D group did not know the condition ($p = 0.006$). No differences were observed in the implementation of cadastral surveys ($p = 0.199$) or the method of land registration title ($p = 0.570$) (Table 8).

Table 8. Comparison of forest conditions between the two groups in Kitakata.

	EM Group		D Group		Total		*p*-Value
	n	%	n	%	n	%	
Forest size *							0.007
<1 ha	6	3.4	7	10.3	13	5.3	
1–3 ha	16	9.1	9	13.2	25	10.2	
3–5 ha	10	5.7	5	7.4	15	6.1	
5–10 ha	27	15.3	3	4.4	30	12.3	
10–30 ha	33	18.8	7	10.3	40	16.4	
30–50 ha	15	8.5	3	4.4	18	7.4	
50 ha<	12	6.8	1	1.5	13	5.3	
Unknown	57	32.4	33	48.5	90	36.9	
Total	176	100.0	68	100.0	244	100.0	
Plantation forest size *							0.000
0 ha	3	1.7	2	3.0	5	2.1	
<1 ha	9	5.2	10	15.2	19	8.0	
1–3 ha	29	16.9	8	12.1	37	15.5	
3–5 ha	15	8.7	2	3.0	17	7.1	
5–10 ha	20	11.6	3	4.5	23	9.7	
10–30 ha	35	20.3	3	4.5	38	16.0	
30–50 ha	4	2.3	0	0.0	4	1.7	
50 ha<	2	1.2	0	0.0	2	0.8	
Unknown	55	32.0	38	57.6	93	39.1	
Total	172	100.0	66	100.0	238	100.0	

Table 8. *Cont.*

	EM Group		D Group		Total		*p*-Value
	n	%	n	%	n	%	
Age of plantation forest *							0.006
Maturity	113	63.8	42	57.5	155	62.0	
Immature	45	25.4	16	21.9	61	24.4	
Both	7	4.0	0	0.0	7	2.8	
Unknown	12	6.8	15	20.5	27	10.8	
Total	177	100.0	73	100.0	250	100.0	
Status of cadastral survey							0.199
Completion	104	56.5	49	65.3	153	59.1	
Partially	18	9.8	6	8.0	24	9.3	
Not-yet	52	28.3	13	17.3	65	25.1	
Unknown	10	5.4	7	9.3	17	6.6	
Total	184	100.0	75	100.0	259	100.0	
Registration type							0.570
Sole	153	85.0	65	87.8	218	85.8	
Joint	7	3.9	1	1.4	8	3.1	
Both	20	11.1	8	10.8	28	11	
Total	180	100.0	74	100.0	254	100	

Note: * *p*-Value < 0.05.

3.3.2. Demographic Characteristics of PFOs

No differences were found between the two groups in the PFOs' age distribution (p = 0.863) or gender (p = 0.156). As for the relationship with FOC, the D group was likelier to be unaware of whether they were members (p = 0.004). Regarding succession, the D group was likelier to have no successor (p = 0.000). In addition, a higher proportion of the EM group was in agriculture (p = 0.042) and forestry (p = 0.028) as the primary income sources for the PFOs (Table 9).

Table 9. Comparison of PFOs' characteristics between the two groups in Kitakata.

	EM Group		D Group		Total		*p*-Value
	n	%	n	%	n	%	
Relationship with FOC *							0.004
Member	151	87.8	52	74.3	203	83.9	
Member successor	6	3.5	1	1.4	7	2.9	
Unknown	15	8.7	17	24.3	32	13.2	
Total	172	100.0	70	100.0	242	100.0	
Successors in forest management *							0.000
Existence	134	74.0	31	41.9	165	64.7	
Absence	47	26.0	43	58.1	90	35.3	
Total	181	100.0	74	100.0	255	100.0	
Primary income source							
Agriculture *	55	30.6	13	17.6	68	26.8	0.042
Forestry *	16	8.9	1	1.4	17	6.7	0.028
Independent business	14	7.8	8	10.8	22	8.7	0.465
Salary and wages	46	25.6	24	32.4	70	27.6	0.282
Pension	81	45.0	36	48.6	117	46.1	0.678
Real estate	2	1.1	0	0.0	2	0.8	1.000
Others	3	1.7	1	1.4	4	1.6	1.000
Total	180	100.0	74	100.0	254	100.0	

Note: * *p*-Value < 0.05.

3.3.3. PFOs' Forest Management Behaviors and Attitudes

Both groups had the highest percentage of respondents, in which the principal was the registered owner (p = 0.981). The principal administrator was the owner himself in both the EM (78.6%) and D groups (64.9%). In both groups, the owner recognized the location (p = 0.355) and boundaries (p = 0.051) of the owned forest; the highest percentage of owners themselves managed the forest, whereas a higher percentage of the D group did not ($p = 0.000$). Comparing the frequency of visits to the forest revealed that the EM group tended to visit more frequently ($p = 0.000$) (Table 10). No difference was observed between the two groups in the percentage of PFOs harvested in the past five years ($p = 0.679$). Likewise, no difference was observed in the reasons for harvesting among PFOs who had logged before. Regarding reforest postharvest, 63.4% of the PFOs in the EM group fully reforested, whereas 51.6% of those in the D group did not, indicating that PFOs with low motivation for forest management tended not to reforest ($p = 0.029$).

Table 10. Comparison of forest management behaviors between the two groups in Kitakata.

	EM Group		D Group		Total		*p*-Value
	n	%	n	%	n	%	
Registration name							0.981
Owner	129	74.1	56	75.7	185	74.6	
Previous generation	34	19.5	14	18.9	48	19.4	
Varies depending on site	9	5.2	3	4.1	12	4.8	
Unknown	2	1.1	1	1.4	3	1.2	
Total	174	100.0	74	100.0	248	100.0	
Person or organization aware of location							0.355
Owner	143	78.6	48	64.9	191	74.6	
Relatives	5	2.7	4	5.4	9	3.5	
FOC	7	3.8	4	5.4	11	4.3	
Municipality	8	4.4	6	8.1	14	5.5	
Not recognized	15	8.2	9	12.2	24	9.4	
Others	4	2.2	3	4.1	7	2.7	
Total	182	100.0	74	100.0	256	100.0	
Person or organization aware of boundary							0.051
Owner	134	74.4	40	54.1	174	68.5	
Relatives	7	3.9	5	6.8	12	4.7	
FOC	5	2.8	2	2.7	7	2.8	
Municipality	8	4.4	6	8.1	14	5.5	
Not recognized	20	11.1	17	23.0	37	14.6	
Others	6	3.3	4	5.4	10	3.9	
Total	180	100.0	74	100.0	254	100.0	
Managing person *							0.000
Owner	136	73.9	34	45.3	170	65.6	
Relatives	2	1.1	3	4.0	5	1.9	
FOC	4	2.2	3	4.0	7	2.7	
Private company	0	0.0	1	1.3	1	0.4	
Not managing	37	20.1	32	42.7	69	26.6	
Varies depending on site	2	1.1	1	1.3	3	1.2	
Others	3	1.6	1	1.3	4	1.5	
Total	184	100.0	75	100.0	259	100.0	
Frequency of owned forest visits by owner *							0.000
Every month	9	4.9	0	0.0	9	3.4	
Several times a year	48	25.9	7	9.2	55	21.1	
Once a year	36	19.5	8	10.5	44	16.9	
Once every few years	49	26.5	20	26.3	69	26.4	
Almost never	30	16.2	28	36.8	58	22.2	
Never	7	3.8	6	7.9	13	5.0	
Varies depending on site	6	3.2	7	9.2	13	5.0	
Total	185	100.0	76	100.0	261	100.0	

Note: * *p*-Value < 0.05.

Differences were also observed in the factors that were important in the decision to harvest. A high percentage of the EM group indicated that they expected to make a substantial profit ($p = 0.005$), whereas a high percentage of the D group indicated that they were willing to sell the land with stumpage ($p = 0.000$) (Table 11).

Table 11. Comparison of reasons for harvesting/reforestation and harvesting decisions between the two groups in Kitakata.

	EM Group		D Group		Total		*p*-Value
	n	%	n	%	n	%	
Harvesting experience in the past 5 years							0.679
Yes	86	49.1	34	45.3	120	48.0	
No	89	50.9	41	54.7	130	52.0	
Total	175	100.0	75	100.0	250	100.0	
Reason for harvesting (multiple answer)							
Suitable age of logging	39	45.3	15	44.1	54	45.0	1.000
To earn extra income	16	18.6	4	11.8	20	16.7	0.428
Recommended by FOC	6	7.0	2	5.9	8	6.7	0.453
Recommended by private company	20	23.3	12	35.3	32	26.7	0.251
Increase in timber prices	1	1.2	0	0.0	1	0.8	1.000
Expiration of sharing contract	5	5.8	3	8.8	8	6.7	0.686
Wind damage, disease, and insect damage	3	3.5	1	2.9	4	3.3	1.000
Others	9	10.5	3	8.8	12	10.0	1.000
Total	86	100.0	34	100.0	120	100.0	
Reforest postharvest *							0.029
Totally	52	63.4	11	35.5	63	55.8	
Partially	3	3.7	4	12.9	7	6.2	
Not planted	26	31.7	16	51.6	42	37.2	
Others	1	1.2	0	0.0	1	0.9	
Total	82	100.0	31	100.0	113	100.0	
Intention to harvesting and reforestation *							0.000
Harvesting and reforestation	73	40.8	10	13.2	83	32.5	
Only harvesting	9	5.0	12	15.8	21	8.2	
Undecided	94	52.5	52	68.4	146	57.3	
No place to harvesting	3	1.7	2	2.6	5	2.0	
Total	179	100.0	100	100.0	255	100.0	
Factors to consider in the decision to harvesting (multiple answer)							
Reasonable benefits *	106	59.9	29	40.3	135	54.2	0.005
Sell with land *	9	5.1	20	27.8	29	11.6	0.000
Trust of buyer	51	28.8	24	33.3	75	30.1	0.543
Reforest postharvest *	63	35.6	21	29.2	84	33.7	0.377
Not reforest postharvest	11	6.2	6	8.3	17	6.8	0.583
Neighboring owner also harvest	6	3.4	4	5.6	10	4.0	0.481
Others	8	4.5	6	8.3	14	5.6	0.238
Total	177	100.0	72	100.0	249	100.0	

Note: * *p*-Value < 0.05.

4. Discussion and Conclusion

The NSFM aims to realize efficient and sustainable timber production based on consolidating the forest land of PFOs with low future intentions. In this study, we administered a questionnaire to PFOs in Miyazaki Prefecture, one of the most active areas for timber production in Japan, to examine the socioeconomic factors that affect PFOs' future intentions.

The PFOs' willingness to manage forests varies by region [32]. First, this study compared PFOs' future intentions and the factors that might influence them across regions with different forest ownership sizes. In the small-scale region, 52.7% of PFOs desired to reduce the future management scale, whereas 70.9% of PFOs in the large-scale region desired to maintain or increase the management scale. A comparison of the two regions revealed the problems in private forest management. A common problem in small and large regions was the lack of awareness of forests owned by PFOs themselves. In particular, the fact that many PFOs were unaware of the size of their forests and planted forests indicated that PFOs do not have the sources to understand the value of their forests and consider the direction of future forest management. The importance of successors in forest management was also indicated. Additionally, this study revealed the challenges specific to the small-scale regions. The small-scale regions showed fewer future intentions to manage forests. Many PFOs were unaware of the maturity of their planted forests as well as the size of their forests. This may be attributed to the lack of understanding in many PFOs regarding the economic value of their planted forests, which may be one factor that reduces their willingness to manage their forests. Since many PFOs did not understand the value of their forests, their decision-making process regarding forest management was passive, as evidenced by the reasons for their decision to harvest their forest. Additionally, many PFOs wanted to quit forestry, whereas many did not reforest postharvest (Table 12). These results indicated the need to develop a framework to provide PFOs with enough information to consider future management directions while implementing NSFM. At the same time, since small-scale regions are less willing to manage forests than large-scale regions, aggerating forest management right by municipalities is promising for the sustainable management of forests. However, since the workload of municipalities is excessive [11], the prefectural government should support municipalities in small-scale areas with an emphasis on small-scale regions. PFOs' low willingness to manage their forests may harm timber production, as owner identification and rights identification are particular barriers to timber production in the small-scale region [33,34]. Similar results were obtained for forest and planted forest size, as they influenced the current willingness to manage [13,16,18–20].

Table 12. Inter-regional differences of forest management problems.

Trend in Forest Management Issues in Small-Scale Regions	Forest Management Issues Common to Both Regions
• Low intention for forest management • Unaware of forest maturity • Passive decision-making for harvesting • Low reforestation rate • PFOs want to quit forestry	• Lacked basic knowledge of owned forest condition (forest size, planted forest size) • Few PFOs have future forest management plans • Low future intention of PFOs with no successor

Next, based on the survey results, we examined the factors affecting the future intentions of PFOs and the forest management behaviors of PFOs with low future intentions. The factor affecting future intention in small-scale regions was the presence of successors [24]. No difference was found in the forest size [13,16,18–20], which was considered a factor while analyzing the results, likely because the forest size was biased toward small-scale. In contrast, forest size, planted forest size, and planted forest maturity were the factors of forest condition that influenced the future intentions in large-scale regions. These results suggest that the economic value of forests affects the future intentions of the PFOs in large-scale regions. In addition, the existence of successors, the relationship with the FOC, and forestry's position as an income source were also indicated as factors. Since it is essential to clarify the forest management behaviors of the PFOs with low future intentions to consider forestry policies, we summarized the characteristics of the forest management behaviors and attitudes of the D group. Among the common issues associated with both regions, the D group tended to have scarce forest management and be willing to dispose of their forestlands. Therefore, these PFOs tended not to implement reforestation because they had less emphasis on the reforest postharvest. In small-scale regions, few PFOs had future management directions and were unaware of the location and boundaries of their forests (Table 13). The D group was not considered interested in the economic value of the forest. There could be two possible causes behind the lack of interest in economic value: first, they do not have information about the forests, and second, they must dispose of the forests due to the absence of successors to inherit them.

Under the Forest Management Law, municipalities must conduct surveys of PFOs' intentions, and some have already begun to do so. Many PFOs are likely to respond to this survey without having all the facts they need to decide on their future forest management intentions. A procedural flaw can be identified regarding the intention survey, which encourages PFOs to make decisions without information about their forests. Before the survey, the government must provide an opportunity for the PFOs to know the location and boundaries of the forest, resource status, and other information. In addition, the omission of local forest ownership size from the criteria for allocating FETT to municipalities is a flaw of this system [5]. FETT is used for "expenses related to forest improvement and its promotion, such as thinning, human resource development and securing of bearers, promotion of timber use, and public awareness." The areas requiring enhanced forest improvement are those with low future management intentions. Furthermore, the results of this study suggest that many PFOs in small-scale regions would like to outsource management or relinquish their land. Therefore, the administrative burden of conducting the survey and forest management aggregation in municipalities with small-scale regions is considered high. For municipalities, securing finances is the most critical aspect of operating NSFM [30]. These results suggest that FETT allocation criteria could still be considered based on the regional characteristics of ownership size. Differences were also observed in the factors that influenced the decision of forest PFOs to harvest between the two regions with different ownership sizes.

Table 13. The factors affecting the PFOs' future intention, tendency of D group's forest management behaviors.

	Small-Scale Regions	Large-Scale Regions	Common Issues to Both Regions
Forest condition factors	-	• Owned forest size • Plantation forest size • Recognition of maturity of planted forest	-
PFOs' characteristics factors	• Existence of successor	• Existence of successor • Recognition of relationship with FOC • Position of forestry as income source	• Existence of successor
Trends in management behaviors and attitudes toward owned forests among Group D	• Scarce forest management • Want to dispose of land • Less interest in the economic value of owned forest • Undecided future forest management plan • Do not reforest postharvest • Less emphasis on reforests postharvest • Few frequencies of owned forest visits • Do not recognize location of owned forest • Do not recognize boundary of owned forest	• Scarce forest management • Want to dispose of land • Less interest in the economic value of owned forest • Do not reforest postharvest • Less emphasis on reforests postharvest • Few frequencies of owned forest visits	• Scarce forest management • Want to dispose of land • Less interest in the economic value of owned forest • Do not reforest postharvest • Less emphasis on reforests postharvest • Few frequencies of owned forest visits

This study examined the factors that influence the PFOs' future intentions in small and large regions by categorizing them by their future intentions and comparing the two types. Comparing the two groups revealed that the EM group prioritized economic benefits and the sustainability of the forest resource in their harvesting decisions. In contrast, the D group had more PFOs who wanted to relinquish their land and withdraw from forestry management. The EM group tended to own more forestland, suggesting that the size or economic value of their forest holdings influenced their future willingness to manage their forests. The D group was characterized by less frequent forest visits and a greater proportion of PFOs who lacked basic forest knowledge, such as area, location, and boundaries. These findings suggest that PFOs' lack of knowledge about their forests may result in uninterested in forest management.

The existence of successors is an essential factor in the continuity of forest management [24], in addition to the size, especially planted forest size [18–20]. Therefore, information on the forest owned, forest area, and the availability of successors are factors influencing willingness to future forest management. The forest management behaviors of PFOs with low future willingness to manage revealed issues regarding forest sustainability. In terms of harvesting decisions, the EM group emphasized the economic benefits and sustainability of the resource. However, the D group saw the logging decision as an opportunity to withdraw from forestry management and passively made logging decisions. This suggests that forests owned by PFOs with low future intention goals are a source of the increased abandonment of reforested.

To better reflect effective forest policy through PFOs typologies [14], examining the factors underlying the decisions of typified PFOs is necessary [12]. PFOs who wish to reduce the size of their future management have poorer forest management behaviors and are likelier to abandon the reforested area. Therefore, the method of categorizing PFOs based on their future willingness to manage the forest with resource sustainability and efficiency of operations was considered reasonable. However, promoting the transfer of forests owned by PFOs with a low future intention to forest management is insufficient; measures are also required to increase PFOs' willingness to forest management. Furthermore, PFOs must be given more opportunities to learn enough about their forests to make informed decisions about future management direction. Especially, the NSFM must consider the ways to develop forest information, provide PFOs with opportunities to obtain such information, encourage PFOs who are willing to manage their forests, and strengthen municipal work structures [11].

Therefore, the role of FOCs who have a good understanding of the status of local forests is crucial [26]. In large areas where the economic value of forests is relatively high, strengthening the relationship between PFOs and FOCs may be effective in motivating PFOs to manage the forests. It is expected that PFOs will be more likely to obtain information on their forests from FOCs, which will provide an opportunity for PFOs to recognize the economic value of their forests. In addition, since many PFOs are willing to dispose their lands in small-scale regions, the aggregation of the forest management rights by the municipalities will be required for sustainable forest management. As the workload of municipalities is expected to increase due to this policy, it will be necessary for the prefectural government to support the municipalities with small forest ownership in a focused manner [6,8].

Author Contributions: Conceptualization, N.O. and S.O.; methodology, N.O.; formal analysis, N.O.; investigation, N.O. and S.O.; writing—original draft preparation, N.O.; writing—review and editing, N.O., S.O. and N.T.; supervision, N.T.; project administration, N.O.; funding acquisition, N.O. and N.T. All authors have read and agreed to the published version of the manuscript.

Funding: This research received JSPS KAKENHI Grant Numbers JP19K20509, JP19KK0027 and JP21H03709.

Institutional Review Board Statement: Not applicable.

Informed Consent Statement: Not applicable.

Acknowledgments: The authors wish to thank the respondents for their valuable contribution to the completion of this paper.

Conflicts of Interest: The authors declare no conflict of interest.

Appendix A

Table A1. Comparison of forest management behaviors between two regions.

	Kunitomi		Kitakata		*p*-Value
	n	%	n	%	
Registration name					0.097
Owner	122	78.2	202	72.1	
Previous generation	30	19.2	56	20.0	
Varies depend on site	2	1.3	4	1.4	
Unknown	2	1.3	18	6.4	
Total	156	100.0	280	100.0	
Managing person					0.190
Owner	88	55.0	187	64.0	
Relatives	3	1.9	6	2.1	
FOC	3	1.9	9	3.1	
Private company	0	0.0	1	0.3	
Not managing	64	40.0	82	28.1	
Varies depend on site	0	0.0	3	1.0	
Others	2	1.3	4	1.4	
Total	160	100.0	292	100.0	
Person or organization aware of location					0.573
Owner	121	75.6	209	73.1	
Relatives	3	1.9	9	3.1	
FOC	3	1.9	14	4.9	
Municipality	10	6.3	16	5.6	
Not recognized	17	10.6	31	10.8	
Others	6	0.7	7	2.4	
Total	160	100.0	286	100.0	

	Kunitomi		Kitakata		*p*-Value
	n	%	n	%	
Person or organization aware of boundary					0.149
Owner	98	60.9	192	67.8	
Relatives	5	3.1	13	4.6	
FOC	2	1.2	9	3.2	
Municipality	16	9.9	15	5.3	
Not recognized	34	21.1	43	15.2	
Others	6	3.7	11	3.9	
Total	161	100.0	283	100.0	
Frequency of owned forest visits by owner					0.650
Every month	6	3.8	9	3.0	
Several times a year	29	18.1	58	19.6	
Once a year	35	21.9	55	18.6	
Once every few years	31	19.4	76	25.7	
Almost never	38	23.8	65	22.0	
Never	15	9.4	19	6.4	
Varies depends on site	6	3.8	14	4.7	
Total	160	100.0	296	100.0	

References

1. Forestry Agency. *Annual Report on Trends in Forest and Forestry in Japan Fiscal Year 2021 Japan*; National Forestry Extension Association in Japan: Tokyo, Japan, 2022; pp. 53–116. (In Japanese)
2. Ota, I. The shrinking profitability of small-scale forestry in Japan and some recent policy initiatives to reverse the trend. *Small-Scale For.* **2002**, *1*, 25–37. [CrossRef]
3. Hosaka, K.; Konoshima, M.; Uemura, R.; Aruga, K. Optimizing aggregation of small forest stands for thinning operations: A case study in Nasushiobara, Tochigi Prefecture, Japan. *Small-Scale For.* **2022**, *21*, 369–392. [CrossRef]
4. Forestry Agency. *Annual Report on Trends in Forest and Forestry in Japan Fiscal Year 2017 Japan*; National Forestry Extension Association in Japan: Tokyo, Japan, 2018; pp. 13–36. (In Japanese)
5. Yoshihiro, K. A study of calculating divided benchmarks of special purpose tax for maintaining forest. *Mon. Rev. Local Gov.* **2019**, *45*, 3–20. (In Japanese)
6. Kohsaka, R.; Uchiyama, Y. Forest Environmental Taxes at multi-layer national and prefectural levels. *J. Jpn. For. Soc.* **2019**, *101*, 246–252. (In Japanese) [CrossRef]
7. Uchiyama, Y.; Kohsaka, R. Utilization of Forest Environment Transfer Tax in ordinance-designated cities. *J. Jpn. For. Soc.* **2020**, *102*, 173–179. (In Japanese) [CrossRef]
8. Aikawa, T. Municipality support by prefectures. *For. Econ.* **2020**, *72*, 3–15. (In Japanese)
9. Kohsaka, R.; Uchiyama, Y. Forest Environment Transfer Tax, prefectural forest policy, and support for municipalities. *J. Jpn. For. Soc.* **2021**, *103*, 134–144. (In Japanese) [CrossRef]
10. Kohsaka, R.; Osawa, T.; Uchiyama, Y. Forest Environment Transfer Tax and urban-rural collaboration. *J. Jpn. For. Soc.* **2020**, *102*, 127–132. (In Japanese) [CrossRef]
11. Ishizaki, R.; Kanomata, H.; Sasada, K. Size and expertise of municipal forest administration staff: Results of a survey on actual conditions of municipal forest administration (Conducted in 2020). *J. Jpn. For. Soc.* **2022**, *104*, 214–222. (In Japanese) [CrossRef]
12. Boon, T.E.; Meilby, H.; Thorsen, B.J. An empirically based typology of private forest owners in Denmark: Improving communication between authorities and owners. *Scand. J. For. Res.* **2004**, *19*, 45–55. [CrossRef]
13. Ingemarson, F.; Lindhagen, A.; Eriksson, L. A typology of small-scale private forest owners in Sweden. *Scand. J. For. Res.* **2006**, *21*, 249–259. [CrossRef]
14. Ficko, A.; Lidestav, G.; Dhubháin, Á.N.; Karppinen, H.; Zivojinovic, I.; Westin, K. European private forest owner typologies: A review of methods and use. *For. Pol. Econ.* **2019**, *99*, 21–31. [CrossRef]
15. Ficko, A.; Boncina, A. Probabilistic typology of management decision making in private forest properties. *For. Pol. Econ.* **2013**, *27*, 34–43. [CrossRef]
16. Joshi, O.; Mehmood, S.R. Factors affecting nonindustrial private forest landowners' willingness to supply woody biomass for bioenergy. *Biomass Bioenergy* **2011**, *35*, 186–192. [CrossRef]

17. Sakai, M. Minds of forest owners for management and forest resource policy: For using and sustaining the resources of artificial forest. *J. For. Econ.* **1999**, *45*, 3–8. (In Japanese)

18. Ito, N.; Hiroshima, T.; Shiraishi, N. Analysis of factors affecting forest management performance: A case study of Gifu prefecture. *Jpn. J. For. Plan.* **2005**, *39*, 49–57. (In Japanese) [CrossRef]

19. Haga, D.; Katano, Y. Regional variations of resident and absentee forest owners in forest management activity. *J. For. Econ.* **2020**, *66*, 1–15. (In Japanese)

20. Katano, Y. The underlying conditions of abandoned forests in the underpopulated area. *J. For. Econ.* **2016**, *62*, 21–30. (In Japanese)

21. Hayashi, M.; Noda, I. Factors affecting owners' attitude toward forest operations: An analysis of data from questionnaire survey at Kumamoto, Japan. *J. For. Econ.* **2005**, *51*, 1–9. (In Japanese)

22. Kajima, S.; Uchiyama, Y.; Kohsaka, R. Private forest landowners' awareness of forest boundaries: Case study in Japan. *J. For. Res* **2020**, *25*, 299–307. [CrossRef]

23. Noriko, S.; Premysl, M.; Takafumi, F. Attitudes of small-scale forest owners to their properties in an ageing society: Findings of survey in Yamaguchi prefecture, Japan. *Small-Scale For.* **2006**, *5*, 97–110.

24. Hayashi, M.; Noda, I.; Yamada, Y. Factors affecting owners' attitude toward forest management: An analysis of data from surveys at Oita, Japan. *J. For. Econ.* **2006**, *52*, 1–11. (In Japanese)

25. Hayashi, M.; Oka, H.; Tanaka, W. Decision making of forest owners and social relationships: A view from transaction cost economics. *J. For. Econ.* **2011**, *57*, 9–20. (In Japanese)

26. Sasada, K.; Tsuzuki, N. The actual situation and problems of district committee members and regional organizations that work to encourage members of forest owners' cooperatives. *J. Jpn. For. Soc.* **2021**, *103*, 22–32. (In Japanese) [CrossRef]

27. Yamamoto, M. Regional appearance of abandoning plantation forest practices and future of forest management: Issues of reforestation in Hokkaido. *For. Econ.* **2001**, *54*, 16–20. (In Japanese)

28. Fujikake, I. Intensifying logging and replantation abandonment on non-national plantation forests in Miyazaki Prefecture. *J. For. Econ.* **2007**, *53*, 12–23. (In Japanese)

29. Kohroki, K. A case study on non-reforestation and forest holding structure in South Kyushu, Miyazaki Prefecture. *J. For. Econ.* **2007**, *53*, 24–35. (In Japanese)

30. Kushiro, T.; Ito, K. A study on the New Forest Management System. *J. For. Econ.* **2021**, *67*, 1–10.

31. Department of Environment and Forestry Miyazaki Prefecture. *Miyazaki Prefecture Forestry Statistics Handbook 2022*; Department of Environment and Forestry Miyazaki Prefecture: Miyazaki, Japan, 2022.

32. Mochida, H.; Shiga, K. The development of logging industry and the trend of forest owners in Miyazaki prefecture. *T. Jpn. For. Soc.* **1998**, *109*, 79–82. (In Japanese)

33. Toyama, K.; Shiraishi, N. Questionnaire survey on the perceptions of logging contractors in Miyazaki Prefecture for the aggregation of private forests. *Kyushu J. For. Res.* **2010**, *63*, 5–8. (In Japanese)

34. Onda, N.; Chinen, Y.; Owake, T.; Okuyama, Y. Influence of small-scale forest ownership on roundwood production and stumpage trade. *For. Econ.* **2021**, *73*, 1–11.

35. Curent population. Available online: http://www.town.kunitomi.miyazaki.jp/main/welcome/page000568.html (accessed on 19 October 2022).

36. 2015 Census of Agriculture and Forestry in Japan Report and Data on the Results. Available online: https://www.e-stat.go.jp/stat-search/files?page=1&layout=datalist&toukei=00500209&tstat=000001032920&cycle=0&tclass1=000001077437&tclass2=000001077396&tclass3=000001093235&tclass4=000001093483&stat_infid=000031515119&tclass5val=0 (accessed on 24 November 2022).

37. Ministry of Agriculture, Forestry and Fisheries. *1990 Census of Agriculture and Forestry in Miyazaki Prefecture*; Association of Agriculture and Forestry Statistics: Tokyo, Japan, 1992.

38. Statistics of Nobeoka City. Available online: https://www.city.nobeoka.miyazaki.jp/soshiki/1/1364.html (accessed on 19 October 2022).

 *forests*

Article

Current Status and Challenges for Forest Commons (*Iriai* Forest) Management in Japan: A Focus on Forest Producers' Cooperatives and Authorized Neighborhood Associations

Masahiko Ota

Graduate School of Fisheries and Environmental Sciences, Nagasaki University, 1–14 Bunkyo-machi, Nagasaki 852-8521, Japan; masahikoota@nagasaki-u.ac.jp

Abstract: *Iriai* forests are an example of communal forest management in Japan. Local institutions have never been static in governing *iriai* forests and the external environments of *iriai* forests have changed significantly over time. This study examines the management challenges of forest producers' cooperatives (FPCs) and authorized neighborhood associations (ANAs) as the two most important contemporary forms of *iriai* forest management. Data from nine FPCs and three ANAs in the Fukuoka and Saga prefectures of Kyushu Island are used. Surveyed topics included basic information about FPCs and ANAs, recent management activities, financial conditions, and member perceptions of forest management. Some FPCs suffered from disadvantageous forestry circumstances, including low timber prices, decreased number of members, and tax burdens; at the same time, some FPCs greatly profited from non-forestry income or assets, e.g., by leasing or selling forestland. In most cases, basic forest management operations had been conducted by both FPCs and ANAs, and members had maintained attachment to and responsibility for *iriai* forests and a sense of public contribution. Policy recommendations include making legal settings and administrative supports more compatible with contemporary realities, providing greater financial support for management activities, and pursuing multi-level governance to open the commons to wider society.

Keywords: *seisan shinrin kumiai*; *ninka chien dantai*; common property resource; developed countries; external policy influence; forest management activities

Citation: Ota, M. Current Status and Challenges for Forest Commons (*Iriai* Forest) Management in Japan: A Focus on Forest Producers' Cooperatives and Authorized Neighborhood Associations. *Forests* **2023**, *14*, 572. https://doi.org/10.3390/f14030572

Academic Editor: Andrej Ficko

Received: 22 January 2023
Revised: 10 March 2023
Accepted: 11 March 2023
Published: 13 March 2023

1. Introduction

Common property resource (CPR) researchers have argued that local communities can successfully manage natural resources, such as the commons [1,2]. These scholars have criticized the assumption that the commons are open-access contexts that inevitably lead to resource depletion. Within certain conditions, such as existing local institutions, local communities can appropriately manage resources. This claim also undermines the assumption that privatization is the sole method of successful resource management [3].

Recently, CPRs in developed industrialized countries have gained a focus [4,5]. Unlike designing institutions that govern natural resources in overuse contexts in developing countries, research on CPR management in developed countries after industrialization and urbanization should address underuse conditions. Appropriate human interventions concerning CPRs are disappearing, due to the declining population or aging of farmers, forest owners, and fishermen,—i.e., producers and bearers of ecosystem services—as well as changing institutional and market conditions around CPRs. Mitsumata [6] categorized external impacts on the commons—as non-settlement trends, commodification of commons, private corporation-led development, public works projects, legal system development and revisions, administration/policies, and court decisions—and presented typical positive and negative influences of each impact. Of these, non-settlement trends and commodification of the commons are important in the context of developed countries. With regard to non-settlement trends, an exodus may have the positive effect of the improvement of resource conditions in places where overexploitation has been evident, while it may have the negative effect of underuse of resources, resulting in the absence of proper management inputs. The commodification of the commons may have the positive effect of providing opportunities to sustain and strengthen the commons, while it may have the negative effect of monoculture, resulting in resource underuse and neglect once market value is lost.

Traditional common (*iriai*) forests in Japan are a well known example of successful CPR management [7,8]. These forest management systems emerged in the 17th century. *Iriai* forests are forests and semi-natural grasslands that can be collectively accessed by residents in a specific area with specific rights. The resources procured typically include timber, forage, and firewood. Right-holders develop local rules and regulations to sustainably utilize limited resources, including methods, periods, and quantities of harvest.

Local institutions governing *iriai* forests have never been static and external environments surrounding *iriai* forests have changed over time. In particular, *iriai* forests cannot be explored in isolation from national policy developments [9–16]. Since the start of the modern Meiji period in 1868, successive governments have been averse to the existence of the common sphere, including *iriai* forests, because principles of long-standing CPR management are not likely to fit into the top-down control of state agencies nor into the pursuit of efficiency and maximization of monetary profits by private entities. Thus, since the late 19th century, *iriai* forests have been subject to various institutional changes that abrogate *iriai* forests and replace them with other forms of management.

Several studies have emphasized the changes in *iriai* forests as a result of external conditions. Shimada [14] pointed out that local communities can adapt *iriai* institutions to external influences, such as social and demographic conditions. However, communities have been unable to effectively cope with low timber prices, which have been manifested all over Japan since the 1980s; Shimada mentions this has been an influence that works outside the control of local communities. Saito [10] analyzed how municipality mergers in the 2000s in Japan affected forest management by property wards (*zaisanku*). Property wards play a role that enables local community members to hold and maintain rights to *iriai* resources; this institution has been in place since the late 19th century. Saito reported that new municipalities may not fully understand local historical contexts, and thus the autonomy of property wards could be threatened.

Matsushita [11] overviewed the management problems faced by forest producers' cooperatives (*seisan shinrin kumiai*: hereafter FPC). FPCs are cooperatives established to manage *iriai* forests, generally replacing broad-leaved forests with coniferous forests for more intensive forestry production. The Japanese government strongly promoted the establishment of FPCs for forests managed by *iriai* institutions after the late 1960s. The problems among FPCs identified by Matsushita include a lack of sufficient income due to uneven distribution of tree age structure and low timber prices, aging of FPC members as a labor force, and a lack of forestry experts among FPC members. He also denoted the burden of taxes; FPCs are cooperatives and hence, are subject to paying several taxes, including corporate tax, corporate inhabitant tax, and fixed asset tax. As a result, it is difficult for members to find a clear reason to continue their FPCs and so, the number of dissolutions of FPCs is increasing. At the same time, he presented several FPC cases where innovative activities had been taken.

Yamashita et al. [9] compared institutional characteristics of conventional *iriai* organizations, FPCs, and authorized neighborhood organizations (*ninka chien dantai*: hereafter ANA) against the backdrop of an increasing number of FPCs that had been dissolving themselves and changing their status to ANAs. ANAs are local neighborhood associations (such as wards) that have obtained the status of legal entity and can officially register fixed assets, such as *iriai* forests. ANAs have been a significant alternative after the dissolution of FPCs. Specifically with regard to the status change from FPCs to ANAs, the researchers pointed out that the reduction of transaction costs and the exemption from corporate taxes were the primary reasons motivating the change.

The present study aims to examine the management situations of FPCs and ANAs as contemporary forms of *iriai* forest management. There is little doubt that FPCs and ANAs are the two most important forms of contemporary *iriai* forest management. However, except for a few studies [9,11], FPCs and ANAs rarely appear in English literature, despite their importance; thus, more information and insights are needed. In particular, few studies provide detailed case information on kinds of forest management operations, forestry and non-forestry income sources, and locals' perceptions regarding the future of their *iriai* forests; the present study tries to fill this information gap.

An examination of FPCs facing management difficulties and ANAs as an alternative will facilitate understanding of how external legal, market, and social environments affect CPR management in developed industrialized countries. The cases of FPCs and ANAs can also highlight the pitfalls that policies related to administration and forestry in Japan have run into, along with the will of local communities to go along with their forest commons despite difficulties.

In terms of Japanese studies on FPCs and ANAs, Yamashita [17] presented a comprehensive review of the progress and current status of FPCs, drawing on official statistics. Handa [18] indicated that even though the policy intention of FPCs was the aim of more efficient timber production, most FPCs had continued with conventional *iriai*-type management. Several studies [19,20] reported problems and difficulties faced by FPCs, similar to those that Yamashita et al. [9] and Matsushita [11] presented. Although studies on FPCs are likely to be critical of the present situation, a couple of studies indicated the potential of FPCs, because the forest areas that FPCs own are likely to be greater than individual private forests; hence, they may hold advantages in making management plans and obtaining forest certificates [21,22].

2. Materials and Methods

First, the author consulted secondary literature and official statistics to determine the historical policy developments affecting *iriai* forests and the institutional arrangements of FPCs and ANAs.

Second, the author collected firsthand data from surveys through face-to-face interviews in the Fukuoka and Saga prefectures on Kyushu Island. For the Fukuoka prefecture, the author mailed a letter to all 55 FPCs registered in a database requesting an interview. The author received responses from 11 cases, which included 2 ANAs. For the Saga prefecture, the author purposefully selected 1 ANA; this was a case where former FPC members and residents were not the same and the number of residents in the neighborhood was far greater than the number of former FPC members. The author sought to observe what kinds of problems exist in such a situation. In total, the author obtained 9 FPC cases and 3 ANA cases; they are called FPCs A to I and ANAs A to C to preserve anonymity. The 3 ANAs included no cases of status change from the 9 FPCs—in other words, they were separate from each other. Topics surveyed included FPCs' or ANAs' basic information, recent management activities, and financial conditions. Data were collected in October 2018, February to May 2019, October 2020, and May 2021.

It should be noted that the present study is not based on random sampling methods. Regarding the Fukuoka prefecture, the author was able to obtain information from FPCs to which letters from the author were successfully delivered and who were willing to take interview surveys. The addresses of many FPCs provided in the database were wrong; consequently, these FPCs did not receive the letters from the author. It is probable that FPCs with little substantive activity were not included. Regarding the Saga prefecture, the author only focused on 1 ANA. Thus, the present study should be understood as case studies aimed at naturalistic generalization, a process through which readers gain insight by reflecting on the details and descriptions presented in case studies [23].

Part of the data has been published in Ota [24] and Ota [25]. The present study reorganizes the already-published information in terms of its purpose and adds qualitative information that Ota [24] and Ota [25] did not present.

3. Results

3.1. Historical Developments and Recent Institutions Concerning Iriai Forests

3.1.1. Typical *Iriai* Institutions

Iriai institutions were developed during the Edo period (1603–1867). At that time, resources that could be obtained from forests were crucial for subsistence purposes. In addition to firewood and timber, grass was an important source of fertilizer for rice fields. Particularly after the middle of the Edo period, population growth was evident due to the expansion of rice fields. As a result, overexploitation of forests and consequent forest degradation proliferated. Natural disasters occurred frequently. The Tokugawa Shogunate and domain lords ruling at the time developed several theories and measures to address the overexploitation of mountain forest resources, including advice for utilizing trees or planting trees after harvest [26]. In addition to such government measures, local rules and regulations concerning forest resources were developed to collectively avoid conflict among households or hamlets and to prevent overexploitation [7]. Such sets of rules and regulations governing *iriai* forests can be called *iriai* institutions.

The unit of *iriai* institutions was traditional Japanese hamlets (*mura*). In general, the principles of membership were strict: only people who resided in the hamlet had a right to access the resources of *iriai* forests. Although great variety was seen in rules applied among hamlets, or even in one hamlet across seasons or species, practices that carefully utilized limited forest resources were common. Typically, there were limits on the amount and time of harvest and particular techniques in exploiting resources. Contribution of labor to manage the forest areas—such as annual burning of grassland, cutting specific timber or thatch, and planting or enriching particular species—was an important obligation under *iriai* institutions. Enforcement of rules was substantial, with high degrees of compliance. *Iriai* institutions were largely in line with the design principles of CPRs presented by Ostrom [1].

3.1.2. Policy Changes over Time

Institutional change has occurred in three directions: conversion to national property, conversion to municipal property, and conversion to private property [10]. The conversion to national property was established by the Public/Private Ownership Separation Policy, which began in 1873. Under this policy, forests for which evidence of private ownership was not confirmed were incorporated into national forests. Many *iriai* forests were regarded as forests without private ownership, due to their form of collective use; consequently, they fell under national forests.

The first wave of conversion to municipal property occurred with the Municipal Government Act in 1889. This act consolidated traditional hamlets (*mura*) from the Edo period into modernized municipal units of cities, towns, and villages. Processes of consolidation were expected to involve taking over resources under *iriai*-related institutions, such as forests, grasslands, or ponds, and making them the property (assets) of the new municipalities. Most of the peasants at the time strongly opposed this policy, to prevent their resources for subsistence from being taken away. Their protests made it difficult to implement the policy. As a compromise, the government created a new scheme of property wards (*za-isanku*). Under this scheme, right-holders of *iriai* resources can substantively maintain their rights to use, manage, and dispose of resources under the supervision of the municipalities. Property wards are one of the most common forms of *iriai* forest management, together with FPCs; in 2011, there were 3710 property wards in total, not only for forests, but also for ponds, hot springs, and graveyards [27].

The second wave of conversion to municipal property was the Village-owned Forests Integration Policy, which began in 1910. This aimed to incorporate into municipal property the *iriai* forests that had remained owned by former hamlet-like units (registered under the names of former hamlets such as *ku*, *aza*, or *kumi*). However, failing to deliver the expected achievements, this policy ended in 1939.

In the late 19th century, when the first wave of conversion to municipal property was taking place, Japanese law scholars learned about the laws and legal systems in Western Europe, and the Japanese Civil Code was enacted in 1896. In the Civil Code, rights of common (*iriai ken*) are specified in Articles 263 and 294. Rights of common ensure local community members' use of local forests or other natural resources based on local customs maintained since the pre-modern era. At the time that the Civil Code was enacted, more than 80% of the population lived in rural areas and were dependent on basic natural resources, such as firewood, fodder, grass, water for irrigation, and fishery; thus, ensuring the livelihood security of these people was essential.

According to the court rulings to date, rights of common are a private right. This feature is extremely important; as a private right, rights of common have been very strong [28]. If these rights had been public rights, the actualities of the rights of common might have been changed or manipulated at the discretion of the government. As a result of this definition of the rights of common in the Civil Code, on the one hand, the government tried to bring *iriai* forests into the public sphere through policy instruments, but on the other hand, it ensured that local community members hold collective access to *iriai* forests, with the rights of common held as a private right.

However, confusing issues have arisen related to the rights of common. On the one hand, rights of common can exist irrespective of land ownership status; court rulings have established that rights of common exist on national lands. On the other hand, rights of common cannot be registered; a group of people holding rights of common for an *iriai* forest cannot register the land of their forest as holders of rights of common. Consequently, holders of rights of common have been likely to register their forests in the name of former hamlets (such as *ku*, *aza*, or *kumi*) or in the individual names of all right-holders.

Until the 1950s, forests had been primarily used for subsistence purposes in Japan. Forest areas, including those of *iriai* forests, mainly consisted of broad-leaved trees such as oak, used for fuel and agricultural uses, and grassland, used for foraging. Such conditions were consistent with the livelihoods and lifestyles of local farmers in those times. However, circumstances drastically changed after the 1960s. Modernization of lifestyles, such as the increasing use of fossil fuels, resulted in the decreased importance of forest resources for the subsistence economy. Grass and small branches were replaced with chemical fertilizers, thatched roofs were replaced with iron roofs, and firewood was replaced with petroleum and natural gas.

At the same time, the national policy for forestry also changed. In 1964, the Forestry Basic Act was enacted, which states that the objectives of forestry are to increase timber production and productivity and to enhance the income of forestry workers. The government strongly promoted the planting of coniferous trees, such as Japanese cedar (*Cryptomeria japonica*) and Japanese cypress (*Chamaecyparis obtusa*), for intensive forestry. Existing broad-leaved forests were often replaced with coniferous trees.

In this context, the government tried to promote intensive forestry practices in *iriai* forests. The Act on Advancement of Modernization of Rights in Relation to Forests Subject to Rights of Common of 1966 has been the base of this policy shift [11]. This act was created to establish modern types of property rights related to forests by extinguishing pre-modern rights of common, developed on the assumption that communal types of management and decision-making rooted in rights of common are likely to hinder advanced and efficient utilization of forest resources. Under this act, which provides due processes, rights of common are to be extinguished, and forests under *iriai* institutions are to be converted to private property, owned either by cooperatives or individuals.

By 2014, approximately 580,700 ha of *iriai* forests had been subject to this modernization process (i.e., the extinguishment of rights of common). As of 2011, of the modernized *iriai* forests, 52.4% had come under the management of FPCs, 41.0% had been individualized by being equally divided among right-holders of *iriai* forests, 5.5% had become jointly owned private forests, and 1.0% had come to be managed by agricultural producers' cooperatives [11].

3.2. Institutions of FPCs and ANAs

The institutions of FPCs are rooted in the 1951 Forest Act. In 1978, the Forestry Cooperative Act was enacted and it is the legal foundation of FPCs. In Japan, cooperatives in the forestry sector generally refer to forest owners' cooperatives (*shinrin kumiai*: hereafter FOC), which differ from FPCs. Members of FOCs and FPCs are both forest owners, but FOCs jointly undertake forestry operations for forests owned by members and other entities. FOCs target ordinally individual private forests, and have nothing to do with *iriai* institutions.

An FPC is based on the principle of "consistency of ownership, management, and labor". Cooperative members make monetary or in-kind investments and own forests, and in principle, they manage the forests through their labor contributions [17] By law, an FPC is established for joint forest management for more than five people residing in a particular community, under the approval of the prefectural governor. After approval, members invest in the FPC and register the corporation. An *iriai* forest is jointly owned and managed by an FPC—i.e., the cooperative's members.

Only people who reside in the community or make in-kind contributions to the forests can be cooperative members. More than half of the cooperative members should regularly engage in FPC activities and more than one-third of the people who regularly engage in these activities should be FPC members. When an FPC gains profits, dividends are allocated to members according to the number of days that they work for the FPC. Activities that FPCs can undertake include forest management, cultivation of trees and edible mushrooms, agriculture utilizing forest areas, operations for or management of entrusted forests, and other related activities. Members can engage in not only forestry activities but also the cultivation of mushrooms, fruit, or animal husbandry.

Similarities exist between conventional *iriai* organizations and FPCs. Right-holders of *iriai* forests and FPC members overlap in many cases, and despite the change in legal status, FPC members are likely to have a sense of their forest as the commons rather than a sense of intensive commercial forestry [18]. There are also differences between conventional *iriai* organizations and FPCs in that legally, rights of common no longer exist for forests owned by FPCs. Furthermore, FPCs are likely to receive subsidies from municipalities. FPCs have to engage in various kinds of desk work, such as bookkeeping and organizing an annual general assembly, and they are subject to corporate and corporate inhabitant taxes due to their cooperative status [9].

The number of FPCs drastically increased after 1966, when modernization processes began, exploding from 586 in 1966 to 1494 in 1976, with a peak at 3482 in 1996. Since then, the number has decreased (Figure 1); as of 2020, 2693 FPCs could be found in Japan.

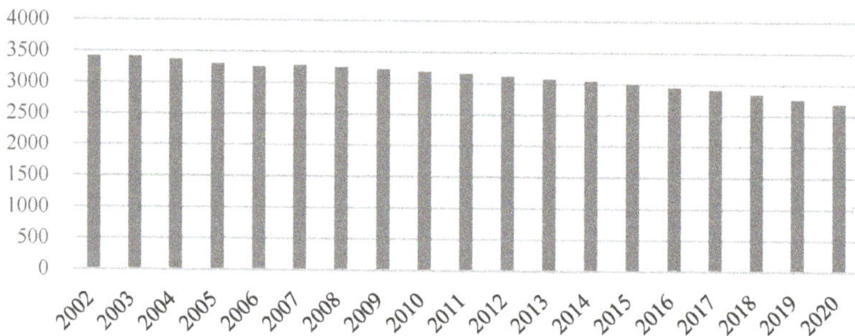

Figure 1. Recent changes in the number of FPCs (created from [29]).

The decreasing number of FPCs in recent years is mostly due to dissolution or status change. Although no official statistics are provided on dissolved FPCs, Table 1 indicates that 30–50 FPCs have been dissolved annually in the last decade. Approximately 72% of the dissolved FPCs have changed their status to ANAs [17].

Table 1. Basic information for the FPCs surveyed.

Pseudonym	A	B	C	D	E	F	G	H	I
Year established	1965	1969	1978	1974	1975	1954	1975	1982	1984
Area owned (ha)	164	36	7	51	61	131	49	93	20
Number of members (at the time of the survey)	72	90	50	32	56	23	60	26	32

(adopted from [24]).

ANAs are defined in the 1991 revised Local Autonomy Law in Japan. Under certain conditions, a local neighborhood association (such as a ward) can obtain the status of legal entity and register fixed assets, such as *iriai* forests, under the name of the neighborhood association. Although the ANA institution was not originally meant to address issues concerning *iriai* forests, it can be used for the sake of such forests. Before 2017, the document procedures were complex and difficult: cooperative members first had to dissolve the FPC through due process—including the liquidation of assets (forests), which may cost around several hundred thousand Japanese yen—and then had to register themselves as an ANA. However, the 2017 revised Forest Act has enabled FPCs to directly change their status to ANAs without dissolution; in this case, members do not have to bear the costs related to liquidation procedures.

One of the most important reasons for becoming an ANA is that ANAs are exempted from corporate tax and corporate inhabitant tax, which FPCs must pay. Given that many FPCs feel the burden of tax payments, tax exemption is a significant advantage [9].

Another distinct point between FPCs and ANAs is that after an ANA is established, its stakeholders are not limited to the former cooperative members. The *iriai* forest becomes an asset of the neighborhood, and so, all residents of the neighborhood become right-holders within the ANA. Decision-making and forest management practices can be affected in cases where the number of residents is greater than the number of former FPC members. This involves a change from a strict membership regime to a much softer and broader community management regime.

3.3. Activities and Management Conditions of the FPCs Surveyed

3.3.1. Overview

Table 1 presents basic information on the surveyed FPCs. With the exception of two cases, the FPCs were established after the Modernization Act in 1966. The average area owned was approximately 68 ha. In terms of forest type, most of the lands that the FPCs owned were planted forests of Japanese cedar. Each FPC's number of members was less than 100 and all FPCs had experienced a decline in membership. A large part of the forests managed by FPCs D, F, and G had been designated as forest reserves, which are important to local environmental conservation and water cultivation.

3.3.2. Tending Activities

Table 2 shows that eight of the nine FPC cases had conducted some activity to tend to planted forests in the few years before the surveys were conducted. Of these eight, six FPCs (A, B, D, E, F, and H) had operations using their own labor. Most of the thinning operations had been entrusted to other forest management bodies, such as FOCs; only FPC A had their own work crews to conduct thinning operations. Four FPCs (B, D, F, and G) received subsidies from the prefecture or cities for their tending operations.

Table 2. Forest management activities among the FPCs surveyed.

Pseudonym	A	B	C	D	E	F	G	H	I
Tending activities	Thinning Improving Cutting Weeding	Thinning Weeding	–	Vine-cutting Weeding	Thinning	Thinning Weeding Maintenance of mountain trails	Thinning	Thinning Weeding	–
Timber production	–	–	–	–	–	Timber from thinning	–	Final harvest Timber from thinning	–
Non-timber forest product production	–	–	–	–	–	–	–	–	–
Non-forestry activities	Land lease	–	Land lease	Land lease	Land lease	–	–	Land lease	–

(adopted from [24]).

3.3.3. Forestry and Non-Forestry Activities

Only FPC H conducted a final harvest of timber; it also sold timber generated from thinning activities. FPC F also conducted thinning and sold timber. No FPC was producing non-timber forest products, such as mushrooms.

Five FPCs (A, C, D, E, and H) engaged in non-forestry activities, all of which involved leasing land to other businesses, such as quarries. As explained in the following section, these FPCs enjoyed large annual incomes. In addition to present incomes, some FPCs held monetary assets that were obtained by selling forestland in the past. Although FPCs B and I had no income, they managed to pay taxes using the assets that they had obtained in the past. FPC D had a large amount of assets due to land selling to a development project.

3.3.4. Case Study Information

This section provides detailed case study information for each FPC, including members' perspectives.

[FPC A]

FPC A was the most active among the surveyed FPCs. It had finished planting Japanese cedar trees by 1995 and conducted tending operations, such as weeding, improving cutting, and thinning. It had working crews for forestry operations, which is rare for an FPC. As of 2019, 18 people worked at the FPC: eight were in their 60s, and six were in their 70s. In addition to the tending operations conducted by work crews, provision of labor, such as the weeding of coastal pine forests, was mandatory for ordinary members.

FPC A held substantial assets, including their own office, because it received annual income from a land lease contract with a company for a quarry, amounting to several million Japanese yen. Thus, FPC A enjoyed stable and substantial money from non-forestry activities, and members could afford to pay the taxes related to the FPC.

In addition to the above-mentioned activities, the members were working on disaster recovery. A landslide occurred in part of their forest in 2018; consequently, they had to remove tree debris to get the roads back to normal conditions.

Even in FPC A, the number of members has decreased from 150 people in 1965 to 91 people in 2009 to 72 people in 2018. The members foresaw that this trend would continue, due to depopulation and population aging.

The president of the FPC told the author that the members were proud of maintaining appropriate forest management, thanks to good income from non-forestry activities, and that they would like to keep the status quo. With increasing societal attention to the role of forests in environmental conservation, they would like to contribute to the public as a forest owner and manager. The president said that in the future, they might want to conduct basic forestry operations for other private (individually owned) forests through contracts.

[FPC B]

FPC B had a history of developing planted forests of Japanese cedar and cypress through a benefit-sharing contract with the city where it is located. Planting started in 1962, before the establishment of the FPC, and the area within its purview is approximately 6 ha. One custodian was hired from the community to supervise this area. The contract with the city emphasizes the importance of the afforestation of mountains and the enrichment of water sources; it is implied that these public functions were of great importance when the FPC was established.

In addition to the planted forests as per the city contract, FPC B had also planted approximately 3 ha of Japanese cedar and cypress. The members conducted tending operations of weeding and thinning, although they had not been able to target all the planted forests that needed tending.

The members received a subsidy from the prefecture for thinning operations, which they entrusted to an FOC. Simple operations, such as weeding, were managed by executive members, which consisted of 15 people in rotation every three years. They intentionally rotated executive members to expose younger members to their *iriai* forest.

FPC B gained no income from forestry and had no constant non-forestry income. In the past, it had sold part of the forestland to the prefecture to construct a check dam; it used the money from this land sale to pay corporate taxes.

The number of cooperative members had decreased from 164 people in 1962 to 90 people in 2018, most of whom were more than 60 years old.

Nevertheless, FPC members had no intention of dissolution or status change. They had managed *iriai* forests in the present form, and so, they would probably continue to in this way. The president told the author that all citizens should bear the management of forests because forests and mountains have public functions, and more support from subsidies was needed because there are limitations to what one cooperative can do.

[FPC C]

The forest area of FPC C is small: 7 ha. Forest accessibility was not bad and the planted Japanese cedar trees were mature. However, the FPC had no immediate plan to fell and sell them, and tending operations were not necessarily required. As a result, FPC C's activities were mostly absent. According to its members' perceptions, the FPC is "on leave" from its activities and operations as a cooperative. At the same time, the members did not want to dissolve or change their status to an ANA. FPC C had a constant income source from leasing land for the placement of telephone poles. It had also sold part of its forests to the city to make roads, through which they received a large amount of money. In this way, the members could afford to pay taxes.

[FPC D]

FPC D was the richest of the surveyed FPCs. It had sold part of its forestland around 40 years ago for a natural park development project, receiving a huge amount of money (more than 100 million Japanese yen). Part of this money had been used for community development, such as renovating a shrine and the community hall. In addition, the FPC had an annual income from leasing land for the placement of sign boards. Consequently, FPC D had no problem paying taxes. They even used their income to enjoy tours to other FPC cases to study management or advanced forestry activities.

The community where FPC D worked had some newcomers. While the FPC had 32 (meaning 32 households) members in 2019, there were more than 140 households in the community. This meant that younger generations lived in the community and participated in tending operations, such as weeding. Even though the number of cooperative members was decreasing, they were optimistic about maintaining this FPC through younger generations, substantive assets, and constant non-forestry income sources.

Part of the forest of FPC D had been designated as a forest reserve for water cultivation. The forests were regarded as important to local environmental conservation.

[FPC E]

FPC E was also a wealthy FPC; it had contracted with a private mining company and leased part of their forestland as a quarry. The FPC received a large annual income from this land leasing—approximately 9.5 million Japanese yen. In addition to corporate taxes, its income tax was massive; however, the FPC had no problem paying it. FPC E also provided dividends to its members.

The number of members was decreasing; however, the rate of decrease was not very severe. The decrease was due to the decline in the population residing in the community. Few FPC members withdrew from membership and the members mostly enjoyed the benefits of cooperative activities.

Exceptionally, in FPC E, members had conducted thinning operations by themselves, with daily allowances; in principle, this was mandatory. According to the president, annual operations fostered a sense of responsibility among members.

[FPC F]

A large part of FPC F's forest had been designated as a forest reserve for water source cultivation. The members had maintained a high level of frequency of tending operations: they conducted weeding and monitoring of forest-area boundaries every year by themselves, and they entrusted thinning to an FOC using a subsidy. The maintenance of trails of the nearby mountain was entrusted to FPC F by the city with a subsidy for this work.

A year before the author's survey, FPC F received forestry income from timber generated from thinning. However, this was exceptional; in the previous decade, there was mostly no other timber income. In the 1960s and 1970s, it enjoyed a huge profit from timber, sometimes as much as 100 million Japanese yen, thanks to increasing demand for utility poles. In the past, the FPC had provided timber for constructing a former community hall to contribute to the community.

The FPC charged 5000 Japanese yen every year as a membership fee. Even with this money, it had a deficit balance due to heavy corporate and fixed asset taxes. The members were spending down the savings that they had earned in the past.

The number of FPC members had decreased from 92 members in 1965 to 23 in 2018. The president told the author that the reason for this was unclear. At the same time, the number of residents in FPC F's community had increased from 120 households several decades ago to more than 600 households recently. This was due to its good accessibility, as people could commute from the community to big cities.

The president said the FPC's management situation was worsening, but the members had no good ideas for breakthroughs. They perceived that having profit-earning activities, either forestry or non-forestry in nature, would be important.

[FPC G]

A large part of the forest of FPC G had been designated as a forest reserve for water cultivation. Its location was near an important source of water for the city.

The number of members had decreased from 92 people in 1975 to 60 in 2018. The president told the author that this decrease may not be due to aging alone: because FPC G had no forestry or non-forestry income, members were charged 10,000 Japanese yen as an annual fee every year, and members who were not willing to pay the fee had withdrawn from the FPC. The money from these fees was used for taxes and administrative costs.

FPC G conducted tending operations, including thinning. It had received subsidies from the prefecture and all operations were entrusted to an FOC using money from the subsidies to pay for it. It also cleaned the walking roads of a nearby natural park, entrusted by the prefecture with a subsidy.

[FPC H]

FPC H was engaged in various management activities. It was the only surveyed FPC to have undertaken a final harvest of timber in recent years, with operations entrusted to an FOC; the profit earned from this operation was several thousand Japanese yen. After the final harvest, the members had wanted to replant cedar and cypress trees; they had little idea of working on broad-leaved forests, such as oak species. The president told the author that if they had been engaged in shiitake mushroom production, which requires oak logs, they might have wanted to replace conifer trees with oak trees. In addition, they had undertaken weeding and thinning operations by themselves. FPC H also received an income from leasing forestland for roads (approximately 750,000 Japanese yen annually).

The number of members had not significantly decreased from 29 people at its inception in 1982 to 26 people in 2018. This was due to its good accessibility to a big city, offering a reasonable distance for commuting. FPC H had no problem paying taxes and continuing its forest management.

[FPC I]

Even though the total area of FPC I's forest was small, it was in coastal areas, thus serving as windbreaks.

FPC I regularly monitored forest-area boundaries, but other than that, it had no tending operations; according to the president, operations were not needed at this phase. There was no income-earning activity: the FPC had sold part of its forest in the past and it drew on the savings earned from these sales to pay taxes. It also charged an annual membership fee of approximately 60,000 Japanese yen; this amount was also used to pay taxes.

Although FPC I's members were attached to their *iriai* forests, they had begun to discuss the possibilities of dissolving the FPC. However, they faced an issue: the number of FPC members (households) was 32, while the number of households in the community was 110. If they dissolved the FPC and became an ANA, all residents in the community would become members of the new ANA. The president told the author that such a change could impose the problems faced by the FPC on other residents; as a result, the members were hesitant to change.

3.4. Activities and Management Conditions of the ANAs Surveyed

3.4.1. Overview

All surveyed ANAs underwent a status change from FPCs in recent years for financial reasons—i.e., burdens of taxation and expected budget shortfall. ANA B had forestry incomes before the status change, but the members felt that such income would not be enough to maintain the status of an FPC.

Even after the status changes, these FPCs continued in their basic forest management activities.

ANAs A and B experienced little issues in terms of membership, because the former FPC members and residents of the communities mostly overlapped. However, for ANA C, where the numbers of FPC members and residents in the community were not the same, difficulties were faced when applying a principle of the ANA, i.e., the forests become an asset for all residents of the community.

3.4.2. Case Study Information

This section provides detailed case study information on each ANA, including members' perspectives.

[ANA A]

ANA A (Fukuoka prefecture), which owns 27 ha, was organized in 2016 after the FPC in Community A was dissolved. The former FPC had been established in 1986. The reason for the dissolution and status change to an ANA was that the cooperative's members lost the motivation to continue intensive forest management activities as an FPC, due to lower timber prices, the aging of cooperative members, and a lack of successors. They decided to choose a less intensive form of management as an ANA.

Even after becoming an ANA, the members continued with basic forest management operations, particularly root-cutting and thinning. Root-cutting was conducted, as a rule, one or two times a year by all residents in the community. Thinning was entrusted to an FOC, paid for by a subsidy from the prefecture.

The former FPC members and the residents of the community mostly overlapped Therefore, their management situation did not drastically change, other than a positive consequence of being free from corporate taxes.

[ANA B]

ANA B (Fukuoka prefecture), which owns 72 ha, was organized in 2020 after the FPC in the community changed its status. The former FPC was established in 1983. During the time of the FPC, members were engaged in the weeding and cleaning of strip roads. This community experienced severe damage from flooding in 2017 and the members had to remove dirt from their forest area. They had undertaken a final harvest of timber, which they entrusted to an FOC. Before the status change, the FPC had a reasonable annual forestry income of approximately 700,000 Japanese yen.

The members decided on the status change at the 2019 annual meeting of the FPC The primary reason was that the FPC would otherwise face bankruptcy. The number of FPC members and residents in the community was declining, due to their remote location Here, when a person's membership ends due to his or her death, and when none of his or her children live in the community, the FPC has to return the amount that the FPC member invested to establish the FPC, i.e., 350,000 Japanese yen. This meant that the more members who died, the more money the FPC had to disburse from their savings. The members were sure that their savings would run out in the near future, despite a certain amount of forestry income. Therefore, they decided to change their organizational status to an ANA The status was changed in 2020, at which point, an agreement was reached among the former FPC members that they would not claim a refund of the invested amount.

The secondary reason for the status change was the burden of taxation; even though the FPC had forestry income, corporate and corporate inhabitant taxes were a burden.

The FPC members felt little concern about changing the organizational status. In this community, residents and FPC members mostly overlapped, and so, issues of membership was not a problem. The president told the authors that they would continue the same level of forestry operations after the status change.

[ANA C]

ANA C (Saga prefecture), owning 23 ha, was organized in 2021 after the FPC in the community changed its status. The former FPC was established in 1990. A large part of the forest had been designated as a forest reserve for water source cultivation. A few years before the status change, the cooperative's members had conducted weeding, using a subsidy from the city. There were no forestry nor non-forestry incomes. The FPC charged annual fees in order to pay taxes.

The members started discussing a status change to an ANA in 2017, due to the burden of taxation. The president told the author that they had a sense of responsibility to manage the *iriai* forest, particularly with regard to disaster prevention and water source cultivation. However, it was unrealistic to continue the FPC situation for future decades.

In the process of deliberation, the FPC faced a problem in that the FPC members and the residents in the community were not the same. As of 2020, the number of FPC members (households) was 26, but the number of households in the community was 80. After an FPC becomes an ANA, the forest becomes an asset for all residents of the community. The FPC members were uncomfortable with this situation, as they had contributed labor and in-kind and monetary investments. Thus, they tried to determine a method by which they could continuously engage in managing the *iriai* forests.

Consequently, the regulations of the newly formed ANA prescribed that a forest division was to be set up consisting of former PFC members. In doing so, the former FPC members tried to maintain responsibility for managing the forests. It was unclear how the benefits from timber production or any other income sources, if any, would be shared in the ANA—in other words, whether benefits would be shared among the former FPC members alone or among all community residents. This issue had remained a gray area.

This arrangement somewhat deviates from the concept and principle of ANAs, which is an organization open to all community residents. At the same time, this arrangement could be regarded as a technique to reconcile the existing legal prescriptions of the ANA setup with local realities.

The former FPC president told the author that the members of the former cooperative were willing to maintain their existing forest management operations, primarily based on the forest division in the ANA. However, they were not sure what other, non-FPC community residents would be involved in or invited to in terms of forest management. At the time of the author's survey, few concrete ideas on collaboration in the community had been developed.

4. Discussion

An examination of the historical developments of external policy influence over *iriai* forests confirmed that FPCs were promoted by the government when timber prices were high in the 1960s. Conventional *iriai* communities, based on rights of common, were considered a hindrance to promoting intensive and efficient forestry. As a result, rights of common were subject to extinguishment through due processes. However, conditions favorable to forestry have disappeared, particularly since the 1980s, given low timber prices and declining FPC membership. The status of a cooperative is generally no longer advantageous; instead, FPCs have the disadvantage of paying corporate taxes even though they have no income. As Shimada [14] indicates, these external factors have worked outside the control of *iriai* communities.

We can observe how policies related to administration and forestry can have negative effects on forest commons. Overall, changes in *iriai* forests after the 1960s have typically been the negative effects of commodification and non-settlement trends [6]. On the one hand, this could be understood as simply a failure of certain policies; on the other hand, this could represent a broader indication that such a policy failure is a consequence of modernization and its simplification of the relations between nature, space, and people [30]. The value of maximizing monetary profits by private entities, which prevailed after the 1960s, should be reconsidered, and revitalizing meaningful human–forest relations in the contemporary context is important.

Through its surveys of FPCs and ANAs, the present study has confirmed several important points. First, some FPCs have suffered from disadvantageous circumstances in forestry, including low timber prices, fewer FPC members, and the burden of corporate taxes. Few FPCs had engaged in forestry production in the few years before the surveys were conducted. These were general trends of FPCs that had been indicated in the previous literature [9,11]. At the same time, some FPCs have enjoyed a large amount of non-forestry incomes or assets, e.g., the leasing or selling of forestland. This point has been less emphasized by previous studies, except for Yamashita [17]. It is noteworthy that there are some wealthy FPCs and that not all FPCs are suffering financially. Therefore, it is necessary that future policy options for FPCs consider their diversity and build on the concrete situations of each FPC. However, the fact that all the confirmed non-forestry activities were forestland leasing or selling indicates that whether or not FPCs have an opportunity to engage in such activities depends on their geographical location (whether the FPC's forest is part of an upcoming project site), regardless of FPCs' management efforts. Thus, it is unrealistic to propose attracting forest development projects in order to lease forestland as a solution to the disadvantageous management circumstances of FPCs.

Second, as previous studies have indicated, becoming an ANA is a reasonable option. As seen in the present results, forest management activities are not likely to drastically change after the status change, as the former FPC members generally have high degrees of attachment to and responsibility for their *iriai* forests. However, as shown in the case of ANA C, difficulties will arise when FPC members and community residents do not overlap.

Third, in most cases of both FPCs and ANAs, basic forest management operations were conducted, at least to some extent. Several received subsidies from the prefecture or city to conduct tending operations and the importance of subsidies was confirmed. However, as mentioned in the Materials and Methods section, the present study did not apply random sampling methods, and thus, the author cannot generalize this in quantitative terms. The qualitative interview results indicate that both FPC and ANA members were likely to feel attached to and responsible for their *iriai* forests. The importance of forests in environmental conservation, water source cultivation, disaster prevention, and climate change mitigation was often emphasized. As the owners and managers of *iriai* forests, they perceived that they had contributed to this public good. It is implied that this sense of pride is one of the important factors maintaining management activities of *iriai* forests in the contemporary context.

It Is noteworthy that local communities are likely to want to persist with forest commons, even when it is difficult. Changing the entity's status to ANA is a creative application of an available institution that originally had nothing to do with forest commons. As can be seen, the existing local will and initiatives should not be overlooked or underestimated. Institutional changes should be encouraged to promote or ease local initiatives that can maintain or revitalize commons management; this lesson can be applied to countries other than Japan.

5. Conclusions

The present study has provided an overview of contemporary *iriai* forest management in Japan, focusing on FPCs and ANAs. It has also presented case studies of several FPCs and ANAs, highlighting the difficulties and struggles that they face as forest commons managers. A simple generalization of the history of Japanese *iriai* forests to the global context is difficult. At the very least, given that forest commons which can entail meaningful human–nature relations have undergone a re-evaluation in the contemporary developed world [6,31], a globally shared question might be how to provide an institutional framework, financial mechanisms, and social understandings that can maintain and revitalize commons.

The author provides three policy recommendations for the Japanese context. First, the legal settings of FPCs and ANAs should be made more compatible with contemporary realities. This has been partially realized through the 2017 revised Forest Act, which enabled easier status change to ANA. However, there is room for further policy modifications in the taxation arrangements of FPCs. In addition, administrative support and consultation opportunities are advisable for FPCs considering a status change to ANA in places where such support has been absent.

Second, greater financial support for management activities is beneficial, particularly for FPCs. Since 2019, the Forest Environment Transfer Tax has been in force in Japan, as a form of payment for ecosystem services [32]. Funds from this tax could be allocated to managers of *iriai* forests. As FPC members feel that they are contributing to the public through forest management, provision of funds from this tax can be seen as reasonable and thereby justified. In the context of global climate change, the ecosystem services provided by managers of *iriai* forests will gain importance.

Third, perusing multi-level governance to open the commons to broader sections of society is key. After becoming ANAs, *iriai* forests become assets of all residents in the community. Given this opportunity, enhanced engagement with people other than former FPC members—e.g., schoolchildren in and outside the community, and environmental volunteers from urban areas—could be considered. Mitsumata and Saito [31] reports cases where new values were created and forest uses were revitalized, as a result of the collaboration of multiple stakeholders. In such a process, the forest composition of *iriai* forests could also be reconsidered; existing planted forests of Japanese cedar and cypress could be gradually turned into mixed forests of conifer and broad-leaved trees. If former FPC members strongly believe that their *iriai* forests should only serve timber production from cedar and cypress trees, changing their thought processes to consider more flexible and diverse uses of forests would also be beneficial.

Funding: This study was financially supported by JSPS KAKENHI under grant number JP21H03709.

Data Availability Statement: The data are not publicly available due to confidentiality reasons.

Conflicts of Interest: The authors declare no conflict of interest.

References

1. Ostrom, E. *Governing the Commons: The Evolution of Institutions for Collective Action*; Cambridge University Press: Cambridge, UK, 1990.
2. Baland, J.; Platteau, J. *Halting Degradation of Natural Resources: Is There a Role for Rural Communities?* Oxford University Press: New York, NY, USA, 1996.
3. McKean, M.A. Common property: What is it, what is it good for, and what makes it work? In *People and Forests: Communities, Institutions, and Governance*; Gibson, C.C., McKean, M.A., Ostrom, E., Eds.; MIT Press: Cambridge, MA, USA, 2000; pp. 27–55.
4. Berge, E.; McKean, M. On the commons of developed industrialized countries. *Int. J. Commons* **2015**, *9*, 469–485. [CrossRef]
5. Miyanaga, K.; Shimada, D. 'The tragedy of the commons' by underuse: Toward a conceptual framework based on ecosystem services and satoyama perspective. *Int. J. Commons* **2018**, *12*, 332–351. [CrossRef]
6. Mitsumata, G. Complementary environmental resource policies in the public, commons and private spheres: An analysis of external impacts on the commons. In *Local Commons and Democratic Environmental Governance*; Murota, T., Takeshita, K., Eds.; United Nations University Press: Tokyo, Japan, 2013; pp. 40–65.
7. McKean, M.A. The Japanese experience with scarcity: Management of traditional common lands. *Environ. Rev.* **1982**, *6*, 63–88. [CrossRef]
8. McKean, M.A. Management of traditional common lands (iriaichi) in Japan. In *Making the Commons Work: Theory, practice, and policy*; Bromley, D.W., Ed.; ICS Press: San Francisco, CA, USA, 1992; pp. 63–98.
9. Yamashita, U.; Balooni, K.; Inoue, M. Effect of instituting "authorized neighborhood associations" on communal (iriai) forest ownership in Japan. *Soc. Nat. Res.* **2009**, *22*, 464–473. [CrossRef]
10. Saito, H. Administrative centralization threatens commons-owning municipal sub-unit: Property Wards (Zaisanku) in Toyota City, Japan. In Proceedings of the 13th Biennial Conference of the International Association for the Study of Commons, Hyderabad, India, 13–17 September 2011.
11. Matsushita, K. Recent problems and new directions for forest producer cooperatives established in common forests in Japan. In *Sustainable Forest Management-Case Studies*; Martin-Garcia, J., Diez, J.J., Eds.; Intechopen: Rijeka, Croatia, 2012; pp. 161–182.
12. Suzuki, T. The custom and legal theory of iriai in Japan: A history of the discourse on the position of the rights of common in the modern legal system. In *Local Commons and Democratic Environmental Governance*; Murota, T., Takeshita, K., Eds.; United Nations University Press: Tokyo, Japan, 2013; pp. 69–86.
13. Muroi, S. History of Japanese common rights: Iriai-ken—Social, judicial and academic overview. In *Land Law and Disputes in Asia: In Search of an Alternative for Development*; Kaneko, Y., Kadomatsu, N., Tamanaha, B.Z., Eds.; Routledge: London, UK, 2021; pp. 182–199.
14. Shimada, D. External impacts on traditional commons and present-day changes: A case study of iriai forests in Yamaguni district, Kyoto, Japan. *Int. J. Commons* **2014**, *8*, 207–235. [CrossRef]
15. Takahashi, T.; Matsushita, K.; de Jong, W. Factors affecting the creation of modern property ownership of forest commons in Japan: An examination of historical, prefectural data. *For. Pol. Econ.* **2017**, *74*, 62–70. [CrossRef]
16. Takahashi, T.; Matsushita, K.; Yoshida, Y.; Senda, T. Impacts of 150 years of modernization policies on the management of common forests in Japan: A statistical analysis of micro census data. *Int. J. Commons* **2019**, *13*, 1021–1034. [CrossRef]
17. Yamashita, U. Seisan shinrin kumiai no ayumi to genkyokumen [The progress and current state of forest producers' cooperatives]. *J. For. Econ.* **2020**, *66*, 26–39. (In Japanese)
18. Handa, R. Seisan shinrin kumiai to iriai rinya no 50 nenshi [A 50-year history of forest producers' cooperatives and iriai forests]. *For. Econ.* **2001**, *54*, 1–13. (In Japanese)
19. Yamashita, U. Kanko kyoyu ni okeru shoyu, shinrinkanri, kenrikankei no jittai [Actual conditions of ownership, forest management, and right relations of customary common ownership]. *For. Econ.* **2014**, *67*, 1–17. (In Japanese)
20. Yamashita, U. Iriai rinya kenkyu no kongo no tenbo [Review of studies on communal (iriai) forests in Japan]. *For. Econ.* **2017**, *70*, 1–21. (In Japanese)
21. Sato, N. Kyoyurin, zaisanku, seisan shinrin kumiai wo gendai ni ikasu [Utilizing common forests, property wards, and forest producers' cooperatives in the contemporary context]. *Gendai Ringyou.* **2012**, *553*, 14–19. (In Japanese)
22. Matsushita, K.; Takahashi, T.; Yoshida, Y.; Senda, T. 2005 nen, 2010 nen noringyo census ni yoru seisan shinrin kumiai no bunseki [A statistical analysis of forest producers' cooperatives using microdata from the 2005 and 2010 World Censuses of Agriculture and Forestry: Forest practices and forest product sales classified by forest holding area]. *J. For Commons.* **2019**, *39*, 60–70. (In Japanese)
23. Melrose, S. Naturalistic generalization. In *Encyclopedia of Case Study Research*; Mills, A.J., Durepos, G., Wiebe, E., Eds.; Sage Publications: Thousand Oaks, CA, USA, 2009; pp. 599–601.
24. Ota, M. Fukuokaken ni okeru seisan shinrin kumiai no katsudo jittai [Activities of production forestry cooperatives in Fukuoka prefecture, Japan]. *Kyushu J. For. Res.* **2021**, *74*, 1–4. (In Japanese)
25. Ota, M. Seisan shinrin kumiai kara ninka chien dantai heno soshiki henko ni kakaru kadai: Sagaken no jirei kara [Problems in the status change from production forestry cooperatives to authorized neighborhood associations: Insights from Saga prefecture, Japan]. *Kyushu J. For. Res.* **2022**, *75*, 109–112. (In Japanese)

26. Kato, M. Kyosei jidai no yama riyo to yama zukuri: Kinsei sanrinsho no ringyo gijutsu [Forest use and management in a symbiosis era]. In *Mori to Hito no Asia: Dento to Kaihatsu no Hazama ni Ikiru [Asia of Forests and People: Living in between Traditions and Development]*; Fukui, K., Tanaka, K., Akimichi, T., Yamada, I., Eds.; Showado: Kyoto, Japan, 1999; pp. 100–130. (In Japanese)

27. Izumi, R.; Saito, H.; Asai, M.; Yamashita, U. *Commons to Chihojichi: Zaisanku no Kako, Genzai, Mirai [Commons and Local Autonomy. Past, Present, and Future of Property Wards]*; J-FIC: Tokyo, Japan, 2011. (In Japanese)

28. Nakao, H. *Rinyaho no Kenkyu [Study of Forest Laws]*; Keiso Shobo: Tokyo, Japan, 1965. (In Japanese)

29. Forestry Agency. Shinrin Kumiai Issei Chosa [National Survey for Forest Owners' Cooperative and Forest Producers' Cooperatives]. 2023. Available online: https://www.maff.go.jp/j/tokei/kouhyou/sinrin_kumiai/ (accessed on 13 January 2023). (In Japanese).

30. Scott, J. *Seeing Like a State: How Certain Schemes to Improve the Human Condition Have Failed*; Yale University Press: New Haven, CT, USA, 1998.

31. Mitsumata, G.; Saito, H. *Mori no Keizaigaku [Economics of the Forest]*; Nippon Hyoron Sha: Tokyo, Japan, 2022. (In Japanese)

32. Kohsaka, R.; Uchiyama, Y. Use of the Forest Environment Transfer Tax for forest data development and exchange: Evidence from all 47 prefectures in Japan. *For. Sci. Technol.* **2022**, *18*, 201–212. [CrossRef]

Article

Economic and Environmental Analysis of Woody Biomass Power Generation Using Forest Residues and Demolition Debris in Japan without Assuming Carbon Neutrality

Masaya Fujino [1],* and Masaya Hashimoto [2]

[1] Faculty of Food and Agricultural Sciences, Fukushima University, Fukushima 9601248, Japan
[2] Mizuho Financial Group, Inc., Tokyo 1008176, Japan
*   Correspondence: fujino@agri.fukushima-u.ac.jp

Abstract: Despite the increasing importance of renewable energy worldwide, the argument that forest biomass power generation is not carbon neutral has been rising. This research used Gifu Biomass Power Co., Ltd. (GBP) in Japan as a case study to investigate this matter. An evaluation was conducted through an input–output analysis on the economic and environmental benefits (i.e., CO_2 reduction) of forest biomass power generation without assuming carbon neutrality. GBP's economic benefits were estimated to be 3452.18 million JPY during the construction period and 114.38 million JPY per year from operations. It was also estimated to generate 21.77 jobs per year in the forestry sector. CO_2 emissions were estimated to increase by 423.02 tons during the construction period and 137,747 tons per year from operations. Although forests may offset CO_2 by absorbing it, woody biomass power generation does not necessarily reduce CO_2 emissions in Gifu Prefecture. The results indicate that woody biomass power generation is effective for the local economy but not necessarily for the global environment. The analysis should include more industrial sectors to clarify the environmental significance of wood biomass power generation without assuming carbon neutrality.

Keywords: CO_2 emission; forestry; Gifu Prefecture; input–output analysis; renewable energy; ripple effect

Citation: Fujino, M.; Hashimoto, M. Economic and Environmental Analysis of Woody Biomass Power Generation Using Forest Residues and Demolition Debris in Japan without Assuming Carbon Neutrality. *Forests* **2023**, *14*, 148. https://doi.org/10.3390/f14010148

Academic Editors: Noriko Sato and Tetsuhiko Yoshimura

Received: 24 December 2022
Accepted: 11 January 2023
Published: 12 January 2023

1. Introduction

1.1. Background

Carbon neutrality, the goal of achieving virtually zero greenhouse gas emissions in the wake of the Paris Agreement adopted in 2015, is now a global trend. A total of 128 countries and self-governing territories have declared their intention to pursue this goal [1]. Climate-related actions are currently being implemented to achieve this. Expectations for renewable energy are growing, and the use of woody biomass is expected to increase.

Meanwhile, the Joint Research Center, designated by the European Commission, published a research report titled "Use of Forest Biomass in EU Energy Production" [2]. The report stated that "The reconstructed Renewable Energy Directive (REDII Directive 2018/2001) envisages zero emissions at the point of biomass combustion. Bioenergy is not accounted for in the energy sector because these emissions are already counted as a change in carbon storage in the land use, land-use change, and forestry (LULUCF) sector (Regulation 2018/841). Therefore, the assumption that bioenergy is 'carbon neutral' within the broader EU climate and energy framework is incorrect" [2] (p. 9).

The above view is quite different from the theory that trees have CO_2 fixed in their trunks until they are cut down; therefore, even if trees are cut down and burned, and CO_2 is generated, they are carbon neutral if they are repeatedly managed and planted. Furthermore, the report points out that forests are being cut down for fuel procurement, which threatens ecosystems and biodiversity. There is a need to balance woody biomass power generation with ecosystems and biodiversity. Thus, the situation surrounding forest biomass is changing drastically.

The Japanese Government has been working to promote renewable energy since the Fukushima Daiichi Nuclear Power Plant accident on 11 March 2011. For example, a feed-in tariff (FIT) system was introduced in July 2012. Renewable energies include solar, wind, hydro, geothermal, and woody biomass, which is a unique energy source. Solar and wind energy are public goods, but woody biomass is a private good. Therefore, when electric power companies purchase this resource, they contribute to the local economy, especially the forestry industry. As such, many forest owners, conduits, and administrators in Japan expect significant economic benefits from woody biomass use. However, power generation using this resource increases CO_2 emissions. Hence, assuming that forest biomass is not carbon neutral, this study determined where and how much economic benefit and where and how much CO_2 emission or reduction can be achieved through forest biomass power generation in Japan. Specifically, this study evaluated the economic and environmental effects of woody biomass power generation during construction and operation through an inter-industry analysis using workplace data from the Gifu Biomass Power Co., Ltd. (GBP), a woody biomass power generation company in Gifu Prefecture. The data used in this study were from the first year of operation (2016). Data for subsequent years were unavailable when this study was conducted, and the projections may differ from actual measurements.

1.2. Current Status of Woody Biomass Power Generation in Japan

Japan is one of the most forested countries in the world. Forests cover approximately 25 million hectares, or 2/3, of the country's land area. Plantation forests account for 40% of this total, or 10 million hectares. The forest stock, mainly planted forests, has increased by approximately 70 million m^3 annually and currently stands at approximately 5.2 billion m^3. Planted forests are coniferous, with cedar and cypress species. Most of these trees were planted after 1945 for building materials, and when the trees reach 30 years of age or more, they are thinned once every ten years or so, with clear cutting occurring after 50 to 60 years. However, steep slopes, intricate contour lines, and heavy precipitation make it difficult to create a road network. Even if overhead wires are used, the small working area causes inefficiencies. Hence, the cost of collecting timber is high, leaving a large amount of timber in the forest.

With the shift to renewable energy being a global trend and Japan having the motive to get rid of nuclear power, thinned wood and forest residues are being used as woody biomass. Woody biomass is burned to generate electricity, with little or no use for the heat produced. To promote the use of woody biomass for power generation, a FIT system was introduced based on the German model (see [3]). The price for electricity generated by woody biomass power plants was determined in 2012 and varied depending on the type of fuel. The price for unused wood was set at 33.6 JPY (including tax); general wood, such as sawmill scraps and palm coconut shells, was at 25.2 JPY; and wood derived from construction materials at 13.65 JPY [4]. Since then, the price has been revised several times. In contrast to government policy that aims to promote the use of unused wood, many woody biomass power plants import and use palm coconut shells from Southeast Asia as raw materials [5].

1.3. Previous Research

The input–output (IO) analysis method is commonly used for analyzing economic spillovers. This method has recently been used to assess greenhouse gas emissions [6–9]. In addition, some studies have used computable general equilibrium (CGE) models, including IO tables [10,11].

Several studies employ IO analysis to evaluate the economic benefits to the forestry sector (see [12]). Others have evaluated the economic impact of woody biomass power generation on the forestry sector [13,14] and the environmental benefits of such power generation [15,16]. Madlener et al. [17] evaluated the economic and environmental benefits of woody biomass power generation. They did not consider the CO_2 emissions attributed to the economic benefits and only evaluated the environmental benefits of alternative energy sources. Given that major electric utilities will reduce energy production if the electricity demand remains constant, we believe that the negative effects of renewable energy production through FIT cannot be ignored. This is because an amount of electricity that matches the demand of electricity must be generated, and it is difficult to store even larger amounts of electricity. Haddad et al.'s study [18] on the bioeconomy in Europe focused primarily on agriculture as a provider of food, feed, fuel, and fiber to bio-based industries. They performed a sensitivity analysis of a 1% increase in forest product input use in the European economies in a CGE framework that considers land use and greenhouse gas emissions by agro-ecological zones.

In Japan, the National Institute for Environmental Studies publishes data on the country's "Embodied Energy and Emission Intensity Data for Japan using input–output tables" (3EID), which can be used for environmental analysis through an IO analysis [19]. Washizu et al. [20] published a Japanese IO table for renewable energy power plants using the investment composition vector. Moriizumi et al. [21] also developed a Japanese IO table for renewable energy. They found that different renewable energy generation technologies likely have large indirect spillover effects on various industries.

Nishiguchi and Tabata [22] examined the social, economic, and CO_2 emission reduction benefits in Japan, including job creation, when generating energy from unused woody biomass (8.58 million tons per year). Hayashi et al. [23] replaced fossil fuels with wood energy and investigated, through an inter-industry analysis, whether this would bring economic benefits to users and local economies. Nakano et al. [24] calculated the production, employment, energy consumption, and CO_2 emissions induced by constructing and operating a power plant fueled by unused woody biomass and curbing energy production using thermal power plants. They also estimated the amount of public benefit gained by conserving the area. Tabata and Okuda [25] analyzed the effectiveness of a woody biomass utilization system in Gifu Prefecture using a life cycle assessment. Japanese forestry processes from the planting stage to woody biomass production were evaluated using process-based ecosystem and forestry cost models and ecological footprint-like indicators. Matsuoka et al. [26] focused on five prefectures: Aomori, Iwate, Miyagi, Akita, and Yamagata, and considered trade among these prefectures. They estimated the annual supply potential of timber and forest biomass resources in Japan, such as small-diameter trees and missing trunk logs, rather than logging residues, from profitable forests that are expected to generate more income than the total cost from planting to final harvest.

Most of these previous studies considered the effects of woody biomass power generation from a social or national perspective. However, a distinct characteristic of woody biomass power generation is its relationship to the local economy. Therefore, it is necessary to analyze the impact of woody biomass power generation on the local economy. This study evaluates the economic and environmental impact of GBP, a woody biomass power generation plant, on the local economy including the impact on Chubu Electric Power Co. Inc. (CEP), which purchases electricity generated from renewable energy sources. Furthermore, instead of assuming carbon neutrality during power generation, we assumed that this process is an emission source and thereby simulated whether woody biomass power generation in Japan absorbs carbon dioxide emissions.

2. Methods

2.1. Input–Output Analysis

The IO analysis method was first introduced by Leontief [27], and the economic effects analyzed comprised direct and indirect effects. The flow of the economic effects from a woody biomass power generation plant is as follows. First, such a plant purchases wood for fuel, creating demand for forest residues from the forestry sector and for demolition debris from the pulp, paper, and wood products sector (direct effect). Moreover, the demand in the forestry sector induces the production of items such as forestry equipment, which are required by this sector (indirect effect).

The IO table shows the relationship between the productive sectors of a given economy in a linear, inter-sectoral model. The relationship between the productive sectors and demand can be expressed as follows:

$$X_i = \sum_{j=1}^{N} X_{ij} + F_i = \sum_{j=1}^{N} a_{ij} X_j + F_i, \tag{1}$$

$$a_{ij} = X_{ij}/X_j, \tag{2}$$

where X_i is the total gross outputs in sector $i = 1, \ldots, N$; a_{ij} are the direct input coefficients that divide X_{ij}, the intermediate demand for sector i from supply sector j by X_j, and F_i is the final demand for products in sector i.

Equation (1) can be rewritten in an abbreviated matrix form as follows:

$$X = (I - A)^{-1} F \tag{3}$$

I is an $N \times N$ identity matrix; A is a matrix of input coefficients; and F is a matrix of the final demand for production. $(I - A)^{-1}$ is known as Leontief's inverse matrix (see [28], p. 21 for details). We used this methodology to examine the economic effects of not only woody biomass power generation but also CEP.

An IO analysis generally requires national IO tables (e.g., [20,21].) However, our analysis focuses on GBP and covers the regional economy. Because the industrial structure of Japan as a whole differs from that of Gifu Prefecture, we use this prefecture's regional IO table. The regional IO table is the same as the national IO table except that the figures are based on the prefecture for which the table was created. Exports and imports are those based on the prefectural level. Therefore, the regional IO table can capture out-of-prefecture demand but not the ripple effects of out-of-prefecture demand.

In the 2005 Gifu Prefecture IO table, three types of tables were created: an integrated major classification table (34 sectors), an integrated medium classification table (108 sectors), and an integrated minor classification table (190 sectors). In the 34-sector table, "Forestry" is grouped with "Arable Agriculture," "Livestock," "Agricultural Services," and "Fisheries" as "Agriculture, Forestry, and Fisheries." Meanwhile, in the table of 108 sectors, the forestry sector is listed with the four sectors mentioned above. Therefore, the 34-sector table for "Agriculture, Forestry, and Fisheries" is subdivided into "Forestry," "Arable Agriculture," "Livestock Production," "Agricultural Services," and "Fisheries" and combined with the 33 sectors other than those in the "Agriculture, Forestry, and Fisheries" sector to create an original 38-sector table. From this table, we derived open inverse matrix coefficients that were then used in the analysis.

2.2. Environmental Effects

CO_2 emissions can be estimated using two approaches: a bottom-up approach based on process analysis (PA) and a top-down approach based on environmental input–output (EIO) analysis. PA accumulates CO_2 emissions from the processes from production to disposal, whereas EIO calculates CO_2 emissions from the entire economic system. Even so, EIO is less accurate than PA. However, the major advantage of an IO analysis approach is that once the model is built, it saves time and labor. Therefore, we adopted EIO to calculate CO_2 emissions.

Sectoral CO_2 emissions were calculated by multiplying the net economic benefit directly by the unit CO_2 emissions. We used the 3EID data from the Japanese IO tables [19]. We assumed that the amount of electricity generated by CEP will decrease as the GBP goes into operation. The economic activity of CEP will be reduced and CO_2 emissions will decrease. Consequently, economic activity by the GBP increases CO_2 emissions.

Direct unit CO_2 emissions vary by sector. The sectoral CO_2 emissions for sector i were calculated as follows:

$$Y_i = PE_i \times u_i - NE_i \times u_i = E_i \times u_i, \tag{4}$$

where Y is sectoral CO_2 emissions; PE is the positive effect; NE is the negative effect; E is the net effect, and u is direct unit CO_2 emissions. Direct unit CO_2 emissions are recorded in the 3EID data. However, given that the Japanese industry is subdivided into 400 sectors, the production value (JPY) and direct CO_2 emissions (t-CO_2) of the 400 sectors were used to calculate a weighted average (JPY/t-CO_2) for each of the 38 sectors, which captured the unit CO_2 emissions.

When fuel wood is burned, CO_2 previously absorbed by the forest is released. Carbon footprint guidance and many published carbon footprint and life cycle assessments (LCA) assume that biomass fuels are carbon neutral (e.g., [29–31]). By contrast, some studies have rejected carbon neutrality for several reasons. For example, Rabl et al. [31], based on the polluter pays principle and the Kyoto protocol that greenhouse gas (positive or negative) contributions should be allocated to those responsible, explicitly stated that emissions and removals occur at each stage of the life cycle of CO_2 counting; Johnson [32] proposed a "change in carbon stocks" to capture the state of CO_2 more accurately; Bright and Strømman [33] investigated biofuels production from Scandinavian forest resources and their road transport. Mäkipää et al. [34] compared the differences in carbon emissions from various harvesting methods. They suggested that using forest residues for energy production leads to a net increase in carbon emissions. CO_2 emissions that occur when electricity is generated from forest residues depend on the residue type, boiler, and power generation efficiency. For additional discussion, see Helin et al. [35].

Considering these factors, the CO_2 emissions generated during wood chip combustion were not deemed to be carbon neutral. In this study, the CO_2 emissions from wood chip incineration used the weight of O_2 stocked in the chips. The O_2 weight in the chips varies depending on the tree species. However, we do not know the blending ratio of each tree species. Following Hashimoto and Moriguchi [36], moisture was assumed to account for 10% of the chip weight and was converted to dry weight. The weight of carbon in the chips was calculated by the weight of carbon multiplied by 0.5. Then, to convert the carbon weight to carbon dioxide weight, the weight was multiplied by 44/12 using the following formula:

$$CO_2 \text{ weight in chips} = \text{chip weight} \times 0.9 \times 0.5 \times 44/12 \tag{5}$$

Thereafter, to consider the net CO_2 emissions of GBP, CO_2 emissions and removals were explicitly counted at each stage of the life cycle. CO_2 emissions are also emitted when harvesting wood and transporting wood to chip makers and GBPs. Because CO_2 emissions from these sources are unknown, the IO table and 3EID were used to estimate CO_2 emissions (timber harvesting is included in the forestry sector, and transport is included in the transportation sector).

3. Materials

3.1. Gifu Prefecture

Gifu Prefecture is in central Japan (Figure 1) and had a population of 2 million in 2016. The prefecture covers an area of approximately 1 million hectares, of which around 0.86 million hectares are forested. The main tree species are cedar and cypress, and with lumber production reaching 370,000 m^3 per year, forestry is one of the most important industries in this prefecture.

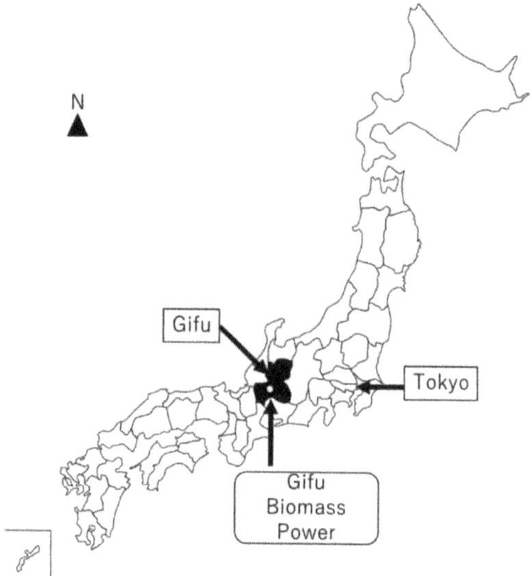

Figure 1. Map of Gifu Biomass Power Co., Ltd.; Gifu Prefecture, where Gifu Biomass Power Co., Ltd. Is located, is in the center of Japan.

There are three woody biomass power generation facilities in the prefecture: One is fueled by wood left over from demolition, another by wood left over from sawmilling, and the third, GBP, uses both woods from forest land and those left over from demolition. Each company consumes a portion of the electricity it generates.

3.2. Gifu Biomass Power Co., Ltd.

Woody biomass power plants are generally located in rural and coastal areas. When forest residues are used, they are expected to be in rural areas, and when imported palm kernel shells are used, they are expected to be in coastal areas.

Mizuho City is in the southern part of Gifu Prefecture and is part of an urban area. The Nagara River and Ibigawa River flow into the surrounding areas, and National Route 21, a major road in Gifu Prefecture, passes through Mizuho City. These conditions are suitable for woody biomass power generation.

The GBP Group consists of two companies: a power generation company (GBP) and a chip company, Biomass Energy Tokai Co., Ltd. (BET); GBP and BET are located at the same site. BET purchases wood residues, demolition residues, and wood chips from the forest, whereas GBP generates 6.25 MWh of electricity, of which 5.40 MWh is sent to CEP. BET and GBP both purchase wood chips from CEP. The plant began commercial operations in December 2014.

The equipment for the power plant was ordered from an out-of-prefecture plant manufacturer for 2.144 billion JPY, while the land preparation and other construction work was ordered from a company within the prefecture for 600 million JPY. The main operating costs are fuel purchases and ash disposal. GBP purchases 86,000 tons of wood chips annually from wood chip companies, including BET (Figure 2). A total of 51,600 tons of thinned wood chips are purchased annually at an average purchase price of 11,300 JPY/ton, and 34,400 tons of demolition wood chips are purchased annually at an average purchase price of 6000 JPY/ton. Demolition debris includes roots and branches of trees cut down for construction work and does not include mill residues. Wood chips derived from forestry are purchased at a ratio of 80% from within the prefecture and 20% from outside the prefecture for construction-derived wood chips. The forestry company belongs to the forestry sector, while the wood chip company belongs to the pulp, paper, and wood products sectors. Fuel purchases amount to 748.2 million JPY from within Gifu Prefecture and 41.28 million JPY outside Gifu Prefecture (Table 1). The amount of ash processed is 5000 tons, and the cost of requesting an out-of-prefecture industrial waste disposal company to process the ash is 75 million JPY (including transportation costs).

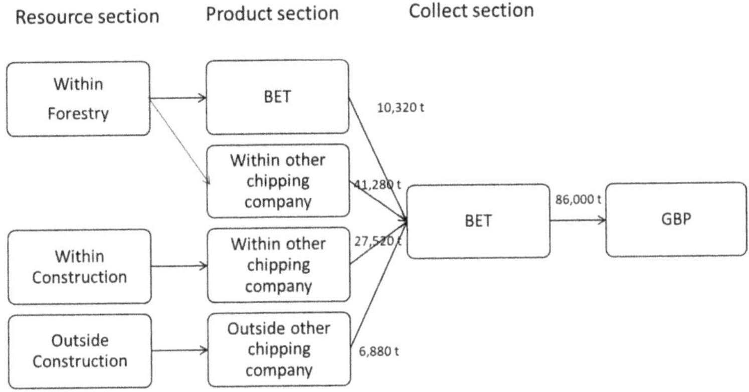

Figure 2. The flow of materials. Gifu Biomass Power Co., Ltd. (GBP) purchases 86,000 tons of wood chips annually from wood chip companies, including Biomass Energy Tokai Co., Ltd. (BET).

Table 1. Fuel purchase amounts (purchase price).

Wood Chips	Purchases (t)	Average Purchase Price Excluding Transfer Fee (JPY/t)	Fuel Purchase Amounts from Gifu (Million JPY)	Fuel Purchase Amounts from Outside of Gifu (Million JPY)
Materials from forestry	**51,600**	**11,300**	**583.08**	**0.00**
to BET	10,320	-	116.62	0.00
to other chipping companies within Gifu	41,280	-	466.46	0.00
Materials from construction	**34,400**	**6000**	**165.12**	**41.28**
to other chipping companies within Gifu	27,520	-	165.12	0
to other chipping companies outside Gifu	6880	-	0	41.28
Amount	**86,000**	**-**	**748.20**	**41.28**

Note: Source: Gifu Biomass Power Co., Ltd.

The negative economic impact of the GBP is the economic benefit resulting from the reduction in energy production by CEP. Since electricity demand cannot be saved if it remains constant, CEP purchases electricity from renewable energy sources and reduces the amount of electricity it generates. Therefore, using CEP's 2014 financial report, we assumed that CEP's expenditures would decrease by 0.032%, which is the ratio of electricity purchased from GBP (43 million kWh) to CEP's electricity production (134,515 million kWh).

4. Results

4.1. Economic Impact of GBP

We estimated the economic benefits of the construction and operations of the biomass power plant in Gifu. Construction is a one-time event, whereas the operations are continuous. The ripple effect during the construction process was quantified at 1553.34 million JPY within the prefecture (Table S1) and 1898.84 million JPY outside the prefecture (Table S2), for a total of 3452.18 million JPY. The out-of-prefecture spillover effect was larger than the within-prefecture spillover effect.

In analyzing the ripple effects of operational processes, the input coefficients for the pulp, paper, and wood products sectors were modified because this study identified industries that produce wood chips as raw materials. In addition, the distribution of input coefficients was modified so that the total input coefficients remained unchanged. The original input factors were 0.020256 for forestry; 1.100155 for pulp, paper, and wood products; and 0.008573 for construction (Table 2). These numbers were derived assuming a sector that produces raw materials for wood chips. The analysis results for the spillover from BET were 0.128984 for forestry; 1 for pulp, paper, and wood products; and 0 for construction. Similarly, the analysis of spillovers from other chipping firms that produce wood chips from unutilized wood showed 0.128984 for forestry; 1 for pulp, paper, and wood products; and 0 for construction. From other chipping firms that produce wood chips from general wood, the results were 0 for forestry; 1 for pulp, paper, and wood products; and 0.128984 for construction. These changes were applied when calculating the first-order ripple effect, and the usual coefficients were used when calculating the second-order ripple effect. This modification involved several assumptions that need to be verified.

Table 2. Input coefficients.

Sector	Original	BET	Other Chipping Company	
		Unused Wood	Unused Wood	General Wood
Forestry	0.020256	0.128984	0.128984	0
Pulp, paper, and wood products	1.100155	1	1	1
Construction	0.008573	0	0	0.128984

The positive within- (out-of-) prefecture ripple effects from the operations amounted to 1266.35 (146.23) million JPY/year. The negative ripple effects were −1222.97 (−75.23) million within (outside) the prefecture. The net ripple effect of the operations was 43.37 (71) million JPY/year within (outside) the prefecture, for a total of 11,438 million JPY/year, indicating that even if CEP's power generation decreases, there will be positive economic effects within and outside Gifu Prefecture. Furthermore, the economic effect outside the prefecture was larger than that within the prefecture.

4.2. Environmental Effects

CO_2 emissions were calculated by multiplying the economic benefits by the CO_2 emissions intensity. CO_2 emissions from the construction of the biomass power plant were quantified at 1230.37 t-CO_2 within the prefecture and 320.69 t-CO_2 outside the prefecture, for a total of 1551.06 t-CO_2 (Table 3). Assuming that woody biomass power generation is carbon neutral, the CO_2 emissions during operation would be −14,740.61 t-CO_2/year within the prefecture and −487.29 t-CO_2/year outside the prefecture. However, assuming the project is not carbon neutral, CO_2 emissions from wood chip combustion would increase by 141,900 t-CO_2/year. Subtracting the emissions, the increase within the prefecture would be 127,159.39 t-CO_2/year, and the combined emissions within and outside the prefecture would be 126,672.1 t-CO_2/year. Thus, it can be assumed that the GBP will increase CO_2 emissions. Given that this situation will continue for 20 years, the total CO_2 emissions from the construction and operational processes will amount to 2,534,993.06 t-CO_2. The dismantling process of the power plant was not calculated because of the lack of information.

Table 3. CO$_2$ emission caused by economic effects with carbon neutrality (unit: t-CO$_2$).

Place	Phase	CO$_2$ Emissions
Within	Construction	1230.37
	Operations	−14,740.61
	—from GBP [1]	1874.39
	—from CEP [2]	−16,614.99
Outside	Construction	320.69
	Operations	−487.29
	—from GBP [1]	315.15
	—from CEP [2]	−802.44

[1] Gifu Biomass Power Corporation; [2] Chubu Electric Power Co., Inc.

5. Discussion

5.1. Verification of Economic and Environmental Impacts of GBP

Since this study did not assume carbon neutrality, CO$_2$ emissions were calculated as positive for both the construction and operational processes. CO$_2$ emissions increased within the prefecture and decreased outside the prefecture. CO$_2$ is considered to impact climate change no matter where it is generated. However, it can be said that the source of CO$_2$ has shifted from outside the prefecture to inside the prefecture because of the woody biomass power generation. No direct data were available for CO$_2$ absorption by afforestation or forest growth in the forestry sector attributable to GBP; therefore, this was not calculated. However, forests can offset CO$_2$ absorption. Gifu Prefecture's CO$_2$ emissions were calculated at 15.91 million tons (2018), excluding CO$_2$ derived from wood pellet combustion at woody biomass power plants. Meanwhile, forests absorb 1.32 million tons of CO$_2$ per year (2018) [37]. Although it is possible to consider 140,000 tons/year of CO$_2$ emissions from GBP to be absorbed by forests, it is important to note that woody biomass power generation did not necessarily reduce CO$_2$ emissions in Gifu Prefecture.

Previous studies on woody biomass power generation have estimated the amount of CO$_2$ emissions reduced through such power generation. Japanese woody biomass power producers generally use woody biomass only for power generation, not heat utilization. Thus, there is room for improvement in energy efficiency and economic efficiency. In this regard, GBP considered the possibility of cogeneration (combined heat and power) before the project started, and the decision was made to use only electricity because there was little demand for heat use around GBP, and the costs were expected to be high compared to the income from heat sales. Further studies are needed to determine other costs and the impact of the heat project on the rural economy and the global environment.

Numerous forestry stakeholders in Japan expect significant economic benefits. Notably, as of 2016, Gifu Prefecture had harvested 389,000 m^3 of timber, had a gross forestry output of 874 million JPY, and employed 899 people [38]. Conversely, GBP plans to use 51,600 tons of forest residues from within the prefecture. Thus, the net economic impact of forestry is 51.48 million JPY per year. From the IO table, it can be seen that the employment inducement coefficient attributed to the net economic impact of forestry is 0.03 persons per million JPY, so the net economic impact of the forestry industry would have increased employment by 1.56 persons per year. However, the 51,600 tons of forest residues to be processed into chips is equivalent to 103,200 m^3 of logs, or 28% of the total timber harvested in Gifu Prefecture. This requires a large amount of labor. Therefore, the input factor from the pulp, paper, and wood products sector to the forestry sector is not fully adjusted, and the input factor may be larger. As such, the economic benefits could be even greater; in addition, the CO$_2$ emissions within Gifu Prefecture could be larger than what has been calculated here.

5.2. Implications for the Promotion of Woody Biomass Generation

Fuel combustion results in CO_2 emissions from woody biomass power plants. The raw materials for the fuel in this case study are forest residues and demolition debris. Initially, forest residues were left over after forest harvesting or thinning, rotted over several years, and emitted CO_2. This residue was collected and used to generate electricity, resulting in CO_2 emissions at the woody biomass power plant. In other words, CO_2 emitted by the forestry sector was transferred to the biomass power plant, and CO_2 emissions remained the same without the GBP. Demolition debris can be considered in the same way. Demolition debris should have been disposed of and emitted CO_2. However, CO_2 emissions were assumed to have been emitted when the debris was transferred to a wood biomass power plant. Therefore, it is shortsighted to assume that woody biomass power plants have become a source of CO_2 emissions by not assuming carbon neutrality, which states that CO_2 emissions from woody biomass power generation are absorbed by the forestry sector. More consideration should be given to including more industrial sectors as well. If wood is harvested for use as fuel for woody biomass power generation, woody biomass power plants can appropriately be viewed as a source of CO_2 emissions. These findings suggest that whether forestry is environmentally friendly should be examined. In addition, carbon neutrality and ecosystem, biodiversity, and soil degradation issues must be considered.

Trees grow at different rates depending on species, climate, and land conditions. Therefore, the amount of CO_2 absorbed by forests per hectare also varies. Comparing CO_2 emissions from woody biomass power generation with CO_2 absorption by forests, it should not be assumed that there is only one universal answer to whether a country is carbon neutral. Consideration should be given to each country or region smaller than a country.

6. Conclusions

Using GBP as a case study, this study evaluated, through an inter-industry analysis, the economic and environmental effects of woody biomass power generation without assuming carbon neutrality. From the analysis, the economic impact of GBP was estimated at 1115.39 million JPY per year, and it was thought to generate 1.56 jobs per year in the forestry sector. In addition, CO_2 emissions were estimated to increase by 1551.06 t-CO_2 during the construction period and 126,672.10 t-CO_2 per year during operations. Notably, even though forests may be able to offset CO_2 emissions by absorbing CO_2, woody biomass power generation does not necessarily reduce CO_2 emissions in Gifu Prefecture. It was found that woody biomass power generation is effective for the local economy but not necessarily for the global environment. However, the input coefficients from the pulp, paper, and wood products sector to the forestry sector were not fully adjusted, and the input coefficients may be larger. Therefore, the economic benefits could be even greater, and the CO_2 emissions within Gifu Prefecture could also be greater than the results of the calculations here. However, the GBP used forest residues and demolition debris as fuel; even without the GBP, CO_2 emissions would have remained the same. This suggests that only the location of the CO_2 emissions changed. To clarify the environmental significance of wood biomass power generation without assuming carbon neutrality, more industrial sectors should be included in the analysis.

This study has two limitations. First, the data collection period was short. A longer period of operation would have resulted in changes in prices and power generation, which could have affected the calculations in this study. Second, no analysis has been done on the decommissioning of power plants due to the lack of data. Future studies should address these issues to complete the analysis of the economic and environmental benefits of woody biomass power plants.

Supplementary Materials: The following supporting information can be downloaded at: https://www.mdpi.com/article/10.3390/f14010148/s1, Table S1: Ripple effects in Gifu prefecture (unit: million JPY). Table S2: Ripple effects for areas outside Gifu prefecture (unit: million JPY).

Author Contributions: M.F. analyzed the data and wrote the manuscript. M.H. collected the data. All authors have read and agreed to the published version of the manuscript.

Funding: This research was financially supported by the Japan Society for the Promotion of Science KAKENHI Grant Number JP 20H00435.

Data Availability Statement: Not applicable.

Acknowledgments: Data used in this paper were provided by Gifu Biomass Power Co., Ltd.

Conflicts of Interest: The authors declare no conflict of interest.

References

1. NewClimate Institute; Oxford Net Zero; Energy & Climate Intelligence Unit; Data-Driven EnviroLab. Net Zero Stocktake 2022: Assessing the Status and Trends of Net Zero Target Setting. Available online: https://zerotracker.net/analysis/net-zero-stocktake-2022 (accessed on 4 November 2022).
2. Camia, A.; Giuntoli, J.; Jonsson, K.; Robert, N.; Cazzaniga, N.; Jasinevičius, G.; Valerio, A.; Grassi, G.; Barredo Cano, J.I.; Mubareka, S. *The Use of Woody Biomass for Energy Production in the EU*; EUR 30548 EN; Publications Office of the European Union: Luxembourg, 2020; p. 182. [CrossRef]
3. García-Alvarez, M.T.; Mariz-Pérez, R.M. Analysis of the success of feed-in tariff for renewable energy promotion mechanism in the EU: Lessons from Germany and Spain. *Procedia-Soc. Behav. Sci.* **2012**, *65*, 52–57. [CrossRef]
4. ERIA Study team. Biomass power generation and wood pellets in Japan. In *Forecast of Biomass Demand Potential in Indonesia: Seeking a Business Model for Wood Pellets*; ERIA Research Project Report FY2022; ERIA, Ed.; Economic Research Institute for ASEAN and East Asia: Jakarta, Indonesia, 2022; pp. 29–45.
5. Levinson, R. The Growing Importance of PKS in the Japanese Biomass Market. Available online: https://biomassmagazine.com/articles/16690/the-growing-importance-of-pks-in-the-japanese-biomass-market (accessed on 4 November 2022).
6. Crawford, R.H. Life cycle energy and greenhouse emissions analysis of wind turbines and the effect of size on energy yield. *Renew. Sust. Energ. Rev.* **2009**, *13*, 2653–2660. [CrossRef]
7. Guo, J.; Zou, L.-L.; Wei, Y.-M. Impact of inter-sectoral trade on national and global CO$_2$ emissions: An empirical analysis of China and US. *Energy Policy* **2010**, *38*, 1389–1397. [CrossRef]
8. Lindner, S.; Legault, J.; Guan, D. Disaggregating the electricity sector of China's input–output table for improved environmental life-cycle assessment. *Econ. Syst Res.* **2013**, *25*, 300–320. [CrossRef]
9. Nakano, S.; Arai, S.; Washizu, A. Economic impacts of Japan's renewable energy sector and the feed-in tariff system: Using an input–output table to analyze a next-generation energy system. *Environ. Econ. Policy. Stud.* **2017**, *19*, 555–580. [CrossRef]
10. Coffman, M.; Bernstein, P.; Wee, S.; Schafer, C. Economic and GHG impacts of natural gas for Hawaii. *Environ. Econ. Policy Stud.* **2017**, *19*, 519–536. [CrossRef]
11. Yamazaki, M.; Takeda, S. A computable general equilibrium assessment of Japan's nuclear energy policy and implications for renewable energy. *Environ. Econ. Policy. Stud.* **2017**, *19*, 537–554. [CrossRef]
12. Cristóbal, S.; Ramón, J. Effects on the economy of a decrease in forest resources: An international comparison. *For. Policy Econ.* **2007**, *9*, 647–652. [CrossRef]
13. Nakamura, R.; Ishikawa, Y.; Matsumoto, A. Endogenous correction of regional economic disparities by making use of regional environmental resources (wooden biomass) extension of inter-regional input-output model. *Input-Output Anal.* **2012**, *20*, 228–242. [CrossRef]
14. Hendricks, A.M.; Wagner, J.E.; Volk, T.A.; Newman, D.H. Regional economic impacts of biomass district heating in rural New York. *Biomass Bioenergy* **2016**, *88*, 1–9. [CrossRef]
15. De, S.; Assadi, M. Impact of cofiring biomass with coal in power plants–A techno-economic assessment. *Biomass Bioenergy* **2009**, *33*, 283–293. [CrossRef]
16. Banerjee, S.; Bit, J. Economic growth versus climate balancing: Some reflections on the sustainable management of forest resource in India. *Decision* **2015**, *42*, 127–145. [CrossRef]
17. Madlener, R.; Koller, M. Economic and CO$_2$ mitigation impacts of promoting biomass heating systems: An input–output study for Vorarlberg, Austria. *Energy Policy* **2007**, *35*, 6021–6035. [CrossRef]
18. Haddad, S.; Britz, W.; Börner, J. Economic impacts and land use change from increasing demand for forest products in the European bioeconomy: A general equilibrium based sensitivity analysis. *Forests* **2019**, *10*, 52. [CrossRef]
19. National Institute for Environmental Studies. Embodied Energy and Emission Intensity Data for Japan Using Input-Output Tables. Available online: https://www.cger.nies.go.jp/publications/report/d031/eng/index_e.htm (accessed on 4 November 2022).
20. Washizu, A.; Nakano, S.; Asakura, K.; Takase, K.; Furukawa, T.; Arai, S.; Hayashi, K.; Okuwada, K. *Economic and Environmental Impact Analysis for Construction of Renewable Energy Power Plants by Extended Input-Output Table*; National Institute of Science and Technology Policy: Tokyo, Japan, 2013; p. 56.
21. Moriizumi, Y.; Hondo, H.; Nakano, S. Development and application of renewable energy-focused input-output table. *J. Jpn. Inst. Energy* **2015**, *94*, 1397–1413. [CrossRef]

22. Nishiguchi, S.; Tabata, T. Assessment of social, economic, and environmental aspects of woody biomass energy utilization: Direct burning and wood pellets. *Renew. Sust. Energ. Rev.* **2016**, *57*, 1279–1286. [CrossRef]
23. Hayashi, T.; Sawauchi, D.; Kunii, D. Forest maintenance practices and wood energy alternatives to increase uses of forest resources in a local initiative in Nishiwaga, Iwate, Japan. *Sustainability* **2017**, *9*, 1949. [CrossRef]
24. Nakano, S.; Murano, A.; Washizu, A. Economic and environmental effects of utilizing unused woody biomass. *J. Jpn. Inst. Energy* **2015**, *94*, 522–531. [CrossRef]
25. Tabata, T.; Okuda, T. Life cycle assessment of woody biomass energy utilization: Case study in Gifu Prefecture, Japan. *Energy* **2012**, *45*, 944–951. [CrossRef]
26. Matsuoka, Y.; Shirasawa, H.; Hayashi, U.; Aruga, K. Annual availability of forest biomass resources for woody biomass power generation plants from subcompartments and aggregated forests in Tohoku Region of Japan. *Forests* **2021**, *12*, 71. [CrossRef]
27. Leontief, W.W. Quantitative input and output relations in the economic systems of the United States. *Rev. Econ. Stat.* **1936**, *18*, 105–125. [CrossRef]
28. Miller, R.E.; Blair, P.D. *Input-Output Analysis: Foundations and Extensions*, 2nd ed.; Cambridge University Press: Cambridge, UK, 2009.
29. Penman, J.; Gytarsky, M.; Hiraishi, T.; Krug, T.; Kruger, D.; Pipatti, R.; Buendia, L.; Miwa, K.; Ngara, T.; Tanabe, K. *Good Practice Guidance for Land Use, Land-use Change and Forestry*; Institute for Global Environmental Strategies (IGES) for IPCC: Kanagawa, Japan, 2003.
30. Intergovernmental Panel on Climate Change. 2006 IPCC Guidelines for National Greenhouse Gas Inventories. Vol. 4. Agriculture, Forestry and Other Land Use. Available online: https://www.ipcc-nggip.iges.or.jp/public/2006gl/index.html (accessed on 4 November 2022).
31. Rabl, A.; Benoist, A.; Dron, D.; Peuportier, B.; Spadaro, J.V.; Zoughaib, A. How to account for CO_2 emissions from biomass in an LCA. *Int. J. Life Cycle Assess* **2007**, *12*, 281. [CrossRef]
32. Johnson, E. Goodbye to carbon neutral: Getting biomass footprints right. *Environ. Impact Assess. Rev.* **2009**, *29*, 165–168. [CrossRef]
33. Bright, R.M.; Strømman, A.H. Fuel-mix, fuel efficiency, and transport demand affect prospects for biofuels in Northern Europe. *Environ. Sci. Technol.* **2010**, *44*, 2261–2269. [CrossRef]
34. Mäkipää, R.; Linkosalo, T.; Komarov, A.; Mäkelä, A. Mitigation of climate change with biomass harvesting in Norway spruce stands: Are harvesting practices carbon neutral? *Can. J. For. Res.* **2015**, *45*, 217–225. [CrossRef]
35. Helin, T.; Sokka, L.; Soimakallio, S.; Pingoud, K.; Pajula, T. Approaches for inclusion of forest carbon cycle in life cycle assessment—A review. *GCB Bioenergy* **2013**, *5*, 475–486. [CrossRef]
36. Hashimoto, S.; Moriguchi, Y. *Data Book: Material and Carbon Flow of Harvested Wood in Japan*; CGER-D034-2004; Center for Global Environmental Research: Tsukuba, Japan, 2004; p. 119.
37. Gifu Prefecture. Total Greenhouse Gas Emissions in Gifu Prefecture. Available online: https://www.pref.gifu.lg.jp/page/3643.html (accessed on 4 November 2022).
38. Gifu Prefecture. Gifu Prefecture Forest and Forestry Statistics 2017. Available online: https://www.pref.gifu.lg.jp/uploaded/attachment/4934.pdf (accessed on 4 November 2022).

Article

The Process and Challenges of Resident-Led Reconstruction in a Mountain Community Damaged by the Northern Kyushu Torrential Rain Disaster: A Case Study of the Hiraenoki Community, Asakura City, Fukuoka Prefecture, Japan

Yoshio Harada [1], Ai Ichinose [1], Tatsuya Owake [2], Kazuo Asahiro [3], Noriko Sato [4,*] and Takahiro Fujiwara [4,*]

1 Graduate School of Bioresource and Bioenvironmental Sciences, Kyushu University, Fukuoka 819-0395, Japan
2 Institute for Creative Cities and Regions, University of Hyogo, Kobe 651-2197, Japan
3 Faculty of Design, Kyushu University, Fukuoka 819-0395, Japan
4 Faculty of Agriculture, Kyushu University, Fukuoka 819-0395, Japan
* Correspondence: sato.noriko.842@m.kyushu-u.ac.jp (N.S.); fujiwara.takahiro.218@m.kyushu-u.ac.jp (T.F.)

Abstract: Frequent torrential rainfall disasters have occurred worldwide in recent years. In Japan, the Northern Kyushu Torrential Rainfall disaster in July 2017 caused extensive damage to Fukuoka and Oita prefectures and significantly impacted local landscapes, from which residents derive pride and identity and, hence, are of the utmost importance. Local communities in Japan are also at risk of extinction due to progressive depopulation. This study discusses community revitalization through landscape creation and related challenges based on the case of the Hiraenoki Community affected by the Northern Kyushu Torrential Rainfall disaster. We conducted long-term participant observations over two years and semi-structured interviews with all households in the community. We found that the landscape project revitalized local pride and involved numerous people outside the community, including the prefectural extension center, university, and Non-Profit Organization (NPO), and provided them with an opportunity to connect. On the other hand, our investigation also revealed the danger of obscuring the original purpose of reconstruction activities because of collaboration with outsiders. This case study elucidates the possibilities and challenges of resident-led reconstruction activities in communities that are facing depopulation and aging problems and are working with organizations within and outside the community.

Keywords: local landscape; observation; depopulation; municipal government; local identity; revitalization; outsiders; collaboration; resilience

Citation: Harada, Y.; Ichinose, A.; Owake, T.; Asahiro, K.; Sato, N.; Fujiwara, T. The Process and Challenges of Resident-Led Reconstruction in a Mountain Community Damaged by the Northern Kyushu Torrential Rain Disaster: A Case Study of the Hiraenoki Community, Asakura City, Fukuoka Prefecture, Japan. *Forests* **2023**, *14*, 664. https://doi.org/10.3390/f14040664

Academic Editor: Timothy A. Martin

Received: 1 March 2023
Revised: 11 March 2023
Accepted: 20 March 2023
Published: 23 March 2023

1. Introduction

In recent years, floods and landslides caused by torrential rainfall have occurred frequently worldwide. In Japan, the Northern Kyushu Torrential Rainfall disaster in July 2017 (NKTR), torrential rainfall in July 2018, the East Japan Typhoon in October 2019, and torrential rainfall in July 2020 caused extensive damage to many areas. These are enumerated in the list of meteorological, seismic, and volcanic phenomena named by the Japan Meteorological Agency [1]. According to the Japan Meteorological Agency, the annual number of occurrences of $\geq$80 mm of precipitation per hour and the annual number of days with $\geq$400 mm of precipitation per day increased significantly between 1976 and 2020 [2]. Therefore, the frequency of cases of intense rainfall over a short period is increasing in Japan.

The NKTR was caused by a linear precipitation system that was formed and sustained by the effect of warm and very humid winds flowing into a stationary seasonal rain front in the vicinity of the Tsushima Strait during 5–6 July 2017 [3]. As a result, continued torrential rain occurred, resulting in record heavy rain in northern Kyushu, such as Asakura City and Toho Village in Fukuoka Prefecture and Hita City in Oita Prefecture [3]. The total precipitation for 2 days in northern Kyushu peaked at >500 mm in certain areas, and new observation records (24-h precipitation) were recorded in Asakura City (545.5 mm) and Hita City (370.0 mm) [3]. This record-breaking precipitation caused severe damage in Fukuoka and Oita prefectures, including 40 casualties and 2 missing persons, and more than 1600 houses were completely or partially destroyed and inundated above the floor level [3]. In addition, torrential rain has severely damaged utilities, such as water supplies and electricity, as well as roads, railways, agriculture, and forestry, which are key industries in this region [3]. Slope failures and mudslides are frequent in mountainous areas and cause large amounts of driftwood [3]. As a result of the NKTR disaster, the Japanese government designated Asakura City, Toho Village, Soeda Town in Fukuoka Prefecture, and Hita City in Oita Prefecture as having experienced "severe disaster areas" following the related law.

Disasters caused by torrential rain also significantly impact the local landscape owing to landslides, housing damage, and the associated reconstruction work. The Landscape Law of Japan (promulgated in 2004) states that "good landscapes are indispensable for creating a beautiful and dignified land and an enriched living environment" (Article 2, Clause 1) [4]. According to this law, "good landscapes are formed through harmony between nature, history, culture, etc. and people's daily lives, economic activities, etc. in each region" (Article 2, Clause 2) and "good landscapes are closely related to the unique characteristics of each region" (Article 2, Clause 3) [4]. Consequently, the local character and regional identity of mountain village communities may be affected by damage to local forest landscapes caused by torrential rainfall disasters. Rehabilitating local landscapes in disaster-affected areas is necessary but not sufficient to recreate beautiful landscapes. Rather, rehabilitation should serve as a disaster prevention and mitigation system utilizing ecosystem services while considering the characteristics of the unique industries that constitute the identity of the region.

Local landscapes are essential to residents who derive pride and identity from them. Therefore, efforts aimed at increasing the willingness of victims to reside in the disaster-affected areas and helping their villages survive through landscape restoration can be considered "creative reconstruction" [5]. Currently, Japan is facing concerns regarding increasing torrential rainfall disasters destroying local landscapes. There have been reports of local governments being unable to adequately respond to disaster recovery owing to the downsizing of administrative operations, including personnel cutbacks [6]. Conversely, local communities have previously played the roles of assisting with dealing with matters that cannot be handled by individuals or families alone, maintenance of local culture, general interest coordination, liaison and coordination between the government and residents, and supplemental functions for the government [7]. Therefore, in situations with limited local government capacity, the role of local communities in the disaster recovery process is likely to be significant. The Reconstruction Design Council in Response to the Great East Japan Earthquake also states that "Given the vastness and diversity of the disaster region, we shall make community-focused reconstruction the foundation of efforts towards recovery. The national government should support reconstruction through general guidelines and institutional design" in the Seven Principles for the Reconstruction Framework [8]. Therefore, the Japanese government is emphasizing the role of local communities in disaster recovery.

Certain recovery activities that do not rely on local governments were reported. For example, in the case of the 2016 Kumamoto earthquake, reconstruction activities centered on community development councils were undertaken in rural villages, which account for most of the affected area [9]. In the case of the NKTR, an agricultural cooperative collaborates with the Non-Profit Organization (NPO) to accept volunteers and plays a role in connecting stakeholders, thus contributing to the recovery of local agriculture [10]. However, depopulation and aging are serious concerns in Japan owing to the rapid population decline, particularly in rural and mountainous areas; moreover, local communities are at risk of extinction [8]. This trend is even more pronounced in communities where the population has been reduced by natural disasters.

Many studies on disasters and resilience were conducted outside Japan. For example, Eakin et al. investigated the linkages between household vulnerability and resilience through the case of torrential rains associated with Hurricane Stan which devastated farm systems in southern Mexico in 2005 [11]. The authors argued that policy interventions not only enable individual survival but also enhance resilience at local, community, and landscape scales, helping to provide local strategies and knowledge on risk management [11]. In Pakistan, which is considered highly vulnerable to climate change impacts, Memon and Ahmed indicated that the lack of multi-sectoral productive economic opportunities had a negative impact on the resilience of rural households and women, and that female-headed households were more vulnerable than male-headed households [12]. Sun et al. identified the disaster types historically faced by rural settlements in Xinjiang, China, and divided the landscape carrier based on the evolution of these settlements [13]. The authors also proposed the resilience mechanism of adaptation to disasters for rural communities in Xinjiang based on the experience of disaster resilience and adaptation in traditional rural settlements [13].

Some European countries are also facing depopulation problems. MacDonald et al. reported a decline in traditional labor-intensive practices and abandonment of marginal agricultural land in many areas, particularly in mountain areas [14]. Lasanta et al. also mentioned that farmland abandonment had a far greater impact on mountain areas because of rural depopulation as well as biophysical constraints and that it would continue in the following decades [15]. Westhoek et al. carried out a scenario study (termed EURURALIS) to stimulate the strategic discussion among national and European Union policymakers on the future of rural areas in Europe and the role of policy instruments [16].

Therefore, it appears that the case of resident-led reconstruction in a mountain community damaged by a natural disaster in Japan can provide valuable insights and information for further discussions on the resilience of communities affected by natural disasters and depopulation. There are many studies focusing on community recovery processes in Japan; however, the majority are cases of communities affected by earthquakes such as the Great East Japan Earthquake [17,18] and few studies have been conducted on torrential rain disasters. However, the potential destruction of local landscapes by torrential rain disasters in Japan is concerning. Given the limited capacity of local governments, the role of local communities in disaster recovery is regarded as crucial. In addition, local communities in rural areas are at risk of extinction. How is the restoration of local landscapes in rural areas affected by torrential rainfall disasters? How have these landscapes affected the revitalization of local communities? What challenges do these landscapes face? To the best of our knowledge, no previous study has addressed these questions.

Therefore, this study aimed to discuss community revitalization through landscape creation and related challenges in the Hiraenoki Community, Asakura City, Fukuoka Prefecture, an area affected by the NKTR.

2. Materials and Methods

2.1. Study Site

We conducted field research in the Hiraenoki Community, Asakura City, Fukuoka Prefecture, Japan. The present "Hiraenoki" Community was formed by merging two communities ("Daira" and "Enoki" communities) during the Meiji period. Although the two communities have become less distinct, residents remain aware of the distinction. Currently, 20 households live in the Hiraenoki Community, 1 of which is an immigrant from outside the community. In addition, two households live outside the community as "semi-community residents" and commute to their homes purchased in the Hiraenoki Community.

The Hiraenoki Community is located on a fan-shaped site. The community has no rice paddy fields; rather, most residents cultivate persimmons, the main agricultural crop in the community. Original Shiwa persimmon (a persimmon species in Japan) trees planted in 1926 remain in the community, and residents have been producing persimmons for >100 years [5]. These persimmon orchards are an essential income source for residents. They also form the center of the local landscape and have become a symbol of the community (Figure 1).

Figure 1. Local landscape formed by persimmon orchards in the Hiraenoki Community (obtained by Yoshio Harada [the first author] on 28 November 2019).

The Hiraenoki Community was severely damaged by the NKTR. Although there were no fatalities in the community, some houses were completely or partially destroyed, and the mountain slopes collapsed in numerous areas. When we conducted this study in the community in 2019, i.e., more than two years after the disaster, slope protection construction was ongoing in several places. Roads within the community were severely damaged. The road to the south of the community had been restored, whereas the road to the north had not. Persimmon orchards were also severely affected by the NKTR. For example, landslide damage in certain areas completely inhibited persimmon production. In other cases, roads leading to persimmon orchards were damaged and became impassable, and persimmon orchards were bought to establish facilities for erosion control. As a result of the NKTR, the population decreased from 37 households (83 residents) before the NKTR to 19 households (45 residents) as of January 2020. In addition, some residents felt that the landscape had deteriorated owing to fallen Japanese cedar (*Cryptomeria japonica*) and cypress (*Chamaecyparis obtuse*) trees from landslides in forests, bamboo encroachment, and slope protection construction (Figure 2).

Figure 2. Ongoing road restorations and slope protection constructions in the Hiraenoki Community (obtained by Yoshio Harada [the first author] on 12 December 2019).

2.2. Hiraenoki Reconstruction Committee

After the NKTR disaster in July 2017, volunteers mainly restored and reconstructed the living environment, including the restoration of infrastructure and houses, as well as the removal of soil and sand encroaching on the roads. Government-led work on collapsed slopes and the restoration of rivers and roads was ongoing at the time of the present study. The Hiraenoki Reconstruction Committee (HRC) was based on the "Road Committee," which was formed by several residents before the NKTR. The Road Committee conducted activities such as reporting collapsed roads in the community to the local government and collecting money for road repairs.

When the living environment for residents in the community was secured to a certain extent through restoration work by volunteers and the government, the HRC was established by agreement among the residents in the regular community meeting on 14 April 2019. The establishment of the HRC was based on the residents' sense of crisis regarding the survival of the community, and the committee members' desire to "make the community a place where residents and others are happy to live."

The HRC comprises seven members, i.e., six residents and a community head, and has three main objectives: (1) community development, including early completion of restoration projects, creation of a living environment, and safety measures; (2) maintenance of persimmon orchards in the community for ≥ 10 years; and (3) creation of a local landscape. Landscape creation is currently the main activity of the HRC and mainly involves the maintenance of the "Kunugiyama Observation" scenic overlook in the community. The landscape creation is funded by the community budget.

During the establishment of the HRC process, the opinion arose that "it would be difficult for community residents alone to carry out reconstruction activities". Therefore, the HRC chairperson consulted an acquaintance who was an alderman. The alderman then introduced the Asakura Extension and Guidance Center of Fukuoka Prefecture (hereafter, Asakura Extension Center) and the Kyushu University Disaster Recovery and Reconstruction Support Team, including the Faculty of Agriculture and Faculty of Design, Fukuoka (hereafter, Kyshu University Team), and requested assistance with the construction activities. Consequently, the Asakura Extension Center and Kyushu University Team participated in activities commencing in June 2019.

The main activities of the HRC include information-sharing meetings held once a month with the HRC members, Kyushu University Team, and Asakura Extension Center; landscape creation work with the community residents; preparation of grant applications for HRC activities; coordination of community events; and petitions to the local government regarding the infrastructure in the community.

2.3. Data Collection

We conducted long-term participant observations between June 2019 and January 2022. We participated in activities organized by the HRC, focusing on local landscape creation, hearing the residents, and observing and recording these activities. All regular meetings by the HRC were in-person meetings. In addition, we conducted semi-structured interviews with all the households in the community in January 2022. Question items included (1) awareness of HRC activities; (2) participation in HRC activities; (3) expectations of the HRC; (4) evaluation of HRC activities; (5) challenges faced by HRC; (6) participation in the Hiraenoki Reconstruction Tree Planting Ceremony; (7) visits to the observation site outside of maintenance activities; (8) impression of the landscape at Kunugiyama observation site; (9) impression of the landscape seen from the Kunugiyama observation site; and (10) interaction inside and outside the community after the establishment of the observation site.

3. Results

3.1. Kunugiyama Observation Site Creation Process

3.1.1. Before the Determination of the Proposed Observation Site (by February 2020)

The main HRC activity in 2019 was developing a policy for reconstruction activities, including identifying the current status of the community after the NKTR disaster and projections for the future. In addition to the exchange of opinions among HRC members, the attractiveness of the community was confirmed through interactive meetings with the community's women's club and ex-residents who relocated from the community after the NKTR. Consequently, the aesthetics of the autumn foliage of persimmon trees and the taste of the well water were identified as the most attractive features of the community.

Based on the opinions of the residents, certain proposals were made such as organizing the river running through the community into a playground for children, promotion of traditional events in the community, and utilization of the community hall. Finally, the proposal, which revitalizes the community through landscape creation by establishing an observation site overlooking the persimmon orchards in the community, was adopted at the autumn persimmon foliage viewing event on 28 November 2019, to which the Kyushu University Team and the Asakura Extension Center were invited.

The HRC decided to plant broadleaf trees that would add color to the vegetation in the community during autumn. The Kyushu University Team assisted in determining the species and location of planted trees. Initially, there were two candidate sites for observation: one on the west side of the community near Kannondo Temple and one on the east side near the Kunugiyama Mountain. The candidate site was decided upon by voting after the autumn foliage viewing event on 28 November 2019. The western side was initially selected; however, the eastern side was ultimately selected as the observation site in February 2020 because a cemetery was located on the western side.

3.1.2. Concretization of the Observation Site Creation (by the Hiraenoki Reconstruction Tree-Planting Ceremony)

After the observation site was determined in February 2020, HRC meetings were suspended owing to the COVID-19 pandemic but resumed in July 2020. The species and size of the seedlings to be planted in accordance with the site conditions and the committee's requirements were then determined in a series of HRC meetings.

The proposed observation site was originally a persimmon orchard that was no longer manageable because of the NKTR. The HRC signed a contract with the three landowners to lease land (2183 m^2) for 20 years (from 2020–2040). In addition, it was necessary to clear the remaining persimmon trees before planting broadleaf trees. Therefore, community residents, Kyushu University Team members, and Asakura Extension Center staff mowed the grass and cleared the remaining persimmon trees. They also built roads and installed drainage ditches in the planned area. Simultaneously, the Kyushu University Team surveyed and mapped the site.

In September 2020, based on the survey results and budget, the number of trees to be planted and their locations were determined based on the advice of Kyushu University Team members. In addition, a detailed plan for the observation site, created by a faculty member of the Faculty of Design of Kyushu University (the fourth author of this paper), was availed to the public to share the image of the completed project (Figure 3). In October 2020, stakes were driven into the planting areas, and weed prevention sheets, deer nets, and simple toilets were installed.

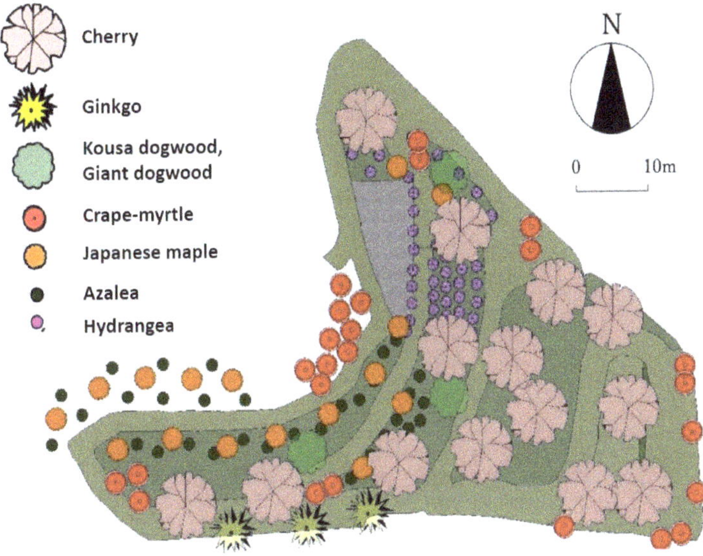

Figure 3. Plan of the observation site (created by Kazuo Asahiro [the fourth author]).

At the HRC meeting in November 2020, the Kyushu University Team introduced NPO Asa-Kuru to the HRC, and they decided that the two organizations would collaborate. It was also decided that the tree planting at the observation site would be held as a "Hilaenoki Reconstruction Tree-Planting Ceremony", inviting residents and ex-residents, staff of Asakura Extension Center and Kyushu University Team, children participating in NPO Asa-Kuru activities, and the media to help disseminate information regarding the project. It was also decided that the ceremony would be held to communicate with the children of the Shiwa School District, where the Hiraenoki Community is located.

In December 2020, HRC planted cherry (*Cerasus* Mill.), crape-myrtle (*Lagerstroemia indica*), and Ginkgo (*Ginkgo biloba* L.) trees and confirmed the location of a signboard before the ceremony (Figure 4). In the subsequent meeting in January, detailed arrangements and roles for the ceremony were determined and a pamphlet was prepared for distribution.

Figure 4. Signboard at the Kunugiyama observation site (obtained by Yoshio Harada [the first author] on 6 March 2021). The signboard notes that this project was led by the HRC and the community residents with support from Kyushu University Team, Asakura Extension Center, and the National Land Afforestation Promotion Organization.

3.1.3. Hiraenoki Reconstruction Tree-Planting Ceremony (6 March 2021)

The Hiraenoki Reconstruction Tree Planting Ceremony was held on 6 March 2021. In total, 110 people participated in the event. Regarding the participation of ex-residents, the HRC sent invitations to 14 people, 4 of whom participated.

Children mainly planted small seedlings, such as hydrangeas (*Hydrangea macrophylla*) and azalea (*Rhododendron* L.), and adults planted medium-sized seedlings, such as Japanese maple (*Acer palmatum* Thunb.). In total, 144 trees, consisting of 9 Someiyoshino cherry (*Prunus yedoensis* Matsumura), 4 Yamazakura cherry (*Cerasus jamasakura*), 4 ginkgo, 2 kousa dogwood (*Cornus kousa*), 1 giant dogwood (*Cornus controversa*), 25 crape-myrtle, 20 Japanese maples, 45 hydrangeas, and 34 azaleas were planted. The planted trees were labeled with the name tag of the person who planted the tree, which was intended to strengthen the connection between the participants and the Hiraenoki Community. At the tree-planting ceremony, children played musical instruments (Kalimba) that were handmade from driftwood generated during the NKTR disaster. The ceremony was reported in newspapers. Subsequently, in April 2021, a group photo of the tree-planting ceremony participants and a project report were distributed. In addition, panels were set up to introduce the past and present of the Hiraenoki Community and the Kunugiyama observation site (Figure 5).

Figure 5. Panel installed at Kunugiyama observation site (created by Kazuo Asahiro [the fourth author]). The panel shows the community residents' thoughts on the Kunugiyama observation as a symbol of reconstruction and their vision of the future of the Hiraenoki community. It also includes a link to a website that introduces the Kunugiyama observation project with a message from HRC and community residents. Please access the website on your smartphone or tablet. Photos from those days, activity scenes, etc. are available.

The observation site project was subsidized (800,000 yen) by the Green and Water Forest Fund of the National Land Afforestation Promotion Organization, and the committee allocated approximately 500,000 yen from their community budget. To minimize costs, all work at the observation site by residents was carried out voluntarily. Consequently, labor costs were reduced by approximately 200,000 yen from the original estimate. In addition, residents utilized their personal connections to obtain saplings at low prices and by division from trees in the community.

During the tree-planting ceremony, we conducted a simple interview with four ex-residents who participated in the tree-planting ceremony. The following three questions were posed: (1) how did you know about the tree-planting ceremony?, (2) do you still own persimmon orchards in the Hiraenoki Community?, and (3) do you want to live in the Hiraenoki Community again?

All four respondents answered that they were informed of the tree-planting ceremony by invitation from the HRC in January 2021. Thus, they were unaware of the Kunugiyama observation site prior to receiving the invitation. In terms of persimmon orchards, three of the four respondents answered that they did not own a persimmon orchard because they had already stopped producing persimmons before the NKTR. One respondent still owned a persimmon orchard in the Hiraenoki Community and visited the community at least once a month to maintain it. When we asked the respondents if they would want to live in the Hiraenoki community again, all four answered "No". Three respondents stated that they could no longer live in the Hiraenoki Community due to the trauma of the disaster experience.

The interview results indicated that it would be very difficult for the community members to reside in the Hiraenoki Community again. On the other hand, we found that their relationship with the Hiraenoki Community had not completely disappeared. For example, an ex-resident visited the Hiraenoki Community to maintain his persimmon orchards. The invitation to the tree-planting ceremony from the HRC also served to connect them with the community.

3.1.4. Activities after the Completion of the Kunugiyama Observation Site (from March 2021)

In June 2021, the HRC, NPO Asa-Kuru, and Kyushu University Teams formed the "Hiraenoki Community Guard Group (*Hiraenoki Satomori Kai*)". As part of these activities, they decided to invite children participating in Asa-Kuru to the Yodo Festival on 23 July 2021, a Shinto ritual held every other year in the Hiraenoki Community, and also to implement recreational activities for them. The Yodo festival is a ritualistic event with fireworks at the Akiba Shrine. The festival was also an opportunity for residents to communicate with each other before the NKTR. At the 2021 Yodo festival, Hiraenoki Community Guard Group staff made preparations in the morning and the children participated in recreation activities in the afternoon, including nature games prepared by the Kyushu University Team, shaved ice, ball scooping, river games by NPO Asa-Kuru, and fireworks prepared by the HRC. The recreational fireworks display was held in front of the community center (Figure 6). Residents also held a firework display at the Akiba Shrine as a Shinto ritual.

Figure 6. Yodo festival in which children participated (obtained by Ai Ichinose [the second author] on 23 July 2021).

Recreational activities with residents and children from the Asa Kuru Community were held at the Reconstruction Autumn Foliage Viewing Event on 28 November 2021 (Figure 7). During the event, a walk around the observation site and a viewing of autumn foliage were held to promote the beauty of autumn foliage in the persimmon garden in the community. The residents served persimmons to the participants and staff from the Asakura Extension Center gave small lectures on persimmon varieties to the children. In addition, a treasure hunt game prepared by the Kyushu University Team and a puppet show by a street performer commissioned by Asa-Kuru were implemented.

Figure 7. Reconstruction Autumn Foliage Viewing Event (obtained by Yoshio Harada [first author] on 28 November 2021).

This event also served as an opportunity to assess the condition of the observation site after the tree-planting ceremony. At the meeting after the event, the participants shared their concerns about the large number of pests that had eaten the leaves of the cherry trees and the fact that the wild boars had broken through the nets. It was also confirmed that four crape-myrtle trees and one Japanese maple tree had died. Therefore, the HRC decided to replant these 5 trees and an additional 16 azaleas and 32 hydrangeas. The replanted tree species are widely found in Japan.

3.2. Results of Semi-Structured Interviews with Community Households

The following opinions on the Hiraenoki Reconstruction Committee were recorded in this study.

3.2.1. Awareness of HRC Activities

Of the 18 respondents, 17 were aware of HRC activities (Table 1). However, of the 10 respondents who stated that they "knew" (excluding the committee members), 7 stated that they only knew about the observation site creation activity. The other three respondents were aware of beautification activities in the community, working with students and children, attending information meetings, and lobbying the local government. The one respondent who answered "don't know" stated that "I was not aware of the activities of the reconstruction committee but was aware of the existence of the observation site".

Table 1. Resident awareness of the Hiraenoki Reconstruction Committee (HRC) activities.

Know	Not Know
17	1

3.2.2. Participation in the Hiraenoki Reconstruction Committee Activities

Of the 11 respondents (excluding the committee members), 9 participated every time they were called upon if possible, for example by mowing the grass at the observation site (Table 2).

Table 2. Participation in HRC activities by residents (excluding committee members).

Participated	Did not Participate
9	2

In addition, 11 residents (non-HRC members) were asked whether they wanted to participate more in HRC activities (Table 3). All the residents said "no" because they had already fully participated in the activities. Other reasons given by the residents who answered "no" included "age", "I can't do anything if I interfere", "participation does not benefit the residents", and "I feel I was forced to participate".

Table 3. Willingness of residents regarding participation in HRC activities (excluding HRC members).

I Want to Participate More	Status Quo Is Good Enough
0	11

In contrast, HRC members were asked if they wanted residents to participate more in committee activities. Three of the seven members answered that they wanted the residents to participate more actively (Table 4). Their reasons were "not enough participation by the population below the early 60 s" and "I feel that the residents' cooperation is still only approximately 80% at the moment". Conversely, the HRC members who answered "I don't think so" had the following opinions: "we can't force them to participate" and "we would like to expand our activity to the outside of the community and cooperate with other communities rather than inside of the community".

Table 4. HRC members perspectives regarding resident participation in HRC activities.

I Want the Residents to Participate More	I Don't Think So
3	4

3.2.3. Expectations of the Hiraenoki Reconstruction Committee

The question was asked in an open-ended format. The most common answer was "maintenance and management of the observation site and its further development" (six respondents). Two respondents answered "early completion of restoration work" and "safety measures for the community and observation site". Other responses were as follows: "maintaining the current situation", "releasing fish into the river", "nursery for fireflies", "organizing abandoned farmland", "creating a lively landscape", "installing a walking trail in the community", "creating a place that can accommodate numerous people", "creating a place for visitors to stop by", "not to be second best", "promotion activities for the observation site", "having more children participate", "calling on the administrative construction office to plant trees for erosion control", "holding social gatherings attended by ex-residents", "arranging for flower viewing", "maintaining and utilizing the Kannondo temple and other facilities in the community effectively", "activities that will properly benefit the community", "activities for the internal residents", and "support for residents' marriages".

3.2.4. Evaluation of the Hiraenoki Reconstruction Committee Activities

Regarding whether the HRC activities aligned with their expectations, seven respondents answered "in line", two answered "not in line", and nine answered "can't say either way" (Table 5). The most common reason for "undecided" was that they could not make a judgment because the results were not yet available. Residents who responded "not in line" felt that the HRC activities were not beneficial to the community and expected the HRC to target the the Hiraenoki Community residents.

Table 5. Evaluation of HRC activities.

In Line	Not in Line	Can't Say Either Way
7	2	9

3.2.5. Challenges of the Hiraenoki Reconstruction Committee

The question was asked in an open-ended format. The most common response was a decrease in the number of bearers owing to depopulation and aging, and the resulting difficulties in long-term maintenance and management" (10 respondents). The second most common response was "There is a difference in views and motivations for reconstruction activities between generations, and relatively young residents do not cooperate well" (7 respondents).

Other opinions expressed by several respondents were that "the observation site has not yet become a gathering place" (4 respondents) and "safety measures have not been taken at the observation site and along the path, posing a risk of injury" (3 respondents). There were also a few comments interpreted as dissatisfaction with the HRC, such as, "The HRC is not open to women's participation", "Insufficient explanation to the residents", and "The HRC is closed and only the leader's opinion is strong".

The following were opinions on the Kunugiyama Observation Site.

3.2.6. Participation in the Hiraenoki Reconstruction Tree Planting Ceremony

Eleven respondents said they "participated and planted trees", three said they "participated but did not plant trees", and four said they "did not participate" (Table 6). The three respondents who answered "participated but did not plant trees" indicated that they were not willing to participate because they felt forced to do so and that they wanted to help the event but felt it was irresponsible to plant trees that they had no way of managing. The four respondents who "did not participate" gave the following reasons: "my schedule did not allow me owing to other commitments", "injury", and "age".

Table 6. Participation in the Hiraenoki Reconstruction Tree Planting Ceremony.

Participated and Planted Trees	Participated But Did Not Plant Trees	Did Not Participate
11	3	4

3.2.7. Visits to the Observation Site Outside of Maintenance Activities

Seven respondents sometimes visited observations outside of the maintenance work. When asked if it was difficult for them to visit the site, eight respondents answered "not hard", seven answered "hard but want to visit", and three answered "hard, so I don't want to visit" (Table 7). This indicated that the majority felt that it was a burden to visit the observation site. Two of the three respondents who answered that they did not want to visit the observation site were relatively young men in their 50s.

Table 7. Opinion regarding the difficulty in visiting the observation site.

Not Hard	Hard, But Want to Visit	Hard, I Don't Want to Visit
8	7	3

3.2.8. Impression of the Landscape at the Kunugiyama Observation Site

When asked if the observation site "feels like a typical Hiraenoki landscape", a photograph was presented to the residents so they could answer based on a common understanding of the question (Figure 8). As a result, eight answered "feels like the typical landscape like Hiraenoki", seven answered that it "does not feel like a landscape like Hiraenoki", and three answered "can't say either way" (Table 8).

Figure 8. Kunugiyama observation site (obtained by Yoshio Harada [the first author] on 28 November 2021).

Table 8. Opinion on the Kunugiyama observation site as the local landscape of the community.

Feels Like the Typical Landscape Like Hiraenoki	Does Not Feel Like Hiraenoki Landscape	Can't Say Either Way
8	7	3

The reasons given for "feels like Hiraenoki landscape" were "because it could be a new symbol of Hiraenoki Community" (5 respondents), "because it is a place of relaxation" (2 respondents), "because it is better to have it than not to have it" (2 respondents), "because it was built by everyone working together" (1 respondent), and "because it is my expersimmon orchard" (1 respondent). In contrast, the reasons given for "not feeling like Hiraenoki Community landscape" were "It is an unfinished project, and no results have been achieved so far" (2 respondents), "It is not very familiar yet in the community" (1 respondent), "There are no fruit trees, which is the industry of Hiraenoki Community" (1 respondent), "It looks like a landscape that can be found anywhere" (1 respondent), and "The original Hiraenoki landscape is not here" (1 respondent). The reasons given for "can't say either way" were "it's my first time here, and I don't know" and "it's an unprecedented landscape."

3.2.9. Impression of the Landscape View from the Kunugiyama Observation Site

The research question "Do you feel that the view from the observation site is the typical landscape of Hiraenoki?" was answered after a photograph was presented to the residents so that they could answer based on a common understanding (Figure 9). Thirteen responded that the view was "landscape like Hiraenoki," four answered that it was "not landscape like Hiraenoki," and one answered, "can't say either way" (Table 9).

Figure 9. Landscape from the Kunugiyama observation site (obtained by Yoshio Harada [the first author] on 28 November 2021).

Table 9. Opinion on the Kunugiyama observation site as the local landscape of the community.

Landscape Like Hiraenoki	Landscape Not Like Hiraenoki	Can't Say Either Way
13	4	1

The reasons given for "feels like Hiraenoki" were "because I can see the autumn leaves of persimmons" (9 respondents), "because I can see the houses, roads, and other residential areas" (4 respondents), "because it is lush green" (1 respondent), and "because I have seen this view for a long time" (1 respondent). Opinions also included "It is sad that there are fewer houses" (1 respondent) and "It would be better if we could see Enoki area as well as Daira area" (1 respondent).

The reasons given for feeling "not Hiraenoki-like" were "We can only see Daira area and not Enoki area" (2 respondents), "The persimmon garden facing east in the morning sun is beautiful, but we can only see the south side from the observation site" (1 respondent), and "It is an unknown view from an unknown location" (1 respondent). One respondent chose "I can't say either way" because "I feel it's just an ordinary view of the countryside."

When asked how they felt about the construction site being visible from the observation site, the most common responses were "I want to improve it as a landscape" (5 respondents), "I have given up on it as inevitable" (5 respondents), and "I feel relieved that it is protected from disasters" (5 respondents).

Other responses were "I feel it will create awareness regarding this community being affected by the disaster" (4 respondents), "I feel that we suffered a major disaster" (3 respondents), "I feel anxious about the disaster" (1 respondent), "I do not want to see it if I could because it reminds me of the disaster" (1 respondent), "The landscape will fade if it turns black" (1 respondent), "I feel the construction is too slow" (1 respondent), "I feel it is just a matter of time because the greenery will return in 10 years" (1 respondent), and " I don't feel anything because I was not affected much by the disaster" (1 respondent).

3.2.10. Interaction Inside and Outside the Community after the Establishment of the Observation Site

As for whether the observation site had ever come up in conversation among the residents in the community, 10 respondents answered "yes" and 8 answered "no." However 6 of the 10 who answered "yes" were HRC members, i.e., more than half of the residents who were not HRC members answered "no" (Table 10).

Table 10. Responses regarding whether the observation site had ever come up in conversation among community residents.

	Yes	No
HRC members	6	1
Non-HRC members	4	7

With regard to whether interaction with the outside community had increased since the establishment of the observation site, 5 respondents answered "increased" and 13 answered "not changed" (Table 11). However, four of the five respondents who answered "increased" were HRC members. Only 1 of the 11 non-HRC residents responded that it had "increased". Therefore, there was a large difference between the perceptions of HRC members and those of non-HRC residents.

Table 11. Opinion on interaction with the outside community.

	Increased	Not Changed
HRC members	4	3
Non-HRC members	1	10

In terms of opinions regarding outsiders coming in and out of the community, 16 responded "agree", 1 responded "disagree", and 1 responded "neither" (Table 12). One respondent who answered "no" stated that he was afraid of the COVID-19 virus being introduced from the outside. The one respondent who answered "neither" stated that he did not feel any particular concern regarding people from the outside coming in and out of the community, but that people had been passing noisily on the road in front of his house for some time, and he would be concerned if such people were to enter the community in the future.

Table 12. Opinion on the Kunugiyama observation site as local landscape of the community.

Agree	Disagree	Neither
16	1	1

4. Discussion

The observation project of the Hiraenoki Community revitalized local pride through landscape creation and involved numerous people outside the community, including the Asakura Extension Center, the Kyushu University Team, NPO Asa-Kuru staff, and children who participated in the Asa-Kuru activities. It provided an opportunity to form connections among people outside the community and residents inside the community. The Kunugiyama observation site is expected to become a "new symbol" for the community and may become an opportunity to recreate the community's identity.

Considering the roles of each entity involved in establishing the observation site, the HRC and cooperative residents played essential roles in planning, preparing project funds, building consensus within the community, and managing the observation site. Conversely, outsiders such as the Kyushu University Team and Asakura Extension Center served as advisors in their respective fields of expertise in the creation of the observation site. They also acted as links between the Asa-Kuru NPO and the Hiraenoki Community. The NPO Asa-Kuru created contact points and communication opportunities between the community and the children. Furthermore, the HRC activities of the observation site led to new developments such as the formation of the "Hiraenoki Community Guard Group (*Hiraenoki Satomori Kai*)" a consultative body of the HRC, a citizens' group (NPO Asa-Kuru), and a university (Kyushu University Team). Therefore, the case of the Hiraenoki Community is regarded as an example as to how interaction with outsiders led to the reaffirmation of the community's appeal and dissemination to the outside community.

In contrast, our interviews revealed that the HRC faced several issues. Firstly, many residents found it challenging to maintain and manage the observation site. In addition, despite considerable participation in the work, awareness among HRC members and other residents in the community differed. For example, perception differed between men and women, among ex-Daira and ex-Enoki community residents, and in various age groups. The communication and sharing of ideas among them were insufficient. Furthermore, certain residents feel an "atmosphere of coercion" around the HRC activities, including the reconstruction tree-planting ceremony. This necessitates the sharing of objectives and issues between the HRC and other residents. Some residents also mentioned "activities targeting residents" as a challenge for the HRC. The HRC's philosophy of landscape creation is to "make the community a place where residents and others are happy to live". However, there is concern that collaboration with outsiders will obscure the original targets of the HRC activities. As in the case in Mashiki Town, which was damaged by the 2016 Kumamoto earthquake [9], the possibility that the landscape creation activities of the Hiraenoki Community can be a means of achieving revitalization of the local community depends on how much participation and understanding can be obtained from the residents in the maintenance activities for the Kunugiyama observation project.

The ex-residents who had moved out of the community were also insufficiently involved in promoting and sharing information regarding reconstruction activities and did not actively participate in the activities. Given the results of interviews with the ex-residents, it would be very difficult for some ex-residents to live in the Hiraenoki Community again. On the other hand, persimmon orchards and the invitation to the tree-planting ceremony from the HRC played roles in connecting ex-residents with the Hiraenoki Community. Therefore, how to utilize the Kunugiyama observation site as a "new communal place" of the Hiraenoki Community to maintain relationships among current residents and ex-residents is an important challenge. It is also essential to achieving the HRC's philosophy to "make the community a place where residents and others are happy to live".

In recent years, there has been increased attention on the "relationship population" in Japan. The Ministry of Internal Affairs and Communications of Japan describes the "relationship population" as a term that refers to people who are involved in a variety of ways with the community, i.e., neither the "settled population" who have moved to the area nor the "exchange population" who have come for sightseeing [19]. On the other hand, Sakuno argued that the "relationship population" should be regarded as one of a number of relationships between urban areas and agricultural and fishing village areas in this new era [20]. In communities facing the challenge of a shortage of residents who keep the community functioning due to population decline and aging, the "relationship population" is expected to become new bearers of the community. In our case study, some ex-residents still had a place attachment and strong network to the Hiraenoki Community even if it would be very difficult for them to live in the Hiraenoki Community again. Therefore, it appeared possible to retain them as a "relationship population" even if they could not be retained as a "settled population". Although our limited data does not allow for further discussion, it appears essential in the reconstruction process to find ways to maintain the "ex-settled population" as the "relationship population" in the post-disaster communities if some residents relocate out of communities after disasters. The reconstruction efforts through observation site creation in the Hiraenoki Community suggest the possibilities and challenges of resident-led landscape creation in the face of aging, depopulation, and reconstruction through cooperation between reconstruction organizations inside and outside the community.

Finally, we were limited in that we could only obtain the opinions of one representative from each household. To understand the perceptions of a wide range of residents in the community, it is necessary to understand the views of other residents, including women. In addition, interviews with ex-residents who have relocated out of the communities after the NKTR are also needed for further discussion. It is also essential to continue participant observation and document resident-led reconstruction activities in the long term.

Author Contributions: Conceptualization, Y.H., A.I., K.A., N.S. and T.F.; methodology, Y.H., A.I., N.S. and T.F.; formal analysis, Y.H.; investigation, Y.H., A.I., N.S. and T.F.; writing—original draft preparation, Y.H.; writing—review and editing, T.O. and T.F.; supervision, K.A., N.S. and T.F.; project administration, N.S.; funding acquisition, N.S. All authors have read and agreed to the published version of the manuscript.

Funding: This work was supported by JSPS KAKENHI Grant Number JP18H04152.

Institutional Review Board Statement: All authors have taken a designated research ethics course required by the universities.

Informed Consent Statement: Informed consent was obtained from all subjects involved in the study.

Data Availability Statement: The data used to support the findings of this study are included within this article.

Acknowledgments: We are deeply grateful to the people of the Hiraenoki Community.

Conflicts of Interest: The authors declare no conflict of interest.

References

1. List of Weather, Earthquake, and Volcanic Phenomena Named by the Japan Meteorological Agency. Available online: https://www.jma.go.jp/jma/kishou/know/meishou/meishou_ichiran.html (accessed on 5 January 2023). (In Japanese).
2. Previous Changes in Torrential Rainfall and Hot Days (Extreme Events). Available online: https://www.data.jma.go.jp/cpdinfo/extreme/extreme_p.html (accessed on 5 January 2023). (In Japanese).
3. White Paper Disaster Management in Japan 2018. Available online: https://www.bousai.go.jp/kaigirep/hakusho/pdf/H30_hakusho_english.pdf (accessed on 14 January 2023).
4. Landscape Law. Available online: https://elaws.e-gov.go.jp/document?lawid=416AC0000000110 (accessed on 14 January 2023). (In Japanese).
5. Asahiro, K.; Sumi, N.; Hashizume, I.; Sato, N.; Fujiwara, T.; Sakuta, K.; Mitani, Y. Recovery support activities through landscape design on mountain village which was damaged by torrential rains: A case of the landscaping of a Kunugiyama Observation spot in Hiraenoki Village, Shiwa, Asakura City, FUKUOKA Prefecture, which was damaged by the torrential rain in July 2017. *Geijutsu Kogaku J. Des.* **2022**, *37*, 17–30. (In Japanese)
6. Miyamoto, T. Problems of disaster recovery in a shrinking society: Overcoming collective denial. *Disaster Kyosei* **2019**, *3*, 11–24. (In Japanese)
7. Yamauchi, Y. The role of communities in an era of low fertility and aging populations: Revitalization of local communities. *Legis. Surv.* **2009**, *288*, 189–195. (In Japanese)
8. Towards Reconstruction "Hope beyond the Disaster". Available online: https://www.cas.go.jp/jp/fukkou/english/pdf/report20110625.pdf (accessed on 14 January 2023).
9. Shibata, Y. The reconstruction process of agricultural villages in the disaster area of the 2016 Kumamoto Earthquake. *J. Rural. Plan. Assoc.* **2019**, *37*, 364–365. (In Japanese) [CrossRef]
10. Noba, R. The role of agricultural cooperatives in the recovery of agriculture in areas affected by torrential rain. *Norin Kinyu* **2021**, *74*, 2–14. (In Japanese)
11. Eakin, H.; Benessaiah, K.; Barrera, J.F.; Cruz-Bello, G.M.; Morales, H. Livelihoods and landscapes at the threshold of change: Disaster and resilience in a Chiapas coffee community. *Reg. Environ. Chang.* **2012**, *12*, 475–488. [CrossRef]
12. Memon, M.H.; Ahmed, R. Multi-topographical landscape: Comparative vulnerability of climate-induced disaster-prone rural area of Pakistan. *Nat. Hazards* **2022**, *111*, 1575–1602. [CrossRef]
13. Sun, Y.; Zhai, B.; Saierjiang, H.; Chang, H. Disaster adaptation evolution and resilience mechanisms of traditional rural settlement landscape in Xinjiang, China. *Int. J. Disaster Risk Reduct.* **2022**, *73*, 102869. [CrossRef]
14. MacDonald, D.; Crabtree, J.R.; Wiesinger, G.; Dax, T.; Stamou, N.; Fleury, P.; Lazpita, J.G.; Gibon, A. Agricultural abandonment in mountain areas of Europe: Environmental consequences and policy response. *J. Environ. Manag.* **2000**, *59*, 47–69. [CrossRef]

15. Lasanta, T.; Arnáez, J.; Pascual, N.; Ruiz-Flaño, P.; Errea, M.P.; Lana-Renault, N. Space–time process and drivers of land abandonment in Europe. *CATENA* **2017**, *149*, 810–823. [CrossRef]
16. Westhoek, H.J.; van den Berg, M.; Bakkes, J.A. Scenario development to explore the future of Europe's rural areas. *Agric. Ecosyst. Environ.* **2006**, *114*, 7–20. [CrossRef]
17. Matanle, P. Post-disaster recovery in ageing and declining communities: The Great East Japan disaster of 11 March 2011. *Geography* **2013**, *98*, 68–76. [CrossRef]
18. Ishikawa, M. A study on community-based reconstruction from Great East Japan earthquake disaster—A case study of Iwanuma City in Miyagi-Pref. *J. Disaster Res.* **2015**, *10*, 807–817. [CrossRef]
19. What Is Relationship Population? Available online: https://www.soumu.go.jp/kankeijinkou/about/index.html (accessed on 25 February 2023). (In Japanese).
20. Sakuno, H. The significance and possibility of relationship population in population decline society in Japan. *Ann. Assoc. Econ. Geogr.* **2019**, *65*, 10–28.

MDPI

St. Alban-Anlage 66

4052 Basel

Switzerland

www.mdpi.com

Forests Editorial Office

E-mail: forests@mdpi.com

www.mdpi.com/journal/forests